Composites Science and Technology

Series Editor

Mohammad Jawaid, Laboratory of Biocomposite Technology, Universiti Putra Malaysia, INTROP, Serdang, Malaysia

This book series publishes cutting edge research monographs comprehensively covering topics in the field of composite science and technology. The books in this series are edited or authored by top researchers and professional across the globe. The series aims at publishing state-of-the-art research and development in areas including, but not limited to:

- Conventional Composites from natural and synthetic fibers
- Advanced Composites from natural and synthetic fibers
- Chemistry and biology of Composites and Biocomposites
- Fatigue damage modelling of Composites and Biocomposites
- Failure Analysis of Composites and Biocomposites
- Structural Health Monitoring of Composites and Biocomposites
- Durability of Composites and Biocomposites
- Biodegradability of Composites and Biocomposites
- Thermal properties of Composites and Biocomposites
- Flammability of Composites and Biocomposites
- Tribology of Composites and Biocomposites
- Applications of Composites and Biocomposites

Review Process

The proposal for each volume is reviewed by the main editor and/or the advisory board. The chapters in each volume are individually reviewed single blind by expert reviewers (at least two reviews per chapter) and the main editor.

Ethics Statement for this series can be found in the Springer standard guidelines here - https://www.springer.com/us/authors-editors/journal-author/journal-author-helpdesk/before-you-start/before-you-start/1330#c14214

Sapuan S. M. · Imran Ahmad
Editors

Composites from the Aquatic Environment

Springer

Editors
Sapuan S. M.
Advanced Engineering and Material Composites Research Centre (AEMC), Department of Mechanical and Manufacturing Engineering
Universiti Putra Malaysia
Selangor, Malaysia

Imran Ahmad
Algae & Biomass Research Lab, Malaysia Japan International Institute of Technology (MJIIT)
Universiti Teknologi Malaysia
Kuala Lumpur, Malaysia

ISSN 2662-1819 ISSN 2662-1827 (electronic)
Composites Science and Technology
ISBN 978-981-19-5326-2 ISBN 978-981-19-5327-9 (eBook)
https://doi.org/10.1007/978-981-19-5327-9

This Springer imprint is published by the registered company Springer Nature Singapore Pte Ltd.
The registered company address is: 152 Beach Road, #21-01/04 Gateway East, Singapore 189721, Singapore

Contents

About the Editors

Sapuan S. M. is a professor of composite materials at Universiti Putra Malaysia. He earned his B.Eng degree in Mechanical Engineering from University of Newcastle, Australia in 1990, MSc from Loughborough University, UK in 1994 and Ph.D. from De Montfort University, UK in 1998. His research interests include natural fiber composites, biocomposites, materials selection and concurrent engineering. To date he has authored or co-authored more than 1521 publications (800 journal papers, 16 authored books, 25 edited books,153 chapters in books and 597 conference proceedings/seminar papers/presentation (32 of which are plenary and keynote lectures and 66 of which are invited lectures). S.M. Sapuan was the recipient of Rotary Research Gold Medal Award 2012, Khwarizmi International Award (KIA). In 2013 he was awarded with 5 Star Role Model Supervisor award by UPM. S.M. Sapuan was recognized as the first Malaysian to be conferred Fellowship by the US-based Society of Automotive Engineers International (FSAE) in 2015. He was the 2015/2016 recipient of SEARCA Regional Professorial Chair. He also received Citation of Excellence Award from Emerald, UK, SAE Malaysia the Best Journal Paper Award, IEEE/TMU Endeavour Research Promotion Award, Best Paper Award by Chinese Defence Ordnance, Malaysia's Research Star Award (MRSA),Top Research Scientists Malaysia Award and Professor of Eminence Award from AMU, India and recently listed in top 2% world scientist by Stanford University, USA. Recently he was elected Fellow of Academy of Science Malaysia.

Imran Ahmad has completed his Ph.D. at the Malaysia-Japan International Institute of Technology, Universiti Teknologi Malaysia, His work was based on the treatment of restaurant wastewater containing FOG using microalgae. Ahmad specializes in the anaerobic treatment of industrial wastewaters, especially landfill leachate. He has published his work in peer-reviewed journals, 6 book chapters, and participated and presented at about 26 international conferences. Ahmad recently published a book with IGI Publications titled "Handbook of Research on Algae as a Sustainable Solution for Food, Energy, and the Environment". Ahmad received best presentation award in 6 International conferences. He received Algae Industry Incubation Consortium (AIIC), Japan grants award and My Membrane award, Malaysia in 2021. He was also the recipient of Young Scholars grant awarded by School of Post Graduate Studies, UTM. Ahmad have teaching experience of 8 years in AICTE approved Engineering Colleges in India. He is having student membership for International Society for applied Phycology (ISAP), International Water Association (IWA), Northeast Algal Society (NEAS), Ecological Society of America (ESA), British Phycological Society, and Northeast Algal Society (NEAS). Ahmad was appointed as the secretary for the International Conference on Environmental Science and Green Technology (ICEGT 2022).

Zooming in to the Composites from the Aquatic Environment

S. M. Sapuan, Imran Ahmad, and J. Tarique

1 Introduction

Composite materials have been used as materials in many industries due to several important reasons such as light weight, high stiffness, and strength properties, aesthetically pleasing, corrosion resistance and part consolidation. The reinforcement phase of the composites is mainly made from synthetic fibres such as glass, carbon, and aramid fibres, although in the recent years, a lot of effort have been intensified to use natural fibres in polymer matrix composites.

There has been an increase in interest in the research and implementation of composites during the last few decades, most likely due to their well-known advantages over traditional materials [1]. In general, combining two or more types of materials results in a composite material with superior properties to its neat precursors. These characteristics can be classified into two main categories: (i) the adaptability of the material in terms of its composition (e.g. polymers, metals, and ceramics); and (ii) the capability of designing a material with a variety of shapes and dimensions [2, 3]. Currently, a large number of studies emphasise the use of the biopolymers chitin and chitosan to create composites with a variety of sizes, shapes, and morphologies, as well as composites for a variety of applications [4–6]. The overwhelming number

S. M. Sapuan · J. Tarique (✉)
Advanced Engineering Materials and Composites Research Centre (AEMC), Department of Mechanical and Manufacturing Engineering, Universiti Putra Malaysia, 43400 UPM Serdang, Selangor, Malaysia
e-mail: tarique5496@gmail.com

S. M. Sapuan
Laboratory of Biocomposite Techology, Institute of Tropical Forestry and Forest Products (INTROP), Universiti Putra Malaysia, 43400 UPM Serdang, Selangor, Malaysia

I. Ahmad
Algae and Biomass Research Laboratory, Malaysia-Japan International Institute of Technology, Universiti Teknologi Malaysia, Jalan Sultan Yahya Petra, 54100, Kuala Lumpur, Malaysia

Sapuan S. M. and I. Ahmad (eds.), *Composites from the Aquatic Environment*, Composites Science and Technology, https://doi.org/10.1007/978-981-19-5327-9_9

of these investigations are devoted to the development of polymer composites, which are multiphase materials composed of a polymeric matrix and fillers. In this sense, the reactive functional groups on the chitin and chitosan backbone provide an infinite number of possibilities for forming a polymeric matrix with tunable shape and dimension for example chemical modification, graft reactions, ionic interactions, and others [7, 8]. For instance, pure chitosan's industrial applicability is limited by its low mechanical strength, a limitation that can be overcome by employing cellulose as reinforcement. Thus, the development of multifunctional, biodegradable composites through the combination of biopolymers is a growing topic with an emphasis on synthesising a diverse range of materials for a variety of applications, including films, foams, fibres, filters, and nanoparticles. Therefore, the purpose of this review paper is to provide the latest information regarding chitin and chitosan which include chitin and chitosan polymer, interaction between water and composites based on chitin/chitosan, chitin and chitosan composites in packaging, and chitin and chitosan composites in biomedical.

The research work on composites utilizing fibres or/and matrices from marine and the aquatic environments is very limited and therefore, it is considered in this book. There is a need to compile information related to these topics. The aquatic environments include oceans; groundwater; glaciers and ice caps; snow and ice; and lakes and rivers [9] and composites from the aquatic environments means the composites with one or both constituent materials are derived from aquatic sources. The examples include fibres and fillers from marine and aquatic plants and animals such as seaweeds, chitosan, water hyacinth, nipah, *Cyperus digitatus*, *Cyperus halpan*, *Cyperus rotundus*, *Scirpus grossus*, *Typha angustifolia* [10], crab shells, etc.

This chapter is an overview of the production and application of aquatic environment-based composites as a promising source for medicinal uses and polymer composites, which is an aspect that has not been widely studied to date. Therefore, there is a need for this review because researchers are interested in the most recent developments in aquatic environment-based materials.

2 Seaweed-Based Composites

Seaweeds are a type of macroalgae that can be found near the coast. Seaweeds are usually regarded one of the members of the algal group, and they come in a variety of sizes. They can range in size from minuscule single cells to sea algae that can grow up to 60 m in length [11]. There are around 10,000 different types of macroalgae, and they account for roughly 10% of all marine life productivity. Seaweed is divided into three families: Chlorophyceae, which is green in color, Rhodophyceae, which is red, and Phaeophyceae, which is brown in color [12]. Seaweeds are macroalgae that live in coastal locations by clinging to rock or any other substrate.

A composite is a material produced by mixing at least two or more materials, which often have different chemical and physical properties [13]. On the other hand, a biocomposite, is a material composed of two or more distinct constituent materials (at least one of which is naturally sourced) that are combined to form a new material that surpasses the constituent materials individually [14]. As a result of increased environmental awareness, concerns about fossil fuel depletion, and a push for more sustainable technologies, composites and biocomposites have gotten a lot of attention in recent years. There have been various investigations on seaweed-based composites using synthetic and natural polymer reinforcement.

Hasan et al. [15] successfully developed and tested biodegradable seaweed films using varying amounts of MCC derived from two different species of bamboo: Lemang Bamboo or L. Bamboo (*Schizostachyum brachycladum*) and Semantan Bamboo or S. Bamboo (*Schizostachyum brachycladum*) (*Gigantochloa scortechinii*). When different volumes and types of MCC were added to pure seaweed films, they noticed that the morphological features of the films were rougher and exhibited ranging waves. In comparison to MCC reinforced seaweed composite films, pure seaweed films have lower tensile strength. The morphology of seaweed composite cracked films reinforced with 3% S. Bamboo MCC, 5% L. Bamboo MCC, and 7% commercial MCC particles is shown in Fig. 1.

Khalil et al. [16] developed seaweed-composite films reinforced with oil palm shell (OPS) nanoparticles. They examined the film's mechanical, physical, and surface structural properties. OPS is a residue of the palm oil mill that is generated after the oil has been extracted or removed from the fruit [17]. It was noticed that the

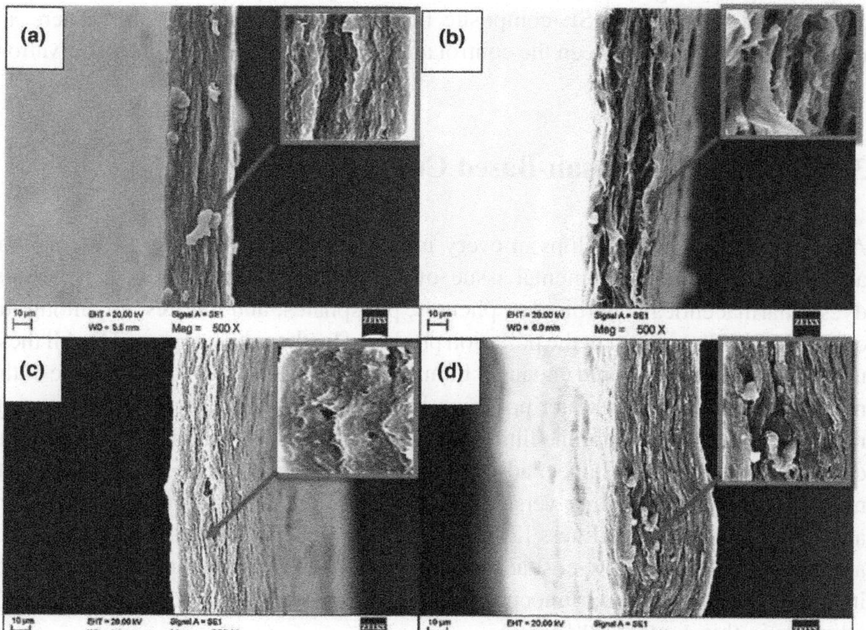

Fig. 1 The SEM images of seaweed composite cracked films including % S. Bamboo MCC, 5% L. Bamboo MCC, and 7% commercial MCC particles [15]

characteristics of seaweed-based films were altered dramatically when composite films were reinforced with OPS nanoparticles.

Khalil et al. [16] also developed a seaweed biocomposite film reinforced with pulp fibers from empty fruit bunches (EFB). EFB fiber is typically produced during the oil extraction mills' process and is classified as a waste product of the palm oil industry [18]. When EFB pulp fibers are utilized as reinforcement, the physical, mechanical, as well as morphological properties of seaweed biocomposite films improve significantly. Because of the strong compatibility between seaweed and EFB pulp fiber, the tensile strength of seaweed-EFB composite films is enhanced. Furthermore, as the EFB pulp fiber concentration in the seaweed-EFB composite films grew, the contact angle decreased. The primary goal of contact angle analysis is to measure the surface hydrophobicity and wettability of materials [19]. Figure 2 depicts the stationary drops for the water contact angle of seaweed incorporated EFB pulp films with varying EFB pulp concentration. Because of its good mechanical characteristics and acceptable hydrophilicity, seaweed composite film reinforced with EFB pulp fibre has a promising future as a packaging material.

Kanmani and Rhim et al. [20] produced antimicrobial films containing carrageenan and grapefruit seed extract (GSE). GSE is primarily extracted from grapefruit pulp, germs, and peel. The active composite films were generated by combining GSE with carrageenan as an antibacterial agent, and the fabricated composite films were physically and mechanically interpreted. Figure 3 depicts scanning electron micrographs of cross-sections of carrageenan control films and common carrageenan reinforced GSE composite films with two levels of GSE. There are noticeable variations between the control and composite films based on observation.

3 Chitin and Chitosan-Based Composites

As the world rapidly develops in every important sector, there are huge concerns about the serious environmental issue of water contamination [21, 22]. Metals, dyes, pharmaceuticals, herbicides, phenols, phosphates, and nitrates are among the contaminants discovered in the pollution produced by these businesses [23]. All these unmanaged behaviors could endanger humans and other species [24, 25]. Some of the methods used to address water pollution include adsorption, electrochemical treatment, precipitation, membrane filtration, electrochemical conversion, and microbial degradation [26]. For instance, adsorption has become a popular method due to its many advantages, including versatility, low cost, high quality, rapid regeneration, and environmental friendliness [27]. Furthermore, renewable and natural resources are being exploited as low-cost adsorbents in this adsorption approach. Because of its hidden availability and innocuous nature, bio-adsorbents are attracting a lot of attention in this context [28].

Furthermore, the ability of the chitin and chitosan composites to be effective bio-adsorbents against various types of pollutants is being investigated in this work. Because chitin has a poor solubility level, it has been covered by its derivative,

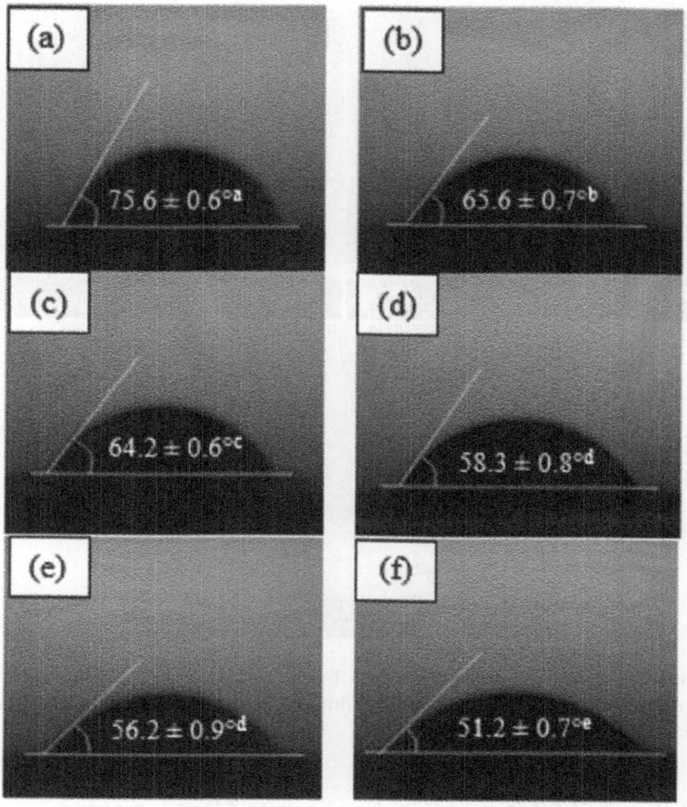

Fig. 2 Stationary drops for water contact angle of SW integrated EFB pulp films with different EFB pulp content [16]

chitosan, which has greater qualities as a soluble biopolymer [29] and rich in—NH_2 and OH groups, which aids in a more efficient adsorption process [30]. However, chitosan's mechanical strength and thermal resistance, as well as its poor stability and acid solubility, has previously hampered its application [27].

According to Ahmed et al. [31] and El Knidri et al. [32], modifying these biopolymers with structural modifications (chitosan/chitin-based composites) can improve their performance as an adsorbent for wastewater contaminants [31, 32]. For example, composites based on chitin and chitosan that use carbonaceous materials such as activated carbon (AC), biochar (BC), carbon nanotubes (CNT), graphene, and graphene oxide (GO) may have good structural stability, improved pore characteristics, and good adsorption capability [33–38].

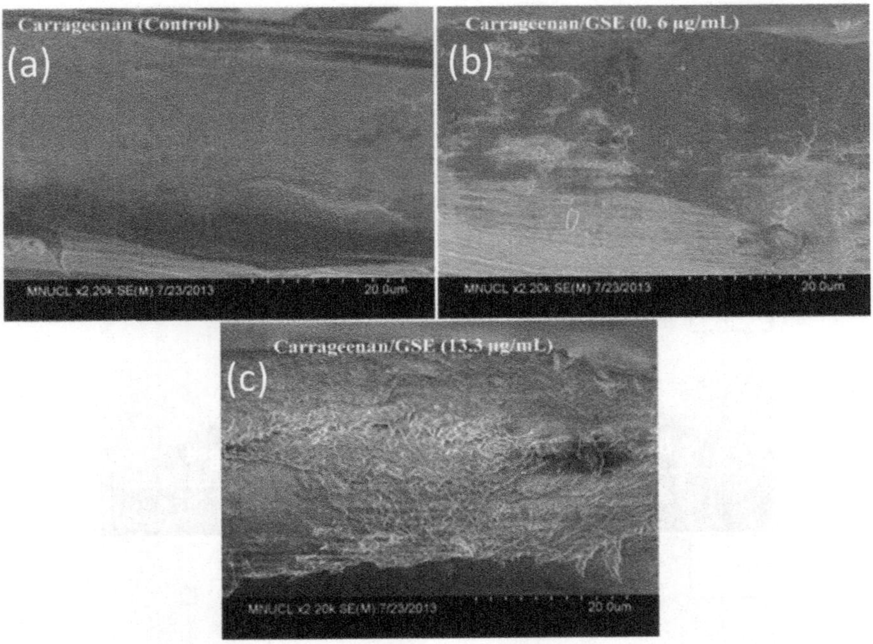

Fig. 3 Scanning electron micrographs photos of cross section of **a** carrageenan control film, **b** carrageenan reinforced with 0.6 μg/mL of GSE film, and **c** carrageenan reinforced with 13.3 μg/mL of GSE film [20]

3.1 Chitin and Chitosan-Carbonaceous Composites

Biopolymer-composites have gained popularity due to their non-toxicity to the environment [39]. As a result, the purpose of this research is to learn more about biopolymer-based composites such as chitin and chitosan. Carbonaceous material, on the other hand, is one of the materials that can be incorporated into the chitin/chitosan structure to improve mechanical and thermochemical capabilities [30, 40]. Furthermore, by improving the efficiency and pore characteristics of biopolymers, carbonaceous chemicals can improve their adsorption capabilities [35]. According to published research, carbonaceous materials that are commonly employed in the construction of adsorbents-based chitin and chitosan composites are graphene oxide, GO (44%) and activated carbon, AC (24%), followed by carbon nanotubes, CNT (19%), biochar, BC (7%), and graphene. This section will elaborate more about the advantages and the properties of these modified chitin and chitosan composites incorporated carbonaceous materials.

3.2 Chitosan/Chitin Biochar Composite

Biochar is a porous carbon that is produced by carbonizing biowaste in a low-oxygen environment [41]. Because of its porous structure and active functional groups, it can be used as a soil conditioner, a catalyst precursor, and a good adsorbent for a range of pollutants [42]. Some studies have shown that chitin and chitosan included composites-based biochar material can be improved [42–45]. Combining natural polymer or biopolymer with biochar is an efficient way to improve the properties of these two materials. The biochar acts as a strong support in this composite because to its favourable structure, which includes a wide surface area and a significant number of active groups, while the CS acts as a complimentary site for pollutant molecules due to its amine and hydroxyl groups [46]. Chitosan and biochar adsorbents have been found to be an effective method for treating both non-organic and organic pollutants [43, 45].

3.3 Chitosan/Chitin Carbon Nanotubes Composite

Carbon nanotubes have been identified as a novel kind of carbonaceous material that has sparked considerable attention since its finding in 1991 [47]. These materials have a huge surface area and outstanding thermochemical properties [48]. CNTs' aggregation behaviour and inadequate structural groups, on the other hand, limit their adsorption application [49]. The simplest way to solve CNT deficiencies is to include biopolymer into them. In terms of $-NH_2$, chitosan provides CNTs with high dispersion and active groups. As a result, a composite like this could be an effective adsorbent for wastewater treatment [50]. Carbon nanotubes incorporated into chitin and chitosan composites are used as effective and high-performance adsorbents [35, 51]. Furthermore, adding carbon nanotubes into biopolymers considerably increases the mechanical properties of biomaterials [52].

3.4 Chitin and Chitosan Nanofibrils Composites Film

As previously stated, when chitosan-based composites are wet, they can change their supramolecular state and take on a hydrophilic character [53]. It can be improved by performing the modification process on chitosan-based composites because only a small amount of chitin nanofibrils can be added to a chitosan matrix [54]; and, by adding 3% of filler (chitin nanofibrils) to composite films increases the durability and modulus of elasticity, all of which is due to the high crystallinity level of chitin nanofibrils [55]. The remarkably similar chemical structures of chitin nanofibrils (filler) and chitosan macromolecules (matrix) reflect the firm connection between the two polymers and their ability to form strong hydrogen bonds. The formation

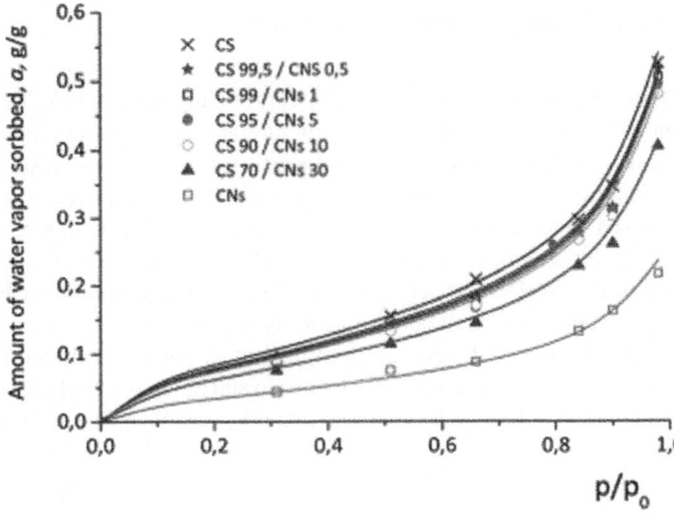

Fig. 4 Isotherms of water vapor sorption determined using the thermal equation of sorption (TES) for raw Chitosan, raw Chitin Nanofibrils, and Chitosan-Chitin Nanofibrils composite films [56]

of stable hydrogen bonds between chitosan and chitin macromolecules, as well as the formation of group structures from chitin nanofibrils, results in decreased molecular flexibility, a higher Young modulus, and a higher tensile stress in composite materials [54, 55]. Figure 4 shows how the amount of chitin nanofibrils in a sample reduces as the amount of chitin nanofibrils in the sample increases. According to the researchers, integrating chitin nanofibrils into the chitosan matrix allows for tighter macromolecule packing, more ordered composite structures, and the creation of liquid crystalline mesophase [56].

4 Applications

Aquatic environment-based polysaccharides are one of nature most abundant biopolymers that can be used as dispersant, scaffold, stabilising, packaging, thickening agent and coating in the food, biomedical and biomass industries due there, excellent film characteristics, high water retaining capacity, biodegradability, and biocompatibility.

4.1 Food Packaging

Seaweed polysaccharides are used in the food industry primarily for their ability to stabilize, emulsify, and develop gels. It is commonly used in the culinary to improve and solidify the structure of jams, jellies, ice creams, and other dairy products. Seaweed-based composite sheets and coatings are also used in food packaging. A product's active packaging is a system. By interacting the product, package, and environment together, you can improve shelf life and safety while maintaining product quality.

One of the active food packaging systems is antimicrobial active packaging. Antimicrobial packaging reduces, impedes, or inhibits the growth of microorganisms by covering them with antimicrobial compounds. Pads, antimicrobial coatings, naturally antimicrobial polymer, and antimicrobial sachets or direct integration into polymer are the four types of antimicrobial packaging [57]. Antimicrobial sachets or pads contain the ingredients in antimicrobials, which are enclosed in a sachet and applied to the packaging. One of the other ways is the direct incorporation of an antibacterial agent into the polymer, which releases the chemical into the packing headspace or onto the food surface.

4.2 Pharmaceutical

Seaweed-based composites could be used for wound dressing, materials, tablet dispersants, bone tissue engineering, cell encapsulation, and scaffolds, among other things [58, 59]. Because polysaccharides derived from seaweed gel quickly, they are frequently employed in drug delivery methods. Several studies have shown that gel formation kinetics have a significant impact on a variety of functional features, including stability, biodegradability, immunological properties, and biocompatibility. The main downside of employing seaweed polymer in drug delivery systems is active component loss due to leaching through the pores of the beads during manufacturing [16]. Many aquatic-based composites have been designed and tested to address the issue of medication delivery applications. Likewise, carrageenan/graphene oxide, carrageenan/gelatin hydrogels, and oxidised alginate/gelatin have been used in bone regeneration and implantation, drug delivery, and hydrogel wound dressing applications, respectively.

5 Conclusions

These biopolymers have proven to be successful in a variety of applications due to their unique film-forming capabilities and outstanding mechanical properties. As a result, seaweed, chitin/chitosan appear to be a very promising renewable resource for the development of biocompatible and environmentally friendly products. Seaweed has been used by humans for its medicinal properties. It is also employed as an emulsifier, gelling agent, and stabilizer in food. The most used polysaccharides from seaweed are alginate, agar, and carrageenan. Several studies aimed at enhancing the composition of chitin and chitosan-based composites are summarized in this review. The basic biopolymers of chitin and chitosan are two of the most abundant polysaccharides, aside from cellulose. Their high availability contributes to the materials' low cost. Because the combining of these biopolymers with other extra materials might offer exceptional physical and chemical applicability, the researchers were encouraged to construct a variety of goods based on these composites based on the particular use.

Furthermore, chitin and chitosan composites containing polysaccharide can be employed as food packaging since their composite films have increased permeability. Finally, chitin and chitosan-inorganic compounds are frequently used in biomedical applications due to their mechanical qualities and biocompatibility. As a result, the advancement of these biopolymer composites must be encouraged to safeguard the green environment, which has recently become a cause of concern. This type of endeavour assures that future generations inherit a better environment in which to live, as well as a better environment for other living species.

Acknowledgements The authors would like to express special thanks to Universiti Teknologi Malaysia and the Ministry of Education for education support.

References

1. Mohanty, A. K., Misra, M., & Drzal, L. T. (2002). Sustainable bio-composites from renewable resources: Opportunities and challenges in the green materials world. *Journal of Polymers and the Environment, 10*(1), 19–26.
2. Camargo, P. H. C., Satyanarayana, K. G., & Wypych, F. (2009). Nanocomposites: Synthesis, structure, properties and new application opportunities. *Materials Research, 12*(1), 1–39.
3. Faruk, O., Bledzki, A. K., Fink, H. P., & Sain, M. (2012). Biocomposites reinforced with natural fibers: 2000–2010. *Progress in Polymer Science, 37*(11), 1552–1596. https://doi.org/10.1016/j.progpolymsci.2012.04.003
4. Ciechanska, D., Wietecha, J., Kucharska, M., Wrzegniewska-Tosik, K., & Kopania, E. (2014). Biomass as a source of functional polymeric materials. *Polimery, 59*(5), 383–392.
5. Muzzarelli, R. A. A. (2011). Biomedical exploitation of chitin and chitosan via mechano-chemical disassembly, electrospinning, dissolution in imidazolium ionic liquids, and supercritical drying. *Marine Drugs, 9*(9), 1510–1533.
6. Rizvi, R., Cochrane, B., Naguib, H., & Lee, P. C. (2011). Fabrication and characterization of melt-blended polylactide-chitin composites and their foams. *Journal of Cellular Plastics, 47*(3), 283–300.
7. Aranaz, I., et al. (2009). Functional characterization of chitin and chitosan. *Current Chemical Biology, 3*(2), 203–230.
8. Rinaudo, M. (2006). Chitin and chitosan: Properties and applications. *Progress in Polymer Science, 31*(7), 603–632.
9. Rich, V. I., & Maier, R. M. (2015). *Aquatic environments*. Elsevier Inc.

10. Bidin, N., Zakaria, M. H., Bujang, J. S., & Abdul Aziz, N. A. (2015). Suitability of aquatic plant fibers for handmade papermaking. *International Journal of Polymer Science*. https://doi.org/10.1155/2015/165868
11. Jayaraman, J., & Ali, N. (2015). 10 use of seaweed extracts for disease management of vegetable crops. *Sustainable Crop Disease Management using Natural Products*, 160.
12. Raj, T. S., Nishanthi, P., Graff, K. H., & Suji, H. A. (2018). Seaweed extract as a biostimulant and a pathogen controlling agent in plants. *International Journal of Tropical Agriculture, 36*(3), 563–580.
13. Jawaid, M., Thariq, M., & Saba, N. (2018). *Durability and life prediction in biocomposites, fibre-reinforced composites and hybrid composites*. Woodhead Publishing.
14. Okonkwo, P. C., Collins, E., & Okonkwo, E. (2017). Application of biopolymer composites in super capacitor. In *Biopolymer composites in electronics* (pp. 487–503). Elsevier.
15. Hasan, M., et al. (2019). Micro crystalline bamboo cellulose based seaweed biodegradable composite films for sustainable packaging material. *Journal of Polymers and the Environment, 27*(7), 1602–1612.
16. Khalil, H. P. S. A., et al. (2017). Oil palm shell nanofiller in seaweed-based composite film: Mechanical, physical, and morphological properties. *BioResources, 12*(3), 5996–6010.
17. Muthusamy, K., et al. (2021). Durability properties of oil palm shell lightweight aggregate concrete containing fly ash as partial cement replacement. *Materials Today Proceedings, 41*, 56–60.
18. Faizi, M. K., et al. (2017). An overview of the Oil Palm Empty Fruit Bunch (OPEFB) potential as reinforcing fibre in polymer composite for energy absorption applications. In *MATEC web of conferences* (Vol. 90, p. 1064).
19. Rane, L. R., Savadekar, N. R., Kadam, P. G., & Mhaske, S. T. (2014). Preparation and characterization of K-carrageenan/nanosilica biocomposite film. *Journal of Materials, 2014*, 1–8.
20. Kanmani, P., & Rhim, J.-W. (2014). Development and characterization of carrageenan/grapefruit seed extract composite films for active packaging. *International Journal of Biological Macromolecules, 68*, 258–266.
21. Gupta, V. K. (2009). Application of low-cost adsorbents for dye removal–a review. *Journal of Environmental Management, 90*(8), 2313–2342.
22. Zia, Q., Tabassum, M., Gong, H., & Li, J. (2019). A review on Chitosan for the removal of heavy metals ions. *Journal of Fiber Bioengineering and Informatics, 12*(3), 103–128.
23. Reddy, D. H. K., & Lee, S.-M. (2013). Application of magnetic chitosan composites for the removal of toxic metal and dyes from aqueous solutions. *Advances in Colloid and Interface Science, 201*, 68–93.
24. Bhatnagar, A., & Sillanpää, M. (2009). Applications of chitin-and chitosan-derivatives for the detoxification of water and wastewater—a short review. *Advances in Colloid and Interface Science, 152*(1–2), 26–38.
25. Ceyhan, Ö., & Baybaş, D. (2001). Adsorption of some textile dyes by hexadecyltrimethylammonium bentonite. *Turkish Journal of Chemistry, 25*(2), 193–200.
26. Sarode, S., et al. (2019). Overview of wastewater treatment methods with special focus on biopolymer chitin-chitosan. *International Journal of Biological Macromolecules, 121*, 1086–1100.
27. Vakili, M., et al. (2014). Application of chitosan and its derivatives as adsorbents for dye removal from water and wastewater: A review. *Carbohydrate Polymers, 113*, 115–130.
28. Ngo, H. H., Guo, W., Zhang, J., Liang, S., Ton-That, C., & Zhang, X. (2015). Typical low cost biosorbents for adsorptive removal of specific organic pollutants from water. *Bioresource Technology, 182*, 353–363.
29. Muxika, A., Etxabide, A., Uranga, J., Guerrero, P., & De La Caba, K. (2017). Chitosan as a bioactive polymer: Processing, properties and applications. *International Journal of Biological Macromolecules, 105*, 1358–1368.
30. Sharififard, H., Rezvanpanah, E., & Rad, S. H. (2018). A novel natural chitosan/activated carbon/iron bio-nanocomposite: Sonochemical synthesis, characterization, and application for

cadmium removal in batch and continuous adsorption process. *Bioresource Technology, 270*, 562–569.
31. Ahmed, M. J., Hameed, B. H., & Hummadi, E. H. (2020). Review on recent progress in chitosan/chitin-carbonaceous material composites for the adsorption of water pollutants. *Carbohydrate Polymers, 247*, 116690. https://doi.org/10.1016/j.carbpol.2020.116690
32. El Knidri, H., Belaabed, R., Addaou, A., Laajeb, A., & Lahsini, A. (2018). Extraction, chemical modification and characterization of chitin and chitosan. *International Journal of Biological Macromolecules, 120*, 1181–1189.
33. Harussani, M. M., Sapuan, S. M., Khalina, A., Rashid, U., & Tarique, J. (2021). Slow pyrolysis of disinfected COVID-19 non-woven polypropylene (PP) waste. In *International Symposium on Applied Sciences and Engineering ISASE2021* (pp. 310–312).
34. Khakpour, R., & Tahermansouri, H. (2018). Synthesis, characterization and study of sorption parameters of multi-walled carbon nanotubes/chitosan nanocomposite for the removal of picric acid from aqueous solutions. *International Journal of Biological Macromolecules, 109*, 598–610.
35. Salam, M. A., El-Shishtawy, R. M., & Obaid, A. Y. (2014). Synthesis of magnetic multi-walled carbon nanotubes/magnetite/chitin magnetic nanocomposite for the removal of Rose Bengal from real and model solution. *Journal of Industrial and Engineering Chemistry, 20*(5), 3559–3567.
36. Xiang Wu, T., Wang, F., Li Tang, Q., & Hua Zhu, Z. (2010). Arrowroot as a novel substrate for ethanol production by solid state simultaneous saccharification and fermentation. *Biomass and Bioenergy, 34*(8), 1159–1164. https://doi.org/10.1016/j.biombioe.2010.03.002
37. Zhang, C., Chen, Z., Guo, W., Zhu, C., & Zou, Y. (2018). Simple fabrication of Chitosan/Graphene nanoplates composite spheres for efficient adsorption of acid dyes from aqueous solution. *International Journal of Biological Macromolecules, 112*, 1048–1054.
38. Liu, J., Zhou, B., Zhang, H., Ma, J., Mu, B., & Zhang, W. (2019). A novel Biochar modified by Chitosan-Fe/S for tetracycline adsorption and studies on site energy distribution. *Bioresource Technology, 294*, 122152.
39. Patricia, M., & Alicia, F. C. (2011). Fluoride removal from water by chitosan derivatives and composites: A review. *Journal of Fluorine Chemistry, 132*(4), 231–240.
40. Frindy, S., et al. (2017). Chitosan–graphene oxide films and CO_2-dried porous aerogel microspheres: Interfacial interplay and stability. *Carbohydrate Polymers, 167*, 297–305.
41. Han, H., Rafiq, M. K., Zhou, T., Xu, R., Mašek, O., & Li, X. (2019). A critical review of clay-based composites with enhanced adsorption performance for metal and organic pollutants. *Journal of Hazardous Materials, 369*, 780–796.
42. Zhang, Z., Zhu, Z., Shen, B., & Liu, L. (2019). Insights into biochar and hydrochar production and applications: A review. *Energy, 171*, 581–598.
43. Afzal, M. Z., Sun, X.-F., Liu, J., Song, C., Wang, S.-G., & Javed, A. (2018). Enhancement of ciprofloxacin sorption on chitosan/biochar hydrogel beads. *Science of the Total Environment, 639*, 560–569.
44. Nitayaphat, W., & Jintakosol, T. (2015). Removal of silver (I) from aqueous solutions by chitosan/bamboo charcoal composite beads. *Journal of Cleaner Production, 87*, 850–855.
45. Xiao, F., Cheng, J., Cao, W., Yang, C., Chen, J., & Luo, Z. (2019). Removal of heavy metals from aqueous solution using chitosan-combined magnetic biochars. *Journal of Colloid and Interface Science, 540*, 579–584.
46. Zhang, L., Tang, S., He, F., Liu, Y., Mao, W., & Guan, Y. (2019). Highly efficient and selective capture of heavy metals by poly (acrylic acid) grafted chitosan and biochar composite for wastewater treatment. *Chemical Engineering Journal, 378*, 122215.
47. Iijima, S. (1991). Helical microtubules of graphitic carbon. *Nature, 354*(6348), 56–58.
48. Sarkar, B., Mandal, S., Tsang, Y. F., Kumar, P., Kim, K.-H., & Ok, Y. S. (2018). Designer carbon nanotubes for contaminant removal in water and wastewater: A critical review. *Science of the Total Environment, 612*, 561–581.
49. Fiyadh, S. S., et al. (2019). Review on heavy metal adsorption processes by carbon nanotubes. *Journal of Cleaner Production, 230*, 783–793.

50. Parlayıcı, Ş., & Pehlivan, E. (2019). Removal of chromium (VI) from aqueous solution using chitosan doped with carbon nanotubes. *Materials Today Proceedings, 18*, 1978–1985.
51. Huang, Y., Lee, X., Macazo, F. C., Grattieri, M., Cai, R., & Minteer, S. D. (2018). Fast and efficient removal of chromium (VI) anionic species by a reusable chitosan-modified multi-walled carbon nanotube composite. *Chemical Engineering Journal, 339*, 259–267.
52. Zhu, H. Y., Jiang, R., Xiao, L., & Zeng, G. M. (2010). Preparation, characterization, adsorption kinetics and thermodynamics of novel magnetic chitosan enwrapping nanosized γ-Fe_2O_3 and multi-walled carbon nanotubes with enhanced adsorption properties for methyl orange. *Bioresource Technology, 101*(14), 5063–5069.
53. Hoeve, C. A. J. (1980). Water in polymers. In S. P. Rowland (Ed.) *ACS symposium series*, (Vol. 127, pp. 135–146).
54. Yudin, V. E., et al. (2014). Wet spinning of fibers made of chitosan and chitin nanofibrils. *Carbohydrate Polymers, 108*, 176–182.
55. Sriupayo, J., Supaphol, P., Blackwell, J., & Rujiravanit, R. (2005). Preparation and characterization of α-chitin whisker-reinforced chitosan nanocomposite films with or without heat treatment. *Carbohydrate Polymers, 62*(2), 130–136.
56. Dresvyanina, E. N., et al. (2020). Thermodynamics of interaction between water and the composite films based on chitosan and chitin nanofibrils. *Carbohydrate Polymers, 245*, 116552.
57. Biji, K. B., Ravishankar, C. N., Mohan, C. O., & Srinivasa Gopal, T. K. (2015). Smart packaging systems for food applications: A review. *Journal of Food Science and Technology, 52*(10), 6125–6135.
58. Liu, H., et al. (2014). Biomimetic and cell-mediated mineralization of hydroxyapatite by carrageenan functionalized graphene oxide. *ACS Applied Materials & Interfaces, 6*(5), 3132–3140.
59. Balakrishnan, B., Mohanty, M., Umashankar, P. R., & Jayakrishnan, A. (2005). Evaluation of an in situ forming hydrogel wound dressing based on oxidized alginate and gelatin. *Biomaterials, 26*(32), 6335–6342.

A Comprehensive Review Based on Chitin and Chitosan Composites

J. Tarique, S. M. Sapuan, N. F. Aqil, A. Farhan, J. I. Faiz, and S. Shahrizan

1 Introduction

Chitin and chitosan are the world's second most prevalent naturally occurring biopolymers. These polysaccharide biopolymers have a long linear chain-like structure that is connected to the functionalizable surface groups via *-D* glycosidic linkages. These biomaterials exhibit unique physical, chemical, mechanical, and optical properties because of their structural characteristics, which contribute to their tunable and exceptional properties such as low density, high porosity, renewability, natural biodegradability, and environmental friendliness, among others. They also have abundant and low-cost natural polymers, and these two are crucial and structurally related polysaccharides that give plants and certain animals structural stability and protection.

In detail, chitin is the most abundant renewable natural resource after cellulose. It is a homopolymer of *N*-acetyl-D-glucosamine residues linked by osidic -1,4 linkages [1]. Chitin and its derivatives are very promising biomolecules with a wide range of biological functions and a high degree of biocompatibility and biodegradability. As a result, they found widespread use in pharmacy, medicine, agriculture, the food and textile industries, as well as cosmetics and wastewater treatment.

There has been an increase in interest in the research and implementation of composites during the last few decades, most likely due to their well-known advantages over traditional materials [2, 3]. In general, combining two or more types of materials results in a composite material with superior properties to its neat precursors. These characteristics can be classified into two main categories: (i) the adaptability of the material in terms of its composition (e.g. polymers, metals, and

J. Tarique · S. M. Sapuan (✉) · N. F. Aqil · A. Farhan · J. I. Faiz · S. Shahrizan
Advanced Engineering Materials and Composites Research Centre (AEMC), Department of Mechanical and Manufacturing Engineering, Universiti Putra Malaysia, 43400 UPM Serdang, Selangor, Malaysia
e-mail: sapuan@upm.edu.my

Sapuan S. M. and I. Ahmad (eds.), *Composites from the Aquatic Environment*, Composites Science and Technology, https://doi.org/10.1007/978-981-19-5327-9_1

ceramics); and (ii) the capability of designing a material with a variety of shapes and dimensions [4, 5]. Currently, a large number of studies emphasise the use of the biopolymers chitin and chitosan to create composites with a variety of sizes, shapes, and morphologies, as well as composites for a variety of applications [6–8]. The overwhelming number of these investigations are devoted to the development of polymer composites, which are multiphase materials composed of a polymeric matrix and fillers. In this sense, the reactive functional groups on the chitin and chitosan backbone provide an infinite number of possibilities for forming a polymeric matrix with tunable shape and dimension for example chemical modification, graft reactions, ionic interactions, and others [9–11]. For instance, pure chitosan's industrial applicability is limited by its low mechanical strength, a limitation that can be overcome by employing cellulose as reinforcement. Thus, the development of multifunctional, biodegradable composites through the combination of biopolymers is a growing topic with an emphasis on synthesising a diverse range of materials for a variety of applications, including films, foams, fibres, filters, and nanoparticles. Therefore, the purpose of this chapter is to provide the latest information regarding chitin and chitosan which include chitin and chitosan polymer, interaction between water and composites based on chitin/chitosan, chitin and chitosan composites in packaging, and chitin and chitosan composites in biomedical.

2 Chitin and Chitosan Polymer

2.1 The Properties of Chitin and Chitosan Polymer

Chitin and chitosan properties are highly variable depending on the origin, deacetylation, protein concentration, and extraction procedures. Deacetylation is the most critical parameter among the others. Among these distinguishing characteristics among chitin and chitosan, the microstructural properties of the polymers remain identical since they have both N-acetyl-d-glucosamine and d-glucosamine units. Thus, to give an overview of several of the most significant features of both polymers, a summarized regarding the attributes of chemical properties, biological properties, and crystallinity are as below.

2.1.1 Chemical Properties of Chitin and Chitosan Polymer

Majority of the polysaccharides that are found in nature, for example cellulose, dextran, pectin, alginic acid, agar, agarose, and carrageenan, are acidic or neutral in behaviour. Chitin and chitosan are both polysaccharides which exist in the simplest form of polysaccharides. Besides, these two polymers have distinguished features such as polyoxysalt formation, their potential to form optical structural characteristics, chelate metal ions and film [12].

Chitin, like cellulose, is a naturally occurring structural polysaccharide but has different characteristics. Muzzarelli [13] and Zikakis [14] stated that chitosan is the non-toxic N-deacetylated derivative of chitin that is universally approved [13, 14]. Rutherford et al. (1978) claimed that chitin is extremely hydrophobic and insolveable in most organic solvents [15]. This is due to its extremely crystalline form; it has a low solubility. The solvent system composed of lithium chloride and tertiary amides is the most effective at dissolving chitin. To be more precise, *N*, *N*-dimethylacetamide, *N*, *N*-dimethylpropionamide, *N*-methyl-2-pyrrolidinone, and 1,3-dimethyl-2-imidazolidinone are the solvents most frequently used to prepare a 5–7% lithium chloride mixture. Other than that, chitin can also be soluble in a hexafluoroisopropanol, hexafluoroacetone and chloro-alcohols when it is combined with aqueous mixture of mineral acids.

In terms of chitosan, which comes from chitin, is soluble in aqueous acids but indissoluble in natural dissolver and aqueous solutions. This is because the existence of free amino groups is usually included to dissolve formic, acetic, lactic, pyruvic, and oxalic acids. Mineral acids—for example hydrochloric and nitric acids can also be applied to produce chitosan mixture; however, phosphoric and sulphuric acids are incompatible with chitosan dissolution.

2.1.2 Biological Properties of Chitin and Chitosan Polymer

Chitin and chitosan are biological polymers with a variety of unique characteristics. The primary benefits are biocompatibility that are acceptable for biodegradation as a result of the biopolymer's origin [16, 17]. *N*-acetylglucosamine is cleaved enzymatically by lysozyme, resulting in the production of glucosamine and *N*-acetylglucosamine. Chitin and chitosan have antimicrobial characteristics [18, 19]. Also, the products from chitin and chitosan are free from toxic and allergic reactions [20]. Besides, chitin and chitosan, also have hemostatic properties. Chitin's hemostatic action is assumed to be caused by vasoconstriction and movement of erythrocytes, clotting element, and platelets to the area of damage [21]. Additionally, chitin and chitosan promote injury curing and may be synthesised as gels, films, fibres, beads, and reinforcement matrices [22].

2.1.3 Crystallinity of Chitin and Chitosan

Crystallinity is a significant property of polysaccharides; in terms of chitin and chitosan, crystallinity is dependent on the proportion of the various monomers that exist in the chitosan. When the *N*-acetyl-2-amino2-deoxy-d-glucose (*N*-acetyl-D-glucosamine) content of the biopolymer exceeds 50, it is called chitosan. The deacetylation technique used to extract the chitosan may have resulted in a decrease in crystallinity of the chitosan, owing to the loss of substantial intermolecular chemical (hydrogen) bonding found in chitin.

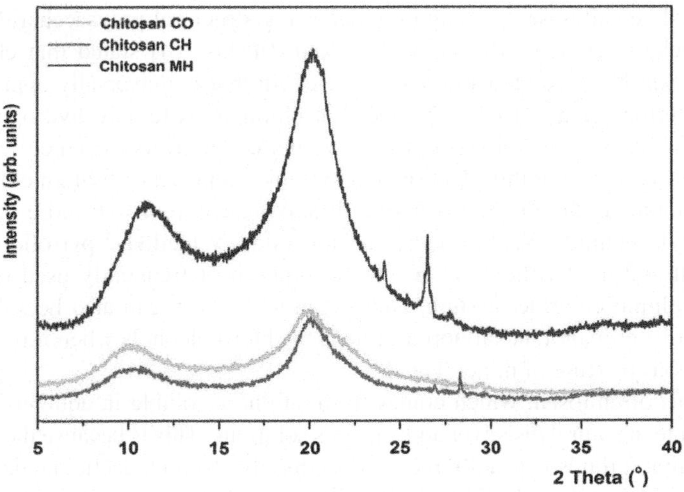

Fig. 1 XRD patterns of a commercial chitosan (green), chitosan developed by conventional heating (red colour), chitosan developed by microwave heating (blue colour) [24]

Chitin, on the other hand, the crystallinity is higher which results in low reactivity, compared to chitosan. Based on Su et al. [23], chitosan was identified by its high refractions at 2θ around 9–10° and 2θ of 20–21°, as well as by its small 2θ values at 26.4°, where the XRD band at 9.9° that is related to a d spacing of 8.92 Å and the refraction at 19.4–20° corresponds to a d spacing of approximately 4.4 Å, sequentially [23]. The crystallinity is influenced by deacetylation at various stages as illustrated in Figs. 1 and 2. Based on Figs. 1 and 2, they discovered that chitosan that has been sold to the public, chitosan that had been drawn out via chemical approach, and chitosan extracted via microwave heating all had distinguished crystallinity level values of 51, 57, and 65%. Additionally, chitin had a notable crystallinity level of 90% and 78% after microwave and conventional heating, sequentially [24].

2.2 Extraction of Chitin and Chitosan Polymer

The methods of extracting chitin and chitosan have varied over years. Commonly people are extracting the chitin and chitosan by using the chemical method as it offers a shorter time of extraction. Despite the various negative consequences of chemical techniques, their rapid removal time makes them the most frequently used commercial treatment option [25]. Besides the chemical methods, there are other methods of extracting chitin which are biological methods. While extracting chitosan, people may consider the conventional heating methods, and the alternative heating methods that use microwave irradiation. All these methods come with a lot of advantages and disadvantages to industries, environment, safety, and health.

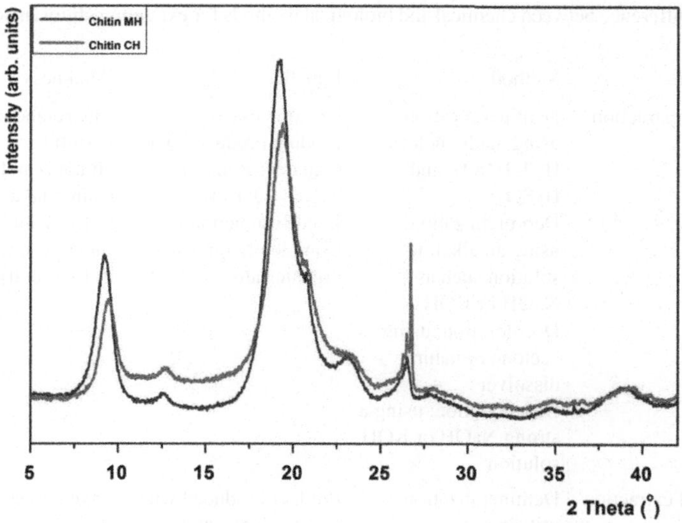

Fig. 2 Chitin extracted by chemical heating (red colour) and microwave heating procedure (blue colour) [24]

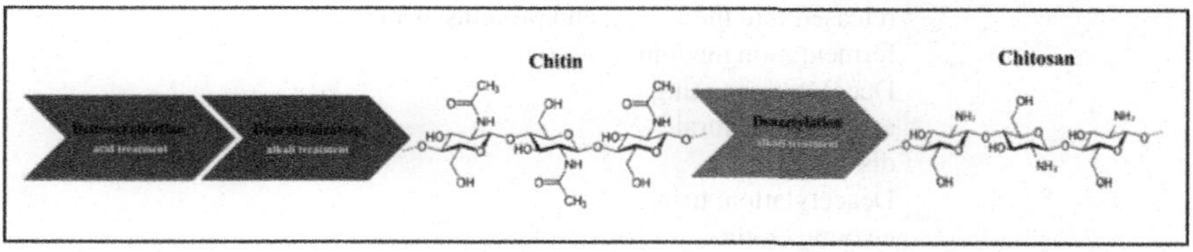

Fig. 3 Process of extracting chitin and chitosan polymer [25]

2.2.1 Chemical and Biological Procedure to Extract Chitin and Chitosan Polymer

Extraction Chitin and chitosan are extracted in two distinct ways: chemically and biologically. Chemical methods require dissolving calcium carbonates and proteins with strong acids and bases which are exceedingly harmful for the environment. To avoid using acidic or alkaline treatments, biological methods are used to obtain chitin and chitosan. In this method, demineralization and deproteinization stages were performed using lactic acid-generating bacteria and bacterial proteases, respectively. While chitin deacetylation is accomplished enzymatically by chitin deacetylase. The summary of the chemical method compared to the biological method is shown in Table 1. However, these chemical applications may result in partial deacetylation of the chitin and hydrolysis of the polymer, resulting in final physiological qualities that are inconsistent [26]. Furthermore, the hydrolyzed protein components become ineffective during this chemical protein removal process.

Table 1 Difference between chemical and biological methods for extracting chitosan polymer [27, 28]

Extraction	Method	Benefit	Weakness
Chemical extraction	Demineralization: using acids such as HCl, HNO_3, and H_2SO_4 Deproteinization: using an alkaline solution such as NaOH or KOH Decoloration: using acetone or natural dissolver Deacetylation: using a strong NaOH or KOH solution	Fast process Product produced with high deacetylation degree (DD%) Used commercially Completely removes organic salts	Environmentally harmful Humans and animals cannot take advantage of the solubilized minerals and proteins as their nutrients
Biological extraction	Demineralization: utilising bacteria-produced lactic acid Deproteinization: using proteases released into the fermentation medium Decoloration: using acetone or natural dissolver Deacetylation: using enzyme chitin deacetylase produced by bacteria	Product produced with an excellent standard Nature friendly Humans and animals can take advantage of the removed minerals and proteins as their nutrients	Take a long time for the process Suit in scope of laboratory studies

2.2.2 Extraction of Chitosan Polymer by Conventional Method and Microwave Irradiation

After the extraction process, normal extraction methods and other possible methods of extraction absorbed by irradiation microwaves [24] take place where they are used to distinguish the naturalness and the grade of result. For a long time, chitin and chitosan polymers have been extracted using a conventional approach, chemically or biologically, involving conventional heating, which takes an amount of time [29], thus, uses a huge amount of energy. Numerous researchers have examined a range of approaches, including electrochemical and microwave radiation, in effort to establish a more efficient, quick, and ecologically friendly approach to extract chitin and chitosan [30, 31].

Microwave irradiation has been popularized as a influential instrument for a faster and effective fusion over the last decade [31], it is rapidly becoming an indispensable method to speed up the synthesis of both organic and inorganic compounds [32], and it can be used efficiently to any sort of chemistry [33]. This innovative method has displaced traditional heating by heating the reaction mass in three dimensions [34], allowing chemical changes to occur in minutes rather than hours or even days [33, 35].

2.2.3 Three General Steps of Extraction Chitin and Chitosan Polymer

Based on Fig. 5, the extraction of chitosan requires three primary steps: demineralization, deproteinization, and deacetylation. An optional action, decolorization, can be add on to remove pigments, primarily Astaxanthin and β-carotene, using a variety of natural and unnatural solvents, including sodium hypochlorite, acetone, and hydrogen peroxide [36–39].

Demineralization: this process is fulfilled in a dilute hydrochloric acid HCl solution and consists of the elimination of calcium carbonate and calcium chloride- the primary unnatural components of crustaceans' exoskeletons. The release of CO_2 gas throughout the digesting response is a fairly good measure of the mineral content. The obtained materials are filtered, neutralised with distilled water, and afterwards dried overnight in an oven.

Deproteinization: proteins are removed by an alkaline treatment with a dilute sodium hydroxide NaOH solution. The composition is filtered, rinsed with deionized water multiple times to eliminate any remaining NaOH, and then dried in an oven all-night. The resulting substance is referred to as pure chitin. Proteins derived from shrimp shell wastes have been shown to be a good source of feedstuffs for animals [40].

Deacetylation: this procedure involves turning chitin to chitosan through the acetyl group removal process. Chitosan is typically prepared by treating it with a concentrated sodium or potassium hydroxide NaOH mixture at a high temperature. Following the reaction, the product is rinsed numerous times with distilled water until it reaches neutrality and then dried overnight in an oven.

2.3 *Modification of Chitin and Chitosan Polymer*

Chitin and chitosan are critical biopolymers with unique chemical and physical characterizations that are used in a variety of implementations including medication transport, antibacterial activity, and other systems [41]. Both chitin and chitosan are the following most prevalent biopolymer, and both possess critical properties such as biocompatibility, biodegradability, high mechanical strength, and nontoxicity [42]. Numerous attempts have been made to modify the physical and chemical properties of this chitin and chitosan polysaccharide. Chitin and chitosan are naturally occurring polysaccharides that perform a variety of chemical and biological functions. These functions are highly dependent on chitin and chitosan's amino groups [43].

Chemical modification of chitin and chitosan is of attraction because it does not alter the basic skeleton of the molecules but retains their original physicochemical and biochemical properties as well as provide new or improved properties [44]. Chemical alteration of chitin and chitosan results in the addition of current characterization that rely on the type of the added group as well as result in a diverse array of derivatives with diverse applications in a variety of sectors. It is widely accepted that the success of chitin and chitosan is due to its low toxicity, biocompatibility, and biodegradability. These two polymers have a tremendous attraction as potential useful biomaterials due to their wide availability and structural diversity [45].

2.3.1 Acetylation of Chitin and Chitosan

Due to their flexibility and solubility, acylated chitin and chitosan derivatives synthesised through *N*-, *O*-, or *N*, *O*-acylation are frequently used. Acetyl is a frequently used O-protecting group, whereas phthaloyl is a frequently used N-protecting group [45]. The N-phthaloylation of chitosan results in chito-derivatives that have more solubility in natural dissolvers and have a chromophore which enables simple monitoring of further modifications [46]. *N*-phthaloyl chitosan may be made by following the procedure to avoid *O*-6 phthaloylation. Chitosan was N-phthaloylated when inside the aqueous acetic acid solution at concentrations ranging from 0 to 10% (v/v).

2.3.2 Quaternization of Chitin and Chitosan

Chitosan is quaternized when the amino group is converted to a positively charged quaternary ammonium group or when a quaternized component is introduced into the polymer backbone [45]. This modification is typically carried out to obtain molecules with extremely specific features, for example enhanced aqueous solubility across a wide pH scope or better bioabsorption. Quaternized chitosan derivatives have been reported and employed in biological applications for drug delivery [47]. As a quaternized derivative, *N*, *N*, *N*-trimethyl chitosan (TMC) has garnered considerable recognition. Indeed, trimethylation of the free amino group is the selected procedure of quaternization by a wide margin. Figure 4 shows a few latest described N; N, N trimethylation approaches.

2.3.3 Oxidative Modification of Chitin

Chitin was oxidatively modified in the presence of $FeCl_3$ by straightly implanting poly(3-hexylthiophene) (P3HT) to its surface. Because of its distinguished standard,

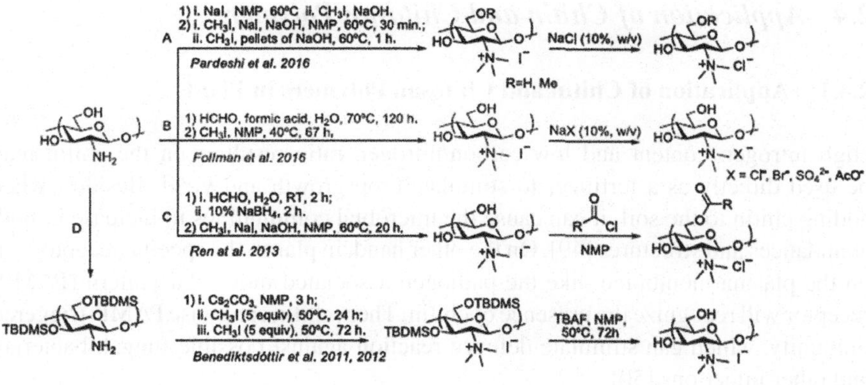

Fig. 4 The latest described *N*; *N*, *N* trimethylation approaches. (Note by, Carvalho et al. [45])

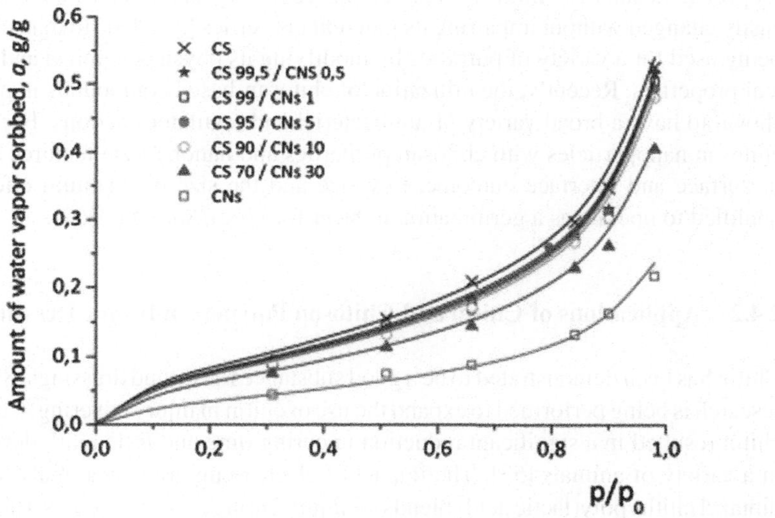

Fig. 5 Isotherms of water vapor sorption by bare Chitosan, bare Chitin nanofibrils and Chitosan-Chitin nanofibrils composite films that are calculated by thermal equation of sorption (TES) [117]

such as good solubility, good capability discharge in a diversity of hues, and less operating voltage, P3HT is employed as a conjugated polymer as well as show improvement of organic light-emitting diode (OLEDS) [48]. Fourier transform infrared (FT-IR) spectroscopy, X-ray diffraction (XRD), UV–vis, fluorescence, transmission electron microscopy (TEM), energy dispersive X-ray (EDX), 1H-nuclear magnetic resonance (H NMR), gel permeation chromatography (GPC), contact angle, and thermo gravimetric analysis were used to characterise the grafted chitin and chitosan (TG) [41].

2.4 *Application of Chitin and Chitosan Polymer*

2.4.1 Application of Chitin and Chitosan Polymers in Plants

High nitrogen content and low carbon/nitrogen ratio resulting in the chitin may be used directly as a fertilizer to stimulate crop growth and yield. Besides, when adding chitin to the soil, it can cause the microbial communities to increase in both abundances and structures [49]. On the other hand, in plants, the specific receptor cell on the plasma membrane, like the pathogen-associated molecular pattern (PAMP) receptor will recognize the presence of chitin. Then, chitin can cause PAMP-triggered immunity, which can stimulate defence reaction against possible fungal, bacterial, and other infections [50].

The biodegradability, biocompatibility, and nontoxicity to people makes chitosan one of the most effective biomaterials in nanotechnology. Additionally, unlike other biopolymers such as chitin, starch, gelatin, cellulose, and glucans, chitosan can be easily changed without impairing its inherent properties [51]. Due to that, chitosan is being used for a variety of purposes by modifying its physicochemical and biophysical properties. Recently, the utilization of chitosan-based compounds in plants was shown to have a broad variety of antibacterial and regulatory actions. For instance, chitosan nanoparticles with chitosan properties and nanoparticle features like those of surface and interface outcome, tiny size and the size of quantum effects were qualified to operate as a germination inducer for *Oryza sativa* L. [52].

2.4.2 Applications of Chitin and Chitosan Polymer in Injury Dressing

Chitin has been demonstrated to be a good substance for wound dressings. Significant research is being performed to expand the use of chitin in injury covering. Curing with chitin resulted in a significant reduction in curing time and little mark development in a variety of animals [53]. The impact of electrospun nonwoven mats containing dibutryl chitin/poly(lactic acid) blends on injury curing of no-hair mice. The outcome indicated that dibutryl chitin effectively enhanced keratinocyte growth [54].

On the other hand, chitosan, a partly deacetylated chitin derivative, has been extensively investigated as an injury covering material. The curing of split-skin graft donor sites was studied using chitosan and a traditional covering. Chitosan was found to promote fast injury to resurface an injury with new epithelium and nerve reconstruction within a vascular dermis [55]. Chitosan aids in wound healing, and the addition of a fundamental fibroblast development element accelerates the rate of regeneration [56]. Besides, the production of a photocrosslinkable chitosan hydrogel for utilization to many types of wounds via brief UV light irradiation [57]. In comparison to untreated controls, the utilization of the photocrosslinkable chitosan hydrogel to full-thickness skin injury on the rear part of mice dramatically stimulated wound shrinking and expedited injury closure and curing [58].

2.4.3 Applications Chitosan Polymer in Metal-Ion Removal

Modified chitosan's high metal ion sorption capabilities are extremely useful for recovering valuable metals or treating contaminated effluents [59]. By introducing additional functional groups towards the chitosan backbone, a vast variety of chitosan derivatives capable of adsorbing metal ions have been produced. To boost sorption particularly for the aim metal, the new moiety is integrated into chitosan to raise the density of the sorption area, modify the pH scope for metal sorption, and replace the sorption area [60]. Chitosan has been shown to be effective at removing Copper (II) ions [61], Zinc (II), and Lead (II) ions in a mixture [62].

3 Interaction Between Water and Composites Based on Chitin/Chitosan

As the world is rapidly developing in every crucial industry, there are major concerns towards the serious issues of the environment which is water pollution [63, 64]. Metals, dyes, medicines, herbicides, phenols, phosphate, and nitrates are among the contaminants found in these industries' pollutants [65]. All of these uncontrolled practices might be dangerous and harmful to humans and other organisms [66, 67]. Adsorption, electrochemical treatment, precipitating, membrane filtration, electrochemical conversion, and microbial degradation. are some of the approaches used to treat this water pollution [68]. For example, adsorption has become a regular approach to be used as it has numerous features such as versatility, relatively inexpensive, excellent quality, rapid regeneration, and environmental friendliness [69]. In addition, renewable and natural resources are being used as it becomes the affordable adsorbents in this adsorption method. Bio-adsorbents are gaining a lot of interest in this context because of their hidden availability and harmless nature [70].

In addition, this study is focusing on the ability of the chitin and chitosan composites to be effective bio-adsorbents towards the various types of contaminants. Since chitin has low solubility level, but it has been covered by its derivative, chitosan which comes with a better properties as it is soluble biopolymer [28, 71] and rich of $-NH_2$ and $-OH$ groups where it helps for a more efficient adsorption process [72]. However, chitosan had been limited by its low mechanical strength and thermal resistance, as well as poor stability and acid solubility [69]. Ahmed et al. [73] and El Knidri et al. [25] stated that the modification of these biopolymers with some additions to their structures (chitosan/chitin-based composites) can improve their ability as an adsorbent for wastewater's pollutants [25, 73]. Example, the composites based on chitin and chitosan that utilized the carbonaceous materials like activated carbon (AC), biochar (BC), carbon nanotubes (CNT), graphene, and graphene oxide (GO) could offer a good stability structure, improved pore characteristics, and good adsorption capability [74–79].

Apart from that, there is a section in this study that discusses the modification of chitosan based composite films which have been strengthened with the chitin nanofibrils. It is important to overcome the problems of chitosan where its molecules have a high hydrophilic nature [80, 81] and even have the instability in supramolecular state when in wet condition which causes mechanical and deformation characteristics to diminish [82].

3.1 Chitin and Chitosan-Carbonaceous Composites

Biopolymer-composites have become widely known with their advantage that is not harmful to the environment [83]. Hence, this study wants to explore more about the composites based on biopolymers called chitin and chitosan. On the other hand, carbonaceous material is one of the elements that can be included into the chitin/chitosan structure to increase their properties in terms of mechanical and thermochemical [72, 84]. Furthermore, by increasing the efficiency and pore properties of the biopolymers, these carbonaceous compounds can increase their adsorption capabilities [76]. Based on the published research, carbonaceous materials which usually used in the fabrication of adsorbents based chitin and chitosan composites are graphene oxide, GO (44%) and activated carbon, AC (24%), followed by carbon nanotubes, CNT (19%), biochar, BC (7%), and graphene (6%) [73]. This section will elaborate more about the advantages and the properties of these modified chitin and chitosan composites incorporated carbonaceous materials.

3.1.1 Chitosan/Chitin-Activated Carbon Composite

Activated carbon exists as the carbonaceous material which has a large surface area and a high adsorption capacity. The qualities of AC, on the other hand, are mostly determined by the raw material utilised and the manufacturing method [85]. Coconut husks, wood, and coal are major resources in the commercial manufacture of this AC [86]. AC is made from a variety of predecessors, including jackfruit extract [87], husk [88], rattan [89], palm date seed [90], and date stones [87, 90]. The major processes in the creation of AC include pyrolysis and activation process. The first stage produces an initial product which is char, then activated to produce AC with a huge surface area [91]. As a result, the cost to produce AC is relatively high. So, fewer quantities of AC is needed for the adsorption process when it is combined with chitosan or chitin, making treatment more cost-effective and environmentally beneficial [92]. However, the presence of micropores prevents the flow of adsorbates with molecular sizes greater than the micropores' size, potentially limiting the use of ACs for big molecule adsorption [93]. The chitosan-activated carbon composite features a high porosity and a strong structure [94].

3.1.2 Chitosan/Chitin Biochar Composite

Biochar can be defined as a porous carbon formed by carbonising biowaste in a low-oxygen environment [95]. Because its structure contains a lot of pores and active functional groups, it may be employed as a soil conditioner, a catalyst prior and an excellent adsorbent for a variety of contaminants [96]. Some investigations have documented the enhancement of chitin and chitosan incorporated composites based biochar material [96–99]. Natural polymer or biopolymer combined with biochar is an effective technique to enhance the properties of these two materials. The biochar serves as a great support in this composite due to its advantageous structure because of its large surface area and a lot of active groups, and in the meantime, the CS provides complementary sites to molecules of pollutants because of its amine and hydroxyl groups [100]. Adsorbents made of chitosan and biochar were shown to be an efficient approach to treat both non-organic and organic contaminants [97, 99]

3.1.3 Chitosan/Chitin Carbon Nanotubes Composite

Carbon nanotubes have been detected as a novel form of carbonaceous material that has gotten a lot of interest since it was first prepared in 1991. [101]. The surface area of these materials is large, and their thermochemical characteristics are excellent [102]. CNTs' aggregation habit and insufficient structural groups, on the other hand, restrict their applicability in adsorption [103]. The easiest solution to address the weaknesses of CNTs is to include biopolymer into them [104]. In terms of $-NH_2$, chitosan offers CNTs an excellent dispersion ability and active groups. As a result, a composite like this might be an ideal adsorbent to treat the wastewater [105]. Composites of chitin and chitosan introduced carbon nanotubes are used as effective and high-performance adsorbents [76, 106]. Furthermore, incorporating carbon nanotubes into biopolymers improves the mechanical characteristics of biomaterials significantly [107].

3.1.4 Chitosan/Chitin Graphene or Graphene Oxide Composite

In terms of thermal, electrical, and mechanical properties, graphene can be said to be the latest findings of carbonaceous material that shows promise. It has a huge specific surface area which makes it an effective adsorbent [108]. In an aqueous solution, however, graphene is easily accumulated, reducing its outer area. Graphene nanoparticles cannot be recovered or reused, and they may pollute the environment, thus limiting its utility in adsorption applications [109]. Graphene oxide (GO) is made from graphite using simple Hummers or more advanced processes. Graphite will be firstly oxidised to become graphite oxide, which is subsequently exfoliated to GO using these processes (simple Hummers and etc.) [110]. Although GO contains several main structural groups, high dispersibility, inclination to agglomerate, and limited recovery restrict its adsorption uses [111]. It is possible to increase the features and performance of GO by combining it into other materials. Graphene oxide or graphene incorporated biopolymer composites, for example, have a favourable form and a high adsorption capability [78, 112, 113]. The adsorption ability of CS-GO composites against palladium metal has been investigated, and it outperforms its individual parts (either CS or GO alone). This might be due to the large surface area of graphene oxide and most active groups of chitosan biopolymers [114]. Hydari et al. [92] reported similar results for the CS-AC composite.

3.2 Chitin and Chitosan Nanofibrils Composites Film

As mentioned previously, chitosan-based composites could change their supramolecular state when in the wet condition and also could come as the hydrophilic nature [82]. It can be improved by doing the modification process on the chitosan-based composites as the minimum quantity of chitin nanofibrils can be added to a chitosan matrix [115]; and also by an addition of 3% of filler (chitin nanofibrils) increases the durability and modulus of elasticity of composite films where all of this due to the high crystallinity level of chitin nanofibrils [116]. The solid connection between chitin nanofibrils (filler) and chitosan macromolecules (matrix) is described by the two polymers near identical chemical structures and their capacity to establish strong hydrogen bonds. The creation of steady hydrogen bonds between chitosan and chitin macromolecules, as well as the production of group structures from chitin nanofibrils, resulting in lower flexibility of molecules, higher Young modulus, and higher tensile stress values in composite materials [115, 116]. It was discovered that when the amount of chitin nanofibrils in the sample grows, the water sorption value decreases which is shown in Fig. 5. It was stated that incorporating chitin nanofibrils into the chitosan matrix allows for tighter macromolecule packing, the production of more ordered composite structures, and the formation of liquid crystalline mesophase [117].

3.2.1 Materials Used and Films Processing

The composite films were made with shrimp chitosan (CS) and chitin nanofibrils (CNs) which were supplied respectively from Biolog Heppe GmbH in Germany and from Mavi Sud s.r.l. in Italy. Apart from that, chitosan has a molecular mass of 1.64 $\times$ 10^5–2 $\times$ 10^5 and a deacetylation degree of 92.4%. Chitin nanofibril concentration in the aqueous system was 20 mg/mL [118]. The composites were then processed when all of the necessary materials had been supplied. The processes that consist of dry casting of polymer solution via a slit die onto a glass substrate and drying at room temperature for 24 h were used to make composite films from composite CS/CNs solutions in 2% acetic acid solution. Then, the samples of chitosan/chitin nanofibril were divided with ratios of 100/0, 99.5/0.5, 99/1, 95/5, 90/10, 70/30, and 0/100, respectively. The composite films according to chitosan-chitin nanofibrils ratios are shown in Table 2. Polymers made about 4% of the total concentration in the solutions

Table 2 The ratios of Chitosan-Chitin nanofibrils in composite films [117]

Sample	Percentage of components		Coefficient for thermal sorption $\alpha \cdot 10^3$, K^{-1}	Constants used in thermal equation of sorption (TES)		Heat of sorption integral, q_i, J/g
	Chitosan (CS)	Chitin Nanofibrils (CNs)		a_0, g/g	E, J/ mol	
1	100	0	−2.8	0.95	276	164
2	99.5	0.5	−2.8	0.90	273	153
3	99	1	−2.8	0.90	264	147
4	95	5	−2.9	0.89	257	142
5	90	10	−2.9	0.86	267	143
6	70	30	−2.9	0.74	263	122
7	0	100	−3.2	0.46	248	75

where it is the ideal concentration for chitosan fibres produced from wet spinning processes [115, 119]. The produced films were exposed for 10 min to a 10% aqueous solution comprising sodium hydroxide and ethanol (1:1), then rinsed and dried in air [118].

3.2.2 Isotherms of Water Vapor Sorption by Composite Films

Water vapour sorption isotherms for chitosan films and composite combined chitin nanofibrils films are shown in Fig. 5. The forms of polymeric sorbents that expand in sorbate vapours are typical of all isotherms. In the range of P/P_0 ~0.4–0.6, it is detected that rising of the isotherm curve for all of the films examined, signalling the beginning of the transition of the polymers' amorphous domains into a state of high elasticity. The water sorption capacity in chitosan is the highest; however, when the number of chitins nanofibrils in a film grows, the sorption value decreases. Chitin is more crystalline than chitosan, which explains the difference [118, 120, 121]. The equilibrium phase for water vapours sorption by the examined films was explored using the thermal equation of sorption (TES) [122–124].

3.2.3 Thermodynamics of the Chitosan and Chitin Nanofibrils Composite Films to the Water System

As chitin concentrations varied from 0.5 to 30%, the chitosan and chitin combination is thermodynamically stable. In composite films, chitosan and chitin produce ordered structures and the macromolecules are packed more densely [118, 125]. The Gibbs energies of mixing, entropies, and enthalpies of mixing pass through minimums in the range of chitin concentrations from 1 to 5%. The greatest interaction between chitosan

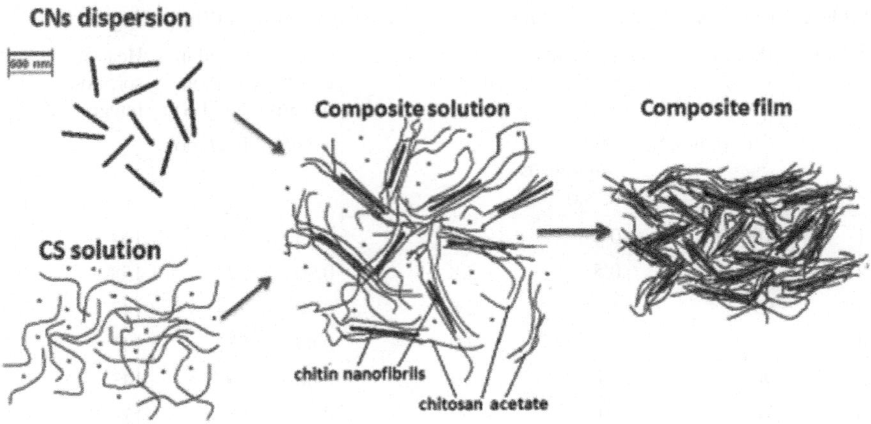

Fig. 6 Formation of supramolecular structure of CS-CNs composite film in the process of film preparation [125]

and chitin was found thermodynamically in this area or range; macromolecules form high ordered structures, and they are grouped together in the most ordered structural formations, which fits with the findings described in [117]. After a significant rise in chitin content (up to more than 5% by weight), the system may split into two phases, with nanofibrils merging into bigger structures. CNs have a large specific surface area and can effectively absorb chitosan from water. As a result, the self-formation of chitosan molecules on the chitin surface, as well as the production of a liquid-crystalline mesophase, as illustrated in Fig. 6 [126], is included, which is unsuitable for water during the absorption of water vapours by the composite material [127]. Diffraction patterns of chitosan films change as chitin nanofibrils are added, and the degree of crystallinity increases [125]. Figure 6 displays a model of supramolecular structure creation in composite films.

3.3 Chitin and Chitosan Composites as Adsorbents

Chitin and chitosan, and their modified composite's structure were being extensively researched as adsorbents to treat wastewater. The increased hydrophilicity conferred by the presence of -OH groups of glucose units might account for chitosan's great capacity for dye adsorption. They are widely used in the fabrication of different adsorbent materials, including as films, membranes, aerogels, hydrogels, composites, and so on, due to the availability of multiple functional groups with high reactivity and the flexibility of the polymer chain [128].

3.3.1 Chitin Based Adsorbent

Due to the presence of reactive $-NH_3$ and –OH functional groups on its structure, the use of second principally available and regenerable polysaccharide biopolymers which is chitin has attracted significant interest from researchers as the importance of green chemistry's growth. Chitin's polysaccharide composition also allows for a variety of changes. In contrast, the dye removal process causes rapid fragmentation of the biopolymer carriers and weak mechano-chemical activity, making it a less recognised material in pollutants removal. Combining them with various types of biopolymers, artificial polymers, carbon-based materials, organic and inorganics, and other materials might improve their shape and adsorption capability. For the creation of chitin blends and composites, many processes such as simple mixing, coprecipitation, spray and freeze spray drying, lyophilization, melt extrusion, solution/solvent casting phase inversion process, electrospinning, and polymerization are used [129]. As a result, various chitin-based adsorbents for the adsorption of dye molecules from contaminated water have been discussed, including chitin pristine adsorbents, chitin nanowhiskers adsorbents, graphene oxide enhanced chitin adsorbents, metal ions attached chitin adsorbents, clay integrating chitin adsorbents and natural/synthetic polymer-based chitin adsorbent [128].

3.3.2 Chitosan Based Adsorbent

Due to the simultaneous existence of reactive functional –OH and $-NH_2$ groups, chitosan can act as a bio-adsorbent for the elimination of cationic and/or anionic dyes [130], which can provide dynamic altering sites for metals and metal oxides. With contaminants, chitosan also may create intermolecular hydrogen bonds [131, 132]. The presence of a significant number of main amino groups accounts for most chitosan's unique features. It also has antibacterial and antifungal properties, may be used to separate proteins, and can bind with a variety of transition metal ions. The amino group, in combination with the hydroxyl group, makes chitosan a highly reactive polysaccharide with high adsorption ability [133]. However, because of its poor mechano-chemical behaviour and the quick fragmentation of biopolymer carriers during the adsorption process, it is less commonly used in environmental remediation. The chitosan biopolymer's mechano-chemical alterations give exceptional structural features as well as unique physicochemical behaviour and applications. Various chitosan blends and composites for textile dye detoxification have been discussed in detail, including chitosan hydrogel adsorbents, graphene oxide tethered chitosan adsorbents, magnetic chitosan adsorbents, clay incorporated chitosan adsorbents, carbon and bio-based chitosan adsorbents, organic moiety doped chitosan adsorbents, inorganics embedded chitosan adsorbents, and polymeric chitosan adsorbents [128].

3.3.3 Regeneration, Desorption and Reusability of Adsorbents

Adsorbents' ability to regenerate and reuse themselves is a key factor in their practical uses. The prospective adsorbent should be able to regenerate quickly and be reused several times while maintaining the same degree of performance [111]. The fatigued adsorbent is cured with a specific solvent, solution, or a mixture of both to break the bond between the adsorbent and the adsorbate and release the adsorbate from the adsorbent surface, a process known as desorption [134]. This has the advantages of lowering total operating costs, recovering adsorbate molecules, and reducing solid by-product waste generation [135]. There are two types of desorption processes; (1) thermal desorption, in which desorption is accomplished by heating. However, the technique necessitated costly pre-treatment, perception issues, the release of hazardous chemicals, unsafe process parameters, and other factors, all of which rendered the approach ecologically unsustainable. (2) Chemical or solvent desorption, in which desorption is done by chemical reactions. Various eluents were used as desorbing agents in this approach, and the process was cost-effective, ecologically friendly, reasonably quick, safe energy, no adsorbent waste, and allowed the agents and adsorbates to be recovered [136]. For the regeneration of the different functionalized adsorbents from the reaction mixture, several desorbing eluents/solvents such as NaOH, KOH, $NaNO_3$, NaH_2PO_4, HCl, HNO_3, H_2SO_4, EDTA, ADPT, and others have been used [134, 137].

3.3.4 Adsorption Application of Chitosan/Chitin-Based Composites

Adsorption is a filter process that involves the deposition of a fluid adsorbate on a solid adsorbent's surface and in the inter pores [136]. This method has been found as an efficient, straightforward, affordable, and environmentally friendly method for treating wastewater [138]. Adsorption performance is mostly determined by the type of adsorbent used and the adsorption circumstances (e.g., temperature, time, pH, concentration, etc.). Therefore, various types of pollutants will be discussed according to the application of chitosan/chitin-based composites adsorbents.

Heavy Metals

Because of their indestructible and poisonous characteristics, heavy metals are considered hazardous pollutants. Batteries, mining, fertiliser, and painting industries produce these pollutants, which can be detected in their wastewaters [139]. Copper, chromium, cadmium, and lead are the most often measured heavy metal ions, according to the data [73]. This might be due to the significant advantages of recovering these metals and avoiding their high harmful levels in water once they are present [86]. High levels of copper, for example, can harm human organs and cause cancer. In addition, cadmium has a negative effect on the human liver. Other than that, chromium is extremely dangerous to humans, because of its carcinogenic nature, causing cancer and even worse which is death [85]. As a result, several investigations have focused on metal ion adsorption on chitin and chitosan introduced carbonaceous material composites [73].

Synthetic Dyes

Because of their widespread use and manufacturing, dyes are typical organic contaminants [95]. Because a lot of molecules in dye are complex and indestructible, they can resist the sunlight exposure through water and have an impact on aquatic systems. Furthermore, dyes are hazardous to people and other living things [86]. Anionic dyes (acid, direct, and reactive dyes), cationic dyes (basic dyes), and non-ionic dyes (dispersed dyes) are the three types of dyes [140]. Methylene blue (MB) is highly investigated dye, according to the statistics [73], due to its severe toxicity and significant colouring impacts on marine systems [95]. MB has the potential to harm the skin, eyes, and brain [86]. Organs such as the kidney, liver, spleen, lung, and eyes are highly sensitive to malachite green (MG). Crystal violet (CV) dye consumption produces a variety of health concerns, including tissue necrosis, skin problems, jaundice, and vomiting. Congo red (CR) can induce DNA mutations in ecological creatures [141]. As a result, various investigations have looked towards removing these dyes with chitin and chitosan-based composites [73].

Other Pollutants

Adsorption of additional pollutants like medicines, phenolic, herbicides, nitrate, and phosphate must also be addressed as other pollutants besides heavy metals and dyes [73]. Medicines were the subject of the most of published research because of their widespread use by people and animals, continual release by pharmaceutical industries, stability, and harmful impact towards environment [142]. Antibiotics like ciprofloxacin and tetracycline, which are commonly used in bacterial infections' treatment, are the most commonly tested medications. Furthermore, these antibiotics' involvement in water might lead to the development of resistant bacteria, posing a health risk for humans and animals [143]. Petrochemicals, plastics, insecticides, leathers, resins, and other industrial wastewaters included in phenolic pollutants. Even little concentrations of these pollutants in water can have an impact on aquatic creatures. They also cause cancer, jaundice, skin illness, and even worse can cause death to human [144]. Weed management using phenylurea herbicides like monuron, linuron, and isoproturon has a deleterious impact on agricultural crop output. These herbicides are dangerous and can cause cancer [145]. Nitrate and phosphate in water encourages aquatic plants and creatures to flourish excessively. As a result, the oxygen concentration of waterfalls, resulting in algae blooms, which has a deleterious influence on marine life. Hence, chitosan/chitin-based composites have been used widely to treat these contaminants or pollutants [73].

4 Chitin and Chitosan Composites in Packaging

4.1 Application of Chitosan as a Coating

Coatings are described as lucid covers generated by the application of covering stuff to a substrate and could be applied straight to the top of meals as coatings that can be eaten or to the surface of wrapping materials to make them function [146, 147].

Chitosan has been extensively investigated in terms of edible coatings to improve the life cycle of food goods, particularly vegetables and fruits. There are some good evaluations on this topic [148–151]. Coatings are putted and formed right onto food products using a liquid film dispersion (applied using paintbrush, fluidized, sprayed, or dipped) or melted mixture [152]. Coatings can help improve the quality of fruits and vegetables by delaying ripening and water loss, as well as cease lipid oxidation and discoloration, and acting as a carrier of food enhancer [9]. They can also aid in the improvement of meat quality by delaying moisture loss, improving product appearance, minimising lipid oxidation and staining, and serving as a channel for food additives. Furthermore, chitosan coatings have high carbon dioxide and oxygen blockade abilities [153], and its inherent antibacterial qualities can inhibit the growth of microorganisms, increasing the shelf life of the coated food synergistically [152–154].

Preservatives embedded in polymeric matrices are a major trend in coverings, with the goal of increasing shelf life and maintaining food. This form of skim (antimicrobial and/or antioxidant coatings) act as replacement for traditional food coatings, which only protect against water damage or loss [155], and the active compound in chitosan coatings may improve the polysaccharide's essential antimicrobial properties, thus also increasing its protectant ability. Natural active compounds have lately become a big step toward more ecologically friendly packaging by being included into biodegradable films or edible coatings [156]. Active ingredients, likes natural products from plants high in phenolic compounds, can improve chitosan's antibacterial and antioxidant capabilities, enhancing the coating's preservation effects and potential to extend the durability of foods [157–159]. These type of covering also appeals to the increasing number of consumers looking for ecologically friendly packaging options.

Nanocoatings, which are made up of ultrathin nanoscale films (less than 100 nm) grown up on facets, are one of the new concepts being studied as nanotechnology advances. This form of plating has the superiority of not changing the material's surface shape while providing chemical and physical purpose, such as changing gases fencing qualities, facet hydrophobicity, to mention a few [146, 159].

Altering the surface of wrapping materials can be accomplished using a variety of methods and processes, which are split into two categories based on the function of the material to be developed: migratory and nonmigratory active packaging. The former can be accomplished through embedding, noncovalent immobilisation, or layer-by-layer toppling, whereas the closing can be accomplished through photografting or covalent immobilisation [146].

The functioning process of the polymeric substrate is the first step in the covalent implanting of active compounds onto chemically inactive polymers. This step can be accomplished using ecofriendly (free from solvent) methods like ionisation of gamma radiation or chilly plasma gas emit; after that, the oxygen-containing groups with enhanced surface is set to combine with the activated hybrids, which can be added by electrospinning, immersion, spreading procedure [160–162].

Immersion was the most efficient approach for depositing chitosan into an active exterior of polylactic acid (PLA) in terms of veneer uniformity and broadness of the sheet set, but electrospraying was the most adaptable [162]. Furthermore, coupling agents (e.g., ethyl-3[3-dimethylaminopropyl] carbodiimide hydrochloride or 1carbonyldiimidazole, *N*-hydroxysuccinimide) can be utilised in the procedure [146, 163, 164]. Chitosan that is attached onto a PLA surface improves the polymer's antifungal and antibacterial characteristics while also providing antioxidant capabilities to the packaging [162]. When high molecular weight chitosan was electrospun, nanofibers formed on the polymer surface, conferring antioxidant action and bring down material demand; however, the dipping approach produced higher antibacterial activity and a more homogenous surface. When contrasted to specimens bundled in immaculate PLA or infomercial plastic content that is polyethylene terephthalate (PET), which displayed variations after 2 h or 2 h, respectively, covered PLA with chitosan retained the physical features and attributes of apple juice for a prolonged set of time (no changes in colour after 48 h storage) [162]. Chitosan increased the antibacterial ability of polyethylene (PE) when layered solely [164] or in conjunction with vitamin E [163], suggesting that it might be used in food packaging or medicine.

4.2 *Application of Chitosan Films for Food Packaging*

4.2.1 Blends and Bilayers of Chitosan and Other Biopolymers

Single chitosan, compared to mixture of chitosan along other polysaccharides, such as pectin, alginate or starch, [165–167], microbial polysaccharides [168, 169], and proteins, likes gelatin [170] and whey proteins [171, 172], have shown enhancement in terms of water solubility and permeability, and better mechanical properties and better performance. Electrostatic forces in between charged -ve side chain class of the other biopolymer and the protonated amino class of chitosan and the generated polyelectrolyte complexes at the working pH. Some writers found issues with the full solubility of one of the polymers in defined conditions, as well as the production of solid phases between polymers, while creating blends [171].

4.2.2 Nanocomposites

One method for dealing with the hydrophilic character of chitosan's inherent flaws, such as low waterproof, poor barrier and automatic properties, is to use nanoscale boosting;(for example metal oxide nanoparticles, nanocellulose, montmorillonite) in chitosan strips, which able to engage physically or chemically with the polymeric chain [158, 173].

Montmorillonite

Montmorillonite or as known as MMT is as arranged in layer silicate clay of mineral found in volcanic rocks (bentonites) and is evaluated as bioplastic reinforcing substance owing to their extensive accessibility, stretching capability, durability in mechanical, and not costly, to name several advantages [174].

Few research employing nanocomposites based on MMT, and chitosan have just been undertaken, and generally, mechanical and barrier characteristics are enhanced when added with MMT in the chitosan layer film [175–179]. The use of nanoclay in biobased films has been discovered to increase not only rigidity, stiffness, and modulus of rupture, added with the water and oxygen blocker [175, 178, 180, 181] found that the chitosan layer having Montmorillonite had a fantastic light block ability, specially at UV wavelengths, functioning very well to fight oxidation from occurring. In addition, these nanocomposites have been shown to have higher antibacterial activity [178, 180, 181]. They were influenced by this and measured the nanocomposites in an able to compose matrix's food, showing their ability as main packing material, able to postpone deterioration to expand shelf life.

Cellulosic Nanofibers and Nanocrystals

Nanosize Fibre cellulose, including such cellulose nanocrystals (CNCs) and cellulose nanofibers (CNFs), are a tempting chitosan's reinforcement to fabricate one composite films of environmentally friendly with enhanced substantial characteristics because of its high liberality with chitosan. Because of hydrogen bonding and electrostatic association, increased interaction between huge length diameter ratios and chitosan particles nanocellulose leads in the creation of a communal network shape, which promotes the crystalline structure of the films [182, 183]. Therefore, chitosan/nanocellulose composites can be used in a variety of water treatment, biomedical and packaging applications [184, 185]. In two separate trials, cellulose nanocrystalline was assigned as a fortify element in composites of chitosan-guar gum [186], starch-chitosan composites [187], and gelatin-chitosan composites [187]. In all experiments, they created a clear, high thermal stability nanocomposite biopolymer-base with improved barrier and mechanical properties. This novel form of harmless, recyclable, safe as well as biodegradable chitosan/nanocellulose films might one day substitute petroleum-based polymers as a food packaging material.

Metal Oxides

ZnO, SiO_2, TiO_2, and MgO nanoscale metal oxides give benefit to chitosan by integrating antibacterial, UV shielding, and magnetic qualities with strengthening ability [188–190]. Zinc oxide (ZnO) is the most extensively used metal oxides in a number of fields due to its antibacterial and photocatalyst properties. Noorbakhsh-Soltani et al. [189] found that ZnO nanoparticles are considered safe for humans and have been used as preservatives, packaging materials, and water treatments, like other metal oxide nanoparticles. From researched by Youssef et al. (2015) [188], ZnO coated film nanoparticles showed antibacterial activity against dangerous food microorganisms. [189] found that HDPE films that are coated with chitosan/ZnO coating gave excellent antimicrobial resilience towards *S. enterica, E. coli*, and *S. aureus*, with pathogen development completely halted after 24 h of incubation. Titanium dioxide (TiO_2) is a prominent artificial nanomaterial in the energy and environmental fields due to its cheap, excellent photocatalytic efficiency, chemically stable, and biocompatible [190]. It has been demonstrated that adding TiO_2 nanopowder to chitosan-based nanocomposite films improves mechanical properties [191, 192]. Youssef et al. [190] created a chitosan/TiO2 film that showed efficient antibacterial fighting against four pathogens, including *Candida albicans*, *Aspergillus Niger*, *E. coli* and *Staphylococcus aureus*. It also allowed cellular substances to leak via damaged membranes. Furthermore, when compared to pure chitosan, [192] discovered chitosan nanocomposites that contain 5 (w/w percent) MgO increased their tensile strength and elastic modulus by 86% and 38%, respectively. The UV protection and absorbent properties of chitosan nanocomposites incorporating MgO nanoparticles were very impressive. Consequently, the chitosan/metal oxides nanocomposite films created have the prospective as a food packaging substance due to their increased physicochemical qualities.

4.2.3 Active Films of Chitosan

The development of oxidation and microbiological are two process that cause the quality of food to deteriorate, resulting in significant changes such as nutritional value loss, texture changes, and the formation of unwanted compounds like off flavors, coloured compounds, and even toxicated substances for human consumptions [179]. As a result, active packaging holds a crucial part in the food sector, minimizing waste and extending the shelf life of products [179]. Bioactive substances such as antibacterial and antioxidant agents, moisture absorbents, nutraceutical compounds and gas scavengers can be added to chitosan films to maximize their inherent properties. Due to consumer concerns about health, current active packaging research is focusing on developing natural preservative systems that contains antioxidant or antimicrobial characteristics and can be used alternatively to artificial additives and chemical preservatives, including those based on essential oils, fruit and plant extracts, nisin and lysozyme [193, 194]. However, due to the expensive cost of application and other limitations such as their strong odor and possible toxicity, the application of natural substances in food preservation is often restricted [195]. As a result, designing contactless active packages between the drug and the food is an incredible opportunity, with benefits such as no flavour transfer, minimal organoleptic alterations, and even distribution of active compounds in the headspace [179].

The active bio compounds identified that confer improved antibacterial and antioxidants powers to films that are edible has been the focus on the subject of chitosan active packaging, scientific research is being conducted [194, 196]. In addition, a more published study to see the impact of these additional natural compounds to the mechanical properties of film [197]. The films recently have been exposed to food matrices variant aiming to investigate their impact on the food's organoleptic qualities as time passes [158, 179]. Lekjing, [198] studied the effects of chitosan/clove oil on the quality and shelf life of cooked pork sausages, finding that the sum of these two ingredients inhibited delayed lipid oxidation, microbial growth, and extended the shelf life of pork sausages that was cooked for more than a week. However, at the outset of the storage period, there were some detrimental effects on odour and taste qualities. Supplemented ginger and essence oil base rosemary could reduce the oxidative processes of poultry meat in similar studies [158, 179] showed that chitosan films incorporating rosemary exhibited good antimicrobial reaction towards Bacillus cereus and Salmonella enterica in in vitro studies. In general, bioactive compounds integrated in chitosan films have shown tremendous potential in lengthening the life expectancy and keeping the standard of food products, as well as reducing postharvest fungus and foodborne bacteria in the food system. In order to get the more efficient bioactive integrated agents, interactions between chitosan towards bioactive chemicals must be understood by doing the extra work. Furthermore, different from compression moulding or extrusion, where the material is subjected to high temperature, the majority of the research employs the casting process for polymer manufacturing, which is a technique not widely adaptable by the packaging sector.

Adding a variety of lipid components to films, such as fatty acids, vegetable oils, natural waxes and resins, increases hydrophobicity and reduces moisture [199]. Water solubility was reduced in chitosan films containing beeswax [200], and permeability of steam was reduced in films containing oleic acid [201], neem oil [202], and essence oil based cinnamon [203] among others.

Chitosan's intrinsic reactive groups, OH and NH_2, allow it to be chemically modified, expanding its application possibilities. The amine functionalization of chitosan's groups using carbonyl compounds results in chitosan-based Schiff bases that are useful for several food packing products. Antibacterial ability of chitosan-based Schiff bases has been established in the form of powders, whiskers, films, and membranes, with interestingly stronger antibacterial activities than bare chitosan [204]. Furthermore, loading metal nanoparticles or metal ions through the covalent coordination link can improve the antibacterial activity of chitosan-based Schiff bases. Some chitosan-based Schiff bases have enhanced the functional qualities of bare chitosan by demonstrating antioxidant capabilities, in this way (Fig. 7). The Schiff bases formed by reacting D-fructose, quercetin quinones, eugenol aldehyde, or carvacrol aldehyde with chitosan [205].

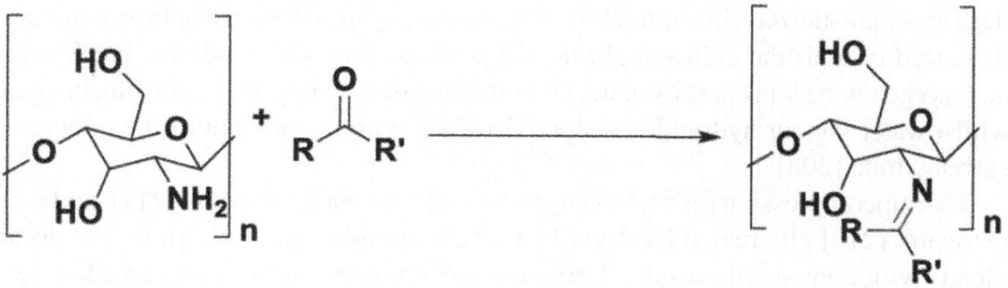

Fig. 7 Imine functionalization for chemical modification of chitosan

4.3 Production

Producers and manufacturers of chitosan and chitin may be found globally. Over the last few decades, chitosan coatings and films have been widely researched, however the majority of information available is restricted to laboratory-scale synthesis using casting operations. The casting method entails diluting the polysaccharide in a specific solvent, which in the case of chitosan is typically an acetic acid solution, and then combining the active compound, the plasticizer, and the desired nanofiller simultaneously on an inert surface to evaporate the solvent and obtain the desired film. As a result, one of the challenges in utilising chitosan is scaling up this laboratory-size procedure to an industrial scale or developing other manufacturing processes to replace the casting methodology.

Plasticizers added (such as glycerol) to chitosan layers before smearing a thermo-mechanical operation (mechanical moulding) resulted in a thermo-plastic substance with acceptable automatic characteristics [205]. This thermomechanical plasticization technique to chitosan film manufacturing could be a better replacement to the old casting method, enabling for larger-scale production of these biodegradable films [206].

4.3.1 Thermoplastic Chitosan Films

Using the thermal–mechanical kneading process, several plasticizers were evaluated in the production of chitosan film [207]. In the presence of water, acetic acid, and the polyol under examination, a variety of non-volatile polyol plasticizers (glycerol, xylitol, and sorbitol) were thermomechanically processed in an internal mixer. Sorbitol (the highest molecular weight polyol tested) produced the best plasticized chitosan with the best thermal, mechanical, and rheological properties, whereas glycerol (the lowest molecular weight polyol tested) produced the worst thermal, mechanical, and rheological properties but had the highest amorphous phase content, which made it easier to process despite its poor properties [207].

Chitosan was recently plasticized in the addition of glycerol and acetic acid solution, and afterwards blended with polyethylene to develop blends with various degrees of plasticized chitosan [208]. The resulting films developed a brown hue and increased haze as the chitosan plasticized content increased, while the mechanical and oxygen barrier properties of the polyethylene films remained largely unchanged, whilst water vapour hydraulic conductivity risen with the amount of the integrated carbohydrate [208].

Thermocompression [208], blown extrusion [209], melt extrusion [211], and melt extrusion [210] all created biodegradable thermoplastic starch-plasticized chitosan blends with comparable results. Extrusion technologies, in this case, enable large-scale manufacturing of plasticized chitosan-based goods, thereby solving the scaleup barrier of chitosan film production.

4.3.2 Deep Eutectic Solvents (DES)-Novel Green Solvents as Plasticizers for Thermomechanical Treatment of Chitosan

DES can also be used to process chitosan. [212] used chitosan 90% deacetylation degree, choline chloride (CC), and citric acid (CA) to make thermocompression moulded films (molar ratio 1:1). The addition of CC and citric acid to chitosan was done individually (not in a liquid combination), and the three-component system was heated to 70 °C for 30 min before adding a 3% acetic acid solution and hot pressing the paste at 120 °C. The water sorption ability of chitosan/choline chloride/CA films was higher than that of chitosan/citric acid films. Additionally, including Choline Chloride through into the chitosan/CA matrix caused a reduction in tensile strength and a little increase in break elongation.

Natural deep eutectic solvents (NADES) generated out of affordable basic substances were used to make thermoplastic chitosan films [212]. Four distinct NADESs based on choline chloride were synthesised using malic acid (MA), lactic acid (LA), citric acid, and glycerol as hydrogen bond donors, while two different chitosan with varying degrees of deacetylation (DD = 76 and 81) served as the polymeric matrix.

Thermocompression moulding was used to create transparent thin chitosan films, and the film properties (mechanical and water resilience) varied depending on the proportion/form. The use of chitosan with less DD and the NADES Choline Chloride/CA and CC/MA resulted in a more comparable surface, fill with less water adsorption, and firmer resistance, but the use of CC/glycerol resulted in a material with worse qualities [212]. Therefore, DES and NADES are compatible green solution materials for use as plasticizers in large-scale chitosan Thermo compressor films with customized characterization.

5 Chitin and Chitosan Composites in Biomedical

Composite materials and applications of chitin and chitosan in applications of biomedical have received considerable attention due to their low sensitivity to foreign

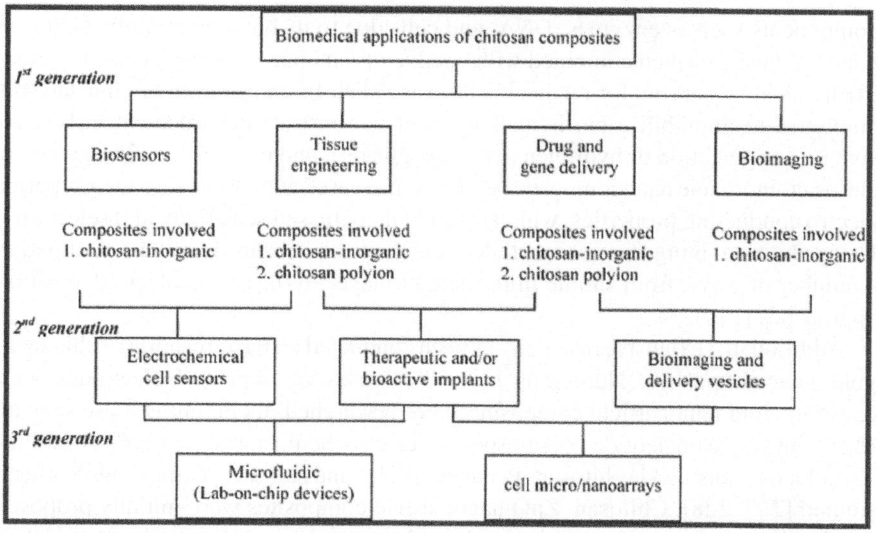

Fig. 8 The evolution of chitosan composites biomedical applications [219]

bodies, intrinsic antibacterial properties, biocompatibility, and biodegradability, as well as their capability to be sculpted into a range of geometries and forms, such as porous structures suited for cell ingrowth and osteoconduction [213]. Due to its biodegradability and biocompatibility, the composite of chitosan including hydroxyapatite is quite popular [214]. Numerous composites with diverse biological functions have been created, including chitosan with alginate, collagen, calcium phosphate, hydroxyapatite, and polysulfone [109, 215–218]. The evolution of chitosan composites in biomedical applications is summarised in Fig. 8.

5.1 Chitosan–Inorganic Materials Composites

One of the most extensively researched chitosan-inorganic materials are various calcium compounds [220]. Due to calcium compounds' inherent brittleness, which results in initial cracking and exfoliation of orthopaedic implants, multiple studies were conducted to develop polymer composites that improve mechanical performance or biocompatibility [221]. Chitosan is a polymer which has garnered interest due to its osteoconductive properties. However, the polymer's lack of adequate mechanical strength precludes it from becoming an implant. As a result, calcium carbonate and calcium phosphate are combined using chitosan to increase mechanical strength, resulting in loads of particle bioresorbable composites for implantation of orthopaedics [219].

A promising use for chitosan–inorganic nanocomposites as a modification of electrodes exists in the field of biosensors. Chitosan is capable of immobilising biological components such as enzymes, DNA, and cells due to its biocompatibility. Chitosan nanocomposites, when combined with conductor nanoparticles, can be used as electrochemical biosensors for medical diagnosis [219]. Initially, chitin or chitosan was employed to immobilise bio ligands in order to construct electrochemical biosensors for the detection of hydrogen peroxide, glucose, and urea. [222–224]. Later on, chitosan–inorganic nanocomposites garnered increased attention due to their superior electro conducting properties, which resulted in increased sensitivity and selectivity. Using chitosan–inorganic nanoparticles composite, the electrode may be modified in a number of ways, from simple film casting to layer-by-layer and electrodeposition approaches [219].

Additionally, extensive research has been conducted on the utilisation of chitosan–gold nanocomposites. Numerous biomedical uses of improved electrodes with chitosan–gold nanoparticle composites were researched, including glucose sensing [225] and single nucleotide polymorphisms electrochemical coding [226]. Additionally, glucose sensing via chitosan–Pt nanoparticles and nanowire composites has been studied [227, 228]. Chitosan–ZnO nanoparticle composites were initially proposed for the detection of hydrogen peroxide and were subsequently modified to detect human IgG. [229, 230]. The glucose and DNA hybridisation detection capabilities of chitosan–ZrO_2 nanocomposites have been explored [231, 232]

Responsiveness and specificity are dependent on the nanomaterial and form of chitosan used. However, by adding a third component, a carbon nanotube, the specificity of these chitosan–inorganic nanocomposite biosensors can be increased. Because the electrode is implemented at a poor applied potential, it reduces the amount of interference caused by other electroactive species detected in bio samples such as ascorbic acid, uric acid, acetaminophen, and many more. Additionally, carbon nanotubes can enhance direct transfer of electrons across electrode layers and redox enzymes, allowing for the identification of redox reactions without the need of reagents [233]. Chitosan, as a cationic polyelectrolyte, disperses carbon nanotubes as a polymeric surfactant, thereby preserving a homogeneous layer of carbon nanotubes on the electrode surface [219]. As a result, the extra component in the chitosan–metal nanocomposites do not impair the precision of the study. Apart from enzymes, antibodies, and DNA, these composite sensors' chitosan matrix can also be used to immobilise living cells for the purpose of detecting cell adhesion [234].

5.2 *Chitosan Polyion Complex Composites*

5.2.1 Chitosan-Protein Composites

Collagen and gelatin are the most studied proteins in conjunction with chitosan. These scaffolds were investigated in a variety of tissue engineering applications such as bone, skin grafting, and nerve regeneration, cartilage repair, and many more. This is because cell adhesion to chitosan alone is inadequate for enhancing early cell attachment and growth, much also if chitosan is employed as a film instead of a three-dimensional porous scaffold [219]. For example, human osteosarcoma (SaOs-2) cells and 3T3 fibroblasts are incapable of adhering to and developing on chitosan membranes [235, 236]. However, they develop nicely on 3D porous chitosan scaffolds [237, 238]. Due to the fact that chitosan enhances angiogenesis and osteogenesis, numerous chitosan–protein composite scaffolds were investigated in nerve regeneration, wound healing, cartilage repair and bone [219].

5.2.2 Chitosan–Glycosaminoglycan Composites

Additionally, chitosan forms biocomposites with a variety of glycosaminoglycans, including heparin, chondroitin sulphate, and hyaluronan. Due to chitosan's positive charge, it reacts to blood proteins, forming thrombus on chitosan layers. As a result, once chitosan is used as a blood contact material, it must be made hem compatible. Sulphation or acylation of chitosan can be used to accomplish this [219]. The other glycosaminoglycan which is chondroitin sulphate, is especially appealing since it improves articular purpose as well as alleviates ache in individuals with arthritis and biodegradable in the large intestine [239, 240]. Chitosan combines with chondroitin sulphate forming a polyion complex that encapsulates the medication for targeted delivery to the colon [241, 242]. Chitosan–glycosaminoglycan–protein composites were also frequently used in the construction of scaffolds such as chitosan–gelatin–hyaluronic acid for the formation of artificial skin [243].

Same as chitosan–protein composites, the composites of chitosan–glycosaminoglycan may improve pathogen adherence as well. The pathogenic fungus, Penicillium marneffei, was known to bind with highly sulphated chitosan and chondroitin sulphate B, heparin, but never to low sulphated chitosan [244]. Heparin–plasma also suppresses human cathelicidin LL-37 peptide's inherent antibacterial activity. Combining chitosan to glycosaminoglycans, on the other hand, improves an antibacterial activity of endogenous antimicrobial peptides in biological liquid that include glycosaminoglycans [245]. These results demonstrate that the inclusion of glycosaminoglycan may jeopardise chitosan's antibacterial action unless another antimicrobial ingredient is included [219].

5.3 Antibacterial Activity in Chitosan

Chitosan is composed of an amorphous copolymer of units of D-glucosamine and *N*-acetyl-D-glucosamine. The two monomeric units which have various ratios result in varying degrees of deacetylation, acetylation patterns, and chitosan molecular weight, all of which have a significant impact on chitosan's antibacterial properties [246, 247]. Due to chitosan's solubility, its antibacterial effectiveness is severely limited. At the moment, chemical modification and the mixture of chitosan with additional antibacterial components are the most often employed ways for improving chitosan's antibacterial activity. Alternatively, it is an effective way for chemically modifying chitosan to increase its solubility and widen its antibacterial spectrum [248]. At the moment, the primary chemical modifications of chitosan are quaternary ammonium salinization, carboxylation, sulfonation, and phosphorylation [248]. In contrast, chitosan can be mixed with other antibacterial materials to generate composites that enhance chitosan's antibacterial action. Other antibacterial materials include, but are not limited to, metal and metal oxide [249].

5.3.1 Wound Dressing

Recently, significant research is being performed to develop novel antibacterial medicines for the treatment of injuries infected with antibacterial resistant pathogens. For a long period of time, silver was seen as an antibacterial agent in the presence of silver sulphadiazine ointments and metal silver. Silver nanoparticles were recognised as a very powerful antibacterial catalyst and were being employed in several therapeutic uses, from silver-based dressings to the equipment of medical silver-coated [250]. Besides, treating patients with deep burns, and wounds are the usage of chitosan. For instance, different α-chitin/nanosilver composite scaffolds have created for use in wound healing [251]. The antibacterial activity of these α-chitin/nanosilver composite scaffolds against *S. aureus* and *E. coli* was shown to be excellent, as it possesses a good blood clotting ability.

Rai et al. [250] described silver nanoparticles as having the following mode of action [250]. Due to their exceptionally huge surface area, silver nanoparticles have superior antibacterial activity compared to other salts. The nanoparticles attach to the cell membrane and also enter the bacterium. Silver nanoparticles interact with both sulphur-containing proteins in the cell membrane and phosphorus-containing molecules such as DNA. When silver nanoparticles enter a bacterial cell, they form a low-molecular-weight zone in the organism's centre, where bacteria concentrate to shield the DNA from the silver ions. The nanoparticles mainly target the respiratory chain, causing cell division and ultimately cell death. The nanoparticles deliver silver ions into the bacteria, boosting the bacteria bactericidal action [250].

5.3.2 Factors Influencing the Antibacterial Activity of Chitosan

Chitosan Concentration

When chitosan's concentration is decreased, it binds to the negatively charged surface of bacteria cells, breaks the bacteria cells' membrane, and produces component leakage, ultimately resulting in the death of bacteria cells. When chitosan concentration is enhanced, chitosan that has been protonated can be wrapped around the bacteria surface to avoid leakage of the component, and positively charged bacteria resist one another to avoid agglutination [252]. Liu et al. [253] conducted research to evaluate the concentration of chitosan impact on antibacterial activity, and a few samples of chitosan were chosen [253]. The experimental findings indicated that all chitosan specimens (20 ppm) together with seven chitosan specimens (50 ppm) (A, B, C, D, E, F, and G) could stimulate *E. coli* growth. To generate an antibacterial low-density polyethylene/chitosan packaging material, 1%, 3%, and 5% of chitosan have been evenly integrated into a polyethylene matrix with a low density. Besides, research by Reesha et al. (2015), antimicrobial efficacy towards *E. coli* was demonstrated by the low-density polyethylene/chitosan film being superior to the virgin low density polyethylene film [254]. Tilapia was stored in cold storage using virgin low-density polyethylene and 1%, 3%, and 5% low density polyethylene/chitosan films. As a result of the effect, tilapia contained in virgin low density polyethylene film has been discarded on the seventh day, but tilapia that has been packed in 1%, 3%, or 5% low density polyethylene/chitosan films could be stored up to 15 days. The researchers stated that the 3% low density polyethylene/chitosan film outperformed the other films evaluated in terms of physical and antibacterial properties and the preservation quality of tilapia steaks can be enhanced when refrigerated.

Temperature

The temperature at which chitosan is stored has a significant effect on its efficacy, which is critical for industrial applications of chitosan. Not only does temperature have a significant effect on chitosan's antibacterial action, but the viscosity or molecular weight of chitosan may also alter with temperature [255]. Besides, temperature significantly affected chitosan's antibacterial action against *E. coli*, with bactericidal activity rising as the temperature climbed between 4 and 37 °C. At lower temperatures, between 4 and 15 °C, the number of *E. coli* declined dramatically in the first five hours and subsequently stabilised. This might be because low temperature affected the pace at which chitosan binds to cells and the number of accessible binding sites on the cell surface, which those two were modified by low temperature [256]. This is further supported when research conducted by Taha and Swailam [257] on chitosan's antimicrobial action against Aeromonas hydrophila came out with the same result with Tsai and Su [256, 257]. The antibacterial properties of water-soluble chitosan also had a significant impact on temperature which explained by the traits of physiological of the bacteria that has been tested or the bacterial-chitosan reaction kinetics; additionally, the antibacterial activity of water-soluble chitosan toward oral bacteria ideal temperature was 37 °C [258].

Degrees of Deacetylation of Chitosan

The degrees of deacetylation of chitosan have a significant effect on its antibacterial capabilities; the greater the degrees of deacetylation of chitosan, the greater the positive charge on chitosan following amino protonation [259]. For instance, when the degree of deacetylation of chitosan increased from 74 to 96%, its antibacterial efficacy against *E. coli* increased progressively [260]. Besides, chitosan with a degree of deacetylation of 95% had a greater antibacterial activity toward *E. coli* compared to chitosan with a degree of deacetylation of 75%, and higher antibacterial activity with the increasing of interaction duration [261].

5.4 Tissue Engineering

Tissue engineering is a critical approach for fixing or replacing biological tissues and organs that have sustained irreversible damage [262]. Because of its nontoxicity, biocompatibility, and biodegradability, chitosan is a substance with great potential for tissue engineering [263]. Additionally, chitosan is structurally similar to glycosaminoglycans, which constitute the majority of the extracellular matrix (ECM) [264] Chitin has been used to create hydrogels, fibrous scaffolds, and porous sponges, all of which have been seeded with the necessary cell types for in vitro or in vivo culture and evaluation [265]. Chitin, rather than chitosan, appears to be a more suitable matrix for regrowth of cartilage tissue, as evidenced by the literature [263]. Additionally, hybrid microspheres of chitin and poly leucine were synthesised by an interfacial polymerization approach based on the ring-opening polymerization of an alpha amino acid *N*-carboxyanhydride. These hybrid microspheres may find value in tissue engineering and drug delivery [266]. Tissue engineering is subdivided into various subfields according to the type of tissue or organ being generated.

5.4.1 Bone

The majority of chitin and chitosan composite materials have been used to construct bone tissue. Bone is made of both organic and inorganic components, the majority of which are collagen and hydroxyapatite. Chitin and chitosan have inherent poor mechanical properties. Thus, chitin can be employed as a bone substitute for bone repair and rebuilding only if its mechanical properties are enhanced with the addition of biomaterials such as hydroxyapatite (HAp), bioactive glass ceramic (BGC), and others [267]. BGC is a term that refers to a collection of osteoconductive silicate-based materials that are employed in bone healing. BGC are synthesised in a variety of ways. Melt quenching and sol–gel are two processes that are frequently utilised [262]. Numerous studies have shown that BGC has an effect on osteoblastic cell differentiation by increasing the levels of distinguishing markers such as ALP, osteocalcin, and osteopontin [268].

The lyophilization procedure was used to create chitin or chitosan/nBGC composite scaffolds. Once the nBGC were distributed equally throughout the adsorbent surface, the composite scaffolds displayed an acceptable degree of porosity. Apart from their potential to become bioactive, the generated nanocomposite scaffolds demonstrated appropriate swelling and degradation properties. The MTT assay, cell attachment and direct contact test, were the tests used to determine the cytocompatibility of the chitin-nBGC and chitosan-nBGC scaffolds [267]. Chitosan–gelatin composite scaffolds with nBGC have been created by combining gelatin with nBGC and the chitosan [269]. With the addition of nBGC, the degradation and swelling properties of the nanocomposite scaffolds decreased, but protein adsorption increased. Investigations into biomineralization demonstrated a considerable increase in mineral deposits on the nanoparticle composite of scaffold as the incubation period increased [267]. The MTT assay, cell adhesion and direct contact test, where the investigations that all demonstrated the nanocomposite scaffolds promote cell adhesion and spread more effectively. These nanocomposite scaffolds have demonstrated efficacy in regenerating alveolar bone [269]. These findings indicated that the composite scaffolds created could be employed in tissue engineering applications.

5.4.2 Tendon and Ligament

Tendons and ligaments are hypocellular tissues, and so their ability to repair following injury is contingent on their anatomical location, extent of tissue loss and vascularity [270, 271]. Autologous treatment of tendons or ligaments tissue engineering comprises construction of scaffold-based constructs that may offer the necessary support and strength that tendon tissue does. Only type I collagen deposition was observed in fibroblasts treated with chitosan/HA composite fibres, emphasising the critical nature of effective cell–scaffold (matrix) interaction in the tissue regeneration cascade [272]. Rotator cuff regeneration employing chitin fabric has revealed that Chitin non-woven fibres assisted in vivo, the regeneration of damaged tendons by the infiltration of cells [273]. Collagen deposition had also been boosted; however, scar tissue formation was seen [273]. High tensile strength is a critical requirement. The quick degradation of chitin in vivo, leads to considerable failure of the weekly regenerated tendon strength as documented in research. When the injection of a composite of chitin and a durable mechanical synthetic polymer; PCL occurs, tendon regeneration is supported [274].

Chitosan can operate as a scaffold for collagen deposition and regeneration of the matrix when fibronectin immobilisation is regulated, due to the fact that it would improve attachment of cells and proliferation, the matrix deposition increasing [275–277]. Chitosan coatings on PLA layers resulted in improved matrix formation. In comparison to PCL, chitosan did not assist cell adhesion or dispersion because of the poor adsorbent of fibre-secrementin. Nevertheless, the matrix setting was substantially boosted in the collagen matrix as it encouraged the gene expression of TGF-431 increasing. Tendon regeneration requires directed proliferation of tenocytes through the length of linked fibres, accompanied by collagen type I deposition.

The scaffold's architecture also has a significant impact. Linked, base etched poly-caprolactone-co-d,l-l-lactide (PCLDLLA) fibres were implanted in noncell-adherent pho-to crosslinked *N*-methacrylatedglycol chitosan (MGC) hydrogel. ECM matrix formation along the PCLDLLA fibres was confirmed by immunohistochemical staining for collagen I, III, decorin, and many more within four weeks. This reports a fresh technique for tissue engineering of ligament by enhancing the scaffold design [278]. Assisted renewal may be incorporated into the scaffold design by establishing alignment channels, either microchannels or aligned microfibrous surfaces.

5.4.3 Cartilage

Cartilage degeneration can be caused by genetic abnormalities, trauma, or disease, and may eventually necessitate surgical intervention, most commonly replacement surgery. Cartilage tissue engineering intends to create a completely healed, functional, and scar-free cartilage that will outperform currently existing therapy techniques for degenerative cartilage. If the cells can contact, mature, and maintain inside the matrix provided, a composite cell scaffold may be an effective tissue engineering structure [262]. Glycosaminoglycans (GAGs) promote cartilage regeneration by providing a stimulating environment. Chitosan induced chondrogenesis due to its structural resemblance to hyaluronic acid (HA) and GAGs found in particular cartilages [262].

Chitosan was frozen and lyophilized to create a microporous scaffold, and the scaffolds were seeded with porcine chondrocytes. While the scaffold kept a spherical form of auricular chondrocytes, the micropores prevented cell penetration. The deposition of the extracellular matrix (ECM) was restricted to the scaffold's perimeter [279]. Injectable chitosan hydrogels and chitosan composites have demonstrated significant benefits on cartilage regeneration [262]. Chitosan is structurally comparable to the ECM found in biological tissue, and hence works as an artificial ECM. Hyaluronic acid functioned as an enhancer to chitosan in building the cartilaginous ECM, hence boosting auricular chondrocyte proliferation in the scaffold lacunae [280]. Starting with direct injection of chitosan into wounded rat cartilage and progressing through various chemical improvement to enhance bonding of chitosan and different ECM mimicking polymers such as chondroitin sulphate and hyaluronic acid, have demonstrated chondrocyte infiltration and accelerated wound healing [281–283].

Fibrous scaffolds were also demonstrated to stimulate the ECM of natural cartilage, resulting in functional effects in chondrogenesis. Chitosan-based composite fibrous scaffolds were constructed, demonstrating the deposition of ECM following chondrocyte culture [262]. While the cationic charge of chondrocytes limits their growth, differentiation of cells can be accomplished by subsequent deposition of ECM [284, 285]. Dynamic cultivation of rabbit auricular chondrocytes resulted in the formation of a full three-dimensional scaffold. The optimal scaffold for cartilage regeneration should combine the spatial structure provided by a porous construct in 3D with the mechanical strength provided by a fibrous mesh. Thus, a mixture of both would have been an appropriate approach to pursue.

6 Conclusion and Future Work

This chapter outlines a number of studies aimed at improving the composition of chitin and chitosan-based composites. Apart from cellulose, the raw biopolymers of chitin and chitosan being two of the most common polysaccharides. Their high availability helps to lower the low cost of the materials. The researchers were motivated to build a variety of products based on these composites based on the unique application because the mixture of these biopolymers with other extra materials might offer remarkable physical and chemical applicability. Incorporating chitin and chitosan hybrids with carbonaceous materials, for example, can improve their availability for treating industrial water pollution. Furthermore, including chitin nanofibrils onto chitosan-matrix composite films could intensify chitosan's hydrophilic character and instability in the supramolecular state. Furthermore, chitin and chitosan composites containing polysaccharide can be used as food packaging since their composite films have enhanced permeability. Finally, because of their mechanical properties and biocompatibility, chitin and chitosan-inorganic compounds are widely used in biomedical applications. As a result, the advancement of these biopolymer composites must be promoted in order to preserve the green environment, which has become a source of concern. This kind of effort ensures that future generations will inherit a better environment to live in, as well as a better environment for other living species.

Acknowledgements The authors would like to express special thanks to Universiti Putra Malaysia and the Ministry of Education for education support.

References

1. Younes, I., Hajji, S., Frachet, V., Rinaudo, M., Jellouli, K., & Nasri, M. (2014). Chitin extraction from shrimp shell using enzymatic treatment. antitumor, antioxidant and antimicrobial activities of Chitosan. *International Journal of Biological Macromolecules, 69*, 489–498. https://doi.org/10.1016/j.ijbiomac.2014.06.013
2. Mohanty, A. K., Misra, M., & Drzal, L. T. (2002). *Journal of Polymers and the Environment, 10*(1/2), 19–26. https://doi.org/10.1023/a:1021013921916
3. Pillai, C. K. S., Paul, W., & Sharma, C. P. (2009). Chitin and chitosan polymers: Chemistry, solubility and fiber formation. *Progress in Polymer Science, 34*(7), 641–678. https://doi.org/10.1016/j.progpolymsci.2009.04.001
4. Camargo, P. H., Satyanarayana, K. G., & Wypych, F. (2009). Nanocomposites: Synthesis, structure, properties and new application opportunities. *Materials Research, 12*(1), 1–39. https://doi.org/10.1590/s1516-14392009000100002
5. Faruk, O., Bledzki, A. K., Fink, H.-P., & Sain, M. (2012). Biocomposites reinforced with natural fibers: 2000–2010. *Progress in Polymer Science, 37*(11), 1552–1596. https://doi.org/10.1016/j.progpolymsci.2012.04.003
6. Ciechanska, D., Wietecha, J., Kucharska, M., Wrzesniewska-Tosik, K., & Kopania, E. (2014). Biomass as a source of functional polymeric materials. *Polimery, 59*(05), 383–392. https://doi.org/10.14314/polimery.2014.383

7. Muzzarelli, R. A. (2011). Biomedical exploitation of chitin and Chitosan via mechano-chemical disassembly, electrospinning, dissolution in imidazolium ionic liquids, and supercritical drying. *Marine Drugs, 9*(9), 1510–1533. https://doi.org/10.3390/md9091510
8. Rizvi, R., Cochrane, B., Naguib, H., & Lee, P. C. (2011). Fabrication and characterization of melt-blended polylactide-chitin composites and their foams. *Journal of Cellular Plastics, 47*(3), 283–300. https://doi.org/10.1177/0021955x11402549
9. Aranaz, I., Mengibar, M., Harris, R., Panos, I., Miralles, B., Acosta, N., Galed, G., & Heras, A. (2009). Functional characterization of chitin and Chitosan. *Current Chemical Biology, 3*(2), 203–230. https://doi.org/10.2174/2212796810903020203
10. Ravi Kumar, M. N. V. (2000). A review of Chitin and Chitosan applications. *Reactive and Functional Polymers, 46*(1), 1–27. https://doi.org/10.1016/s1381-5148(00)00038-9
11. Rinaudo, M. (2006). Chitin and Chitosan: Properties and applications. *Progress in Polymer Science, 31*(7), 603–632. https://doi.org/10.1016/j.progpolymsci.2006.06.001
12. Austin, P. R., Brine, C. J., Castle, J. E., & Zikakis, J. P. (1981). Chitin: New facets of research. *Science, 212*(4496), 749–753. https://doi.org/10.1126/science.7221561
13. Muzzarelli, R. A. A. (1975). Natural chelating polymers. *Analytica Chimica Acta, 80*(1), 200. https://doi.org/10.1016/s0003-2670(01)84996-5
14. Zikakis, J. P. (1984). Preface. *Chitin, Chitosan, and Related Enzymes*, xiii–xv. https://doi.org/10.1016/b978-0-12-780950-2.50005-8
15. Rutherford, F. A., & Austin, P. R. (1978). Marine chitin properties and solvents. In R. A. A. Muzzarelli & E. R. Pariser (Eds.), *Proceedings of the first international conference on Chitin and Chitosan* (pp. 182–192). MIT.
16. Kean, T., & Thanou, M. (2010). Biodegradation, biodistribution and toxicity of chitosan. *Advanced Drug Delivery Reviews, 62*(1), 3–11. https://doi.org/10.1016/j.addr.2009.09.004
17. Muzzarelli, R. A. (1997). Human enzymatic activities related to the therapeutic administration of chitin derivatives. *Cellular and Molecular Life Sciences, 53*(2), 131–140. https://doi.org/10.1007/pl00000584
18. No, H. K., Park, N. Y., Lee, S. H., & Meyers, S. P. (2002). Antibacterial activity of chitosans and chitosan oligomers with different molecular weights. *International Journal of Food Microbiology, 74*(1–2), 65–72. https://doi.org/10.1016/s0168-1605(01)00717-6
19. Lim, S.-H., & Hudson, S. M. (2003). Review of Chitosan and its derivatives as antimicrobial agents and their uses as textile chemicals. *Journal of Macromolecular Science, Part C: Polymer Reviews, 43*(2), 223–269. https://doi.org/10.1081/mc-120020161
20. Rao, S. B., & Sharma, C. P. (1997). Use of Chitosan as a biomaterial: Studies on its safety and hemostatic potential. *Journal of Biomedical Materials Research, 34*(1), 21–28. https://doi.org/10.1002/(sici)1097-4636(199701)34:1%3c21::aid-jbm4%3e3.0.co;2-p
21. Whang, H. S., Kirsch, W., Zhu, Y. H., Yang, C. Z., & Hudson, S. M. (2005). Hemostatic agents derived from chitin and Chitosan. *Journal of Macromolecular Science, Part C: Polymer Reviews, 45*(4), 309–323. https://doi.org/10.1080/15321790500304122
22. Ueno, H., Mori, T., & Fujinaga, T. (2001). Topical formulations and wound healing applications of Chitosan. *Advanced Drug Delivery Reviews, 52*(2), 105–115. https://doi.org/10.1016/s0169-409x(01)00189-2
23. Su, Z., Zhang, M., Lu, Z., Song, S., Zhao, Y., & Hao, Y. (2018). Functionalization of cellulose fiber by in situ growth of zeolitic imidazolate framework-8 (zif-8) nanocrystals for preparing a cellulose-based air filter with gas adsorption ability. *Cellulose, 25*(3), 1997–2008. https://doi.org/10.1007/s10570-018-1696-4
24. El Knidri, H., El Khalfaouy, R., Laajeb, A., Addaou, A., & Lahsini, A. (2016). Eco-friendly extraction and characterization of chitin and chitosan from the shrimp shell waste via microwave irradiation. *Process Safety and Environmental Protection, 104*, 395–405. https://doi.org/10.1016/j.psep.2016.09.020
25. El Knidri, H., Belaabed, R., Addaou, A., Laajeb, A., & Lahsini, A. (2018). Extraction, chemical modification and characterization of Chitin and Chitosan. *International Journal of Biological Macromolecules, 120*, 1181–1189. https://doi.org/10.1016/j.ijbiomac.2018.08.139

26. Chaussard, G., & Domard, A. (2004). New aspects of the extraction of chitin from squid pens. *Biomacromolecules, 5*(2), 559–564. https://doi.org/10.1021/bm034401t
27. Futalan, C. M., Kan, C.-C., Dalida, M. L., Hsien, K.-J., Pascua, C., & Wan, M.-W. (2011). Comparative and competitive adsorption of copper, lead, and nickel using chitosan immobilized on bentonite. *Carbohydrate Polymers, 83*(2), 528–536. https://doi.org/10.1016/j.carbpol.2010.08.013
28. Hamed, I., Özogul, F., & Regenstein, J. M. (2016). Industrial applications of crustacean by-products (chitin, Chitosan, and Chitooligosaccharides): A review. *Trends in Food Science & Technology, 48*, 40–50. https://doi.org/10.1016/j.tifs.2015.11.007
29. Luo, J., Wang, X., Xia, B., & Wu, J. (2010). Preparation and characterization of quaternized chitosan under microwave irradiation. *Journal of Macromolecular Science, Part A, 47*(9), 952–956. https://doi.org/10.1080/10601325.2010.501310
30. Razzaq, T., & Kappe, C. O. (2008). On the energy efficiency of microwave-assisted organic reactions. *Chemsuschem, 1*(1–2), 123–132. https://doi.org/10.1002/cssc.200700036
31. Surati, M. A., Jauhari, S., & Desai, K. R. (2012). A brief review: Microwave assisted organic reaction. *Archives of Applied Science Research*, 645–661.
32. Ajani, O. O., Obafemi, C. A., Ikpo, C. O., Ogunniran, K. O., & Nwinyi, O. C. (2009). Microwave-assisted synthesis and antibacterial activity of some pyrazol-1-ylquinoxalin-2(1H)-one derivatives. *Chemistry of Heterocyclic Compounds, 45*(11), 1370–1378. https://doi.org/10.1007/s10593-010-0435-z
33. Hayes, B. L. (2004). Recent advances in microwave-assisted synthesis. *Aldrichimica Acta, 37*(2), 66–77.
34. Safari, J., Naseh, S., Zarnegar, Z., & Akbari, Z. (2014). Applications of microwave technology to rapid synthesis of substituted imidazoles on silica-supported sbcl3as an efficient heterogeneous catalyst. *Journal of Taibah University for Science, 8*(4), 323–330. https://doi.org/10.1016/j.jtusci.2014.01.007
35. Karthiga, D. G., Sathish, K. K., & Arivalagan, K. (2014). Microwave assisted nanoparticles for drug delivery systems. *International Journal Pharmacy and Pharmaceut, 6*(4), 118–123.
36. Arancibia, M. Y., Alemán, A., Calvo, M. M., López-Caballero, M. E., Montero, P., & Gómez-Guillén, M. C. (2014). Antimicrobial and antioxidant chitosan solutions enriched with active shrimp (*Litopenaeus vannamei*) waste materials. *Food Hydrocolloids, 35*, 710–717. https://doi.org/10.1016/j.foodhyd.2013.08.026
37. Kumari, S., Rath, P., Sri Hari Kumar, A., & Tiwari, T. N. (2015). Extraction and characterization of Chitin and chitosan from fishery waste by chemical method. *Environmental Technology & Innovation, 3*, 77–85. https://doi.org/10.1016/j.eti.2015.01.002
38. Kumirska, J., Czerwicka, M., Kaczyński, Z., Bychowska, A., Brzozowski, K., Thöming, J., & Stepnowski, P. (2010). Application of spectroscopic methods for structural analysis of chitin and Chitosan. *Marine Drugs, 8*(5), 1567–1636. https://doi.org/10.3390/md8051567
39. Srinivasan, H., Kanayairam, V., & Ravichandran, R. (2018). Chitin and chitosan preparation from shrimp shells *Penaeus monodon* and its human ovarian cancer cell line, PA-1. *International Journal of Biological Macromolecules, 107*, 662–667. https://doi.org/10.1016/j.ijbiomac.2017.09.035
40. Benhabiles, M. S., Abdi, N., Drouiche, N., Lounici, H., Pauss, A., Goosen, M. F. A., & Mameri, N. (2013). Protein recovery by ultrafiltration during isolation of chitin from shrimp shells *Parapenaeus longirostris. Food Hydrocolloids, 32*(1), 28–34. https://doi.org/10.1016/j.foodhyd.2012.11.035
41. Hai, T. A., & Sugimoto, R. (2018). Surface modification of chitin and chitosan with poly(3-HEXYLTHIOPHENE) via oxidative polymerization. *Applied Surface Science, 434*, 188–197. https://doi.org/10.1016/j.apsusc.2017.10.197
42. Zargar, V., Asghari, M., & Dashti, A. (2015). A review on chitin and chitosan polymers: Structure, chemistry, solubility, derivatives, and applications. *ChemBioEng Reviews, 2*(3), 204–226. https://doi.org/10.1002/cben.201400025
43. Morimoto, M., Saimoto, H., & Shigemasa, Y. (2002). Control of functions of chitin and chitosan by chemical modification. *Trends in Glycoscience and Glycotechnology, 14*(78), 205–222. https://doi.org/10.4052/tigg.14.205

44. Jayakumar, R., Selvamurugan, N., Nair, S. V., Tokura, S., & Tamura, H. (2008). Preparative methods of phosphorylated chitin and chitosan—an overview. *International Journal of Biological Macromolecules, 43*(3), 221–225. https://doi.org/10.1016/j.ijbiomac.2008.07.004
45. Marques, M. M., Queda, F., Santos, C. V., & Carvalho, L. C. (2016). Selective modification of chitin and chitosan: En route to tailored oligosaccharides. *Chemistry - An Asian Journal, 11*(24), 3468–3481. https://doi.org/10.1002/asia.201601041
46. Ifuku, S., Ikuta, A., Egusa, M., Kaminaka, H., Izawa, H., Morimoto, M., & Saimoto, H. (2013). Preparation of high-strength transparent chitosan film reinforced with surface-deacetylated chitin nanofibers. *Carbohydrate Polymers, 98*(1), 1198–1202. https://doi.org/10.1016/j.carbpol.2013.07.033
47. Mourya, V. K., & Inamdar, N. N. (2008). Trimethyl Chitosan and its applications in drug delivery. *Journal of Materials Science: Materials in Medicine, 20*(5), 1057–1079. https://doi.org/10.1007/s10856-008-3659-z
48. Leclerc, C., Simao-Beaunoir, A.-M., Gauthier, A., Fayad, K., Mamady, H., Beaulieu, C., & Brzezinski, R. (2001). Purification and properties of a β-1,6-glucanase from *streptomyces* sp. EF-14, an actinomycete antagonistic to *phytophthora spp. Applied Microbiology and Biotechnology, 57*(1–2), 117–123. https://doi.org/10.1007/s002530100780
49. Malerba, M., & Cerana, R. (2019). Recent applications of chitin- and chitosan-based polymers in plants. *Polymers, 11*(5), 839. https://doi.org/10.3390/polym11050839
50. Pusztahelyi, T. (2018). Chitin and chitin-related compounds in plant–fungal interactions. *Mycology, 9*(3), 189–201. https://doi.org/10.1080/21501203.2018.1473299
51. Divya, K., & Jisha, M. S. (2017). Chitosan nanoparticles preparation and applications. *Environmental Chemistry Letters, 16*(1), 101–112. https://doi.org/10.1007/s10311-017-0670-y
52. Divya, K., Vijayan, S., Nair, S. J., & Jisha, M. S. (2019). Optimization of chitosan nanoparticle synthesis and its potential application as germination elicitor of *Oryza sativa* L. *International Journal of Biological Macromolecules, 124*, 1053–1059. https://doi.org/10.1016/j.ijbiomac.2018.11.185
53. Muzzarelli, R. A., Mattioli-Belmonte, M., Pugnaloni, A., & Biagini, G. (1999). Biochemistry, histology and clinical uses of Chitins and chitosans in wound healing. *Chitin and Chitinases*, 251–264. https://doi.org/10.1007/978-3-0348-8757-1_18
54. Jang, S. I., Mok, J. Y., Jeon, I. H., Park, K.-H., Nguyen, T. T., Park, J. S., Hwang, H. M., Song, M.-S., Lee, D., & Chai, K. Y. (2012). Effect of electrospun non-woven mats of Dibutyryl Chitin/poly(lactic acid) blends on wound healing in hairless mice. *Molecules, 17*(3), 2992–3007. https://doi.org/10.3390/molecules17032992
55. Singh, R., Shitiz, K., & Singh, A. (2017). Chitin and Chitosan: Biopolymers for wound management. *International Wound Journal, 14*(6), 1276–1289. https://doi.org/10.1111/iwj.12797
56. Mizuno, K., Yamamura, K., Yano, K., Osada, T., Saeki, S., Takimoto, N., Sakurai, T., & Nimura, Y. (2002). Effect of chitosan film containing basic fibroblast growth factor on wound healing in genetically diabetic mice. *Journal of Biomedical Materials Research, 64A*(1), 177–181. https://doi.org/10.1002/jbm.a.10396
57. Ishihara, M., Nakanishi, K., Ono, K., Sato, M., Kikuchi, M., Saito, Y., Yura, H., Matsui, T., Hattori, H., Uenoyama, M., & Kurita, A. (2002). Photocrosslinkable Chitosan as a dressing for wound occlusion and accelerator in healing process. *Biomaterials, 23*(3), 833–840. https://doi.org/10.1016/s0142-9612(01)00189-2
58. Ishihara, M. (2002). Photocrosslinkable chitosan hydrogel as a wound dressing and a biological adhesive. *Trends in Glycoscience and Glycotechnology, 14*(80), 331–341. https://doi.org/10.4052/tigg.14.331
59. Akakuru, O. U., Louis, H., Amos, P. I., Akakuru, O. C., Nosike, E. I., & Ogulewe, E. F. (2018). The chemistry of chitin and chitosan justifying their nanomedical utilities. *Biochemistry & Pharmacology: Open Access, 07*(01). https://doi.org/10.4172/2167-0501.1000241
60. Alves, N. M., & Mano, J. F. (2008). Chitosan derivatives obtained by chemical modifications for biomedical and environmental applications. *International Journal of Biological Macromolecules, 43*(5), 401–414. https://doi.org/10.1016/j.ijbiomac.2008.09.007

61. Wan Ngah, W. S., Endud, C. S., & Mayanar, R. (2002). Removal of copper(ii) ions from aqueous solution onto chitosan and cross-linked chitosan beads. *Reactive and Functional Polymers, 50*(2), 181–190. https://doi.org/10.1016/s1381-5148(01)00113-4
62. Chen, A.-H., Liu, S.-C., Chen, C.-Y., & Chen, C.-Y. (2008). Comparative adsorption of cu(ii), zn(ii), and pb(ii) ions in aqueous solution on the crosslinked chitosan with epichlorohydrin. *Journal of Hazardous Materials, 154*(1–3), 184–191. https://doi.org/10.1016/j.jhazmat.2007.10.009
63. Gupta, V. K., & Suhas. (2009). Application of low-cost adsorbents for dye removal – a review. *Journal of Environmental Management, 90*(8), 2313–2342. https://doi.org/10.1016/j.jenvman.2008.11.017
64. Zhang, L., Zeng, Y., & Cheng, Z. (2016). Removal of heavy metal ions using chitosan and modified chitosan: A review. *Journal of Molecular Liquids, 214*, 175–191. https://doi.org/10.1016/j.molliq.2015.12.013
65. Reddy, D. H., & Lee, S.-M. (2013). Application of magnetic chitosan composites for the removal of toxic metal and dyes from aqueous solutions. *Advances in Colloid and Interface Science, 201–202*, 68–93. https://doi.org/10.1016/j.cis.2013.10.002
66. Bhatnagar, A., & Sillanpää, M. (2009). Applications of chitin- and chitosan-derivatives for the detoxification of water and wastewater— a short review. *Advances in Colloid and Interface Science, 152*(1–2), 26–38. https://doi.org/10.1016/j.cis.2009.09.003
67. Ceyhan, O., & Baybas, D. (2001). Adsorption of some textile dyes by hexadecyltrimethylammonium bentonite. *Turkish Journal of Chemistry, 25*(2), 193–200.
68. Sarode, S., Upadhyay, P., Khosa, M. A., Mak, T., Shakir, A., Song, S., & Ullah, A. (2019). Overview of wastewater treatment methods with special focus on biopolymer chitin-chitosan. *International Journal of Biological Macromolecules, 121*, 1086–1100. https://doi.org/10.1016/j.ijbiomac.2018.10.089
69. Vakili, M., Rafatullah, M., Salamatinia, B., Abdullah, A. Z., Ibrahim, M. H., Tan, K. B., Gholami, Z., & Amouzgar, P. (2014). Application of chitosan and its derivatives as adsorbents for dye removal from water and wastewater: A review. *Carbohydrate Polymers, 113*, 115–130. https://doi.org/10.1016/j.carbpol.2014.07.007
70. Tran, V. S., Ngo, H. H., Guo, W., Zhang, J., Liang, S., Ton-That, C., & Zhang, X. (2015). Typical low cost biosorbents for adsorptive removal of specific organic pollutants from water. *Bioresource Technology, 182*, 353–363. https://doi.org/10.1016/j.biortech.2015.02.003
71. Muxika, A., Etxabide, A., Uranga, J., Guerrero, P., & de la Caba, K. (2017). Chitosan as a bioactive polymer: Processing, properties and applications. *International Journal of Biological Macromolecules, 105*, 1358–1368. https://doi.org/10.1016/j.ijbiomac.2017.07.087
72. Sharififard, H., Shahraki, Z. H., Rezvanpanah, E., & Rad, S. H. (2018). A novel natural Chitosan/activated carbon/iron bio-nanocomposite: Sonochemical Synthesis, characterization, and application for cadmium removal in batch and continuous adsorption process. *Bioresource Technology, 270*, 562–569. https://doi.org/10.1016/j.biortech.2018.09.094
73. Ahmed, M. J., Hameed, B. H., & Hummadi, E. H. (2020). Review on recent progress in Chitosan/chitin-carbonaceous material composites for the adsorption of water pollutants. *Carbohydrate Polymers, 247*, 116690. https://doi.org/10.1016/j.carbpol.2020.116690
74. Karaer, H., & Kaya, İ. (2016). Synthesis, characterization of magnetic chitosan/active charcoal composite and using at the adsorption of methylene blue and reactive blue4. *Microporous and Mesoporous Materials, 232*, 26–38. https://doi.org/10.1016/j.micromeso.2016.06.006
75. Khakpour, R., & Tahermansouri, H. (2018). Synthesis, characterization and study of sorption parameters of multi-walled carbon nanotubes/chitosan nanocomposite for the removal of picric acid from aqueous solutions. *International Journal of Biological Macromolecules, 109*, 598–610. https://doi.org/10.1016/j.ijbiomac.2017.12.105
76. Salam, M. A., El-Shishtawy, R. M., & Obaid, A. Y. (2014). Synthesis of magnetic multi-walled carbon nanotubes/magnetite/chitin magnetic nanocomposite for the removal of Rose Bengal from real and model solution. *Journal of Industrial and Engineering Chemistry, 20*(5), 3559–3567. https://doi.org/10.1016/j.jiec.2013.12.049

77. Wang, F., Yang, B., Wang, H., Song, Q., Tan, F., & Cao, Y. (2016). Removal of ciprofloxacin from aqueous solution by a magnetic chitosan grafted graphene oxide composite. *Journal of Molecular Liquids, 222*, 188–194. https://doi.org/10.1016/j.molliq.2016.07.037
78. Zhang, C., Chen, Z., Guo, W., Zhu, C., & Zou, Y. (2018). Simple fabrication of chitosan/graphene nanoplates composite spheres for efficient adsorption of acid dyes from aqueous solution. *International Journal of Biological Macromolecules, 112*, 1048–1054. https://doi.org/10.1016/j.ijbiomac.2018.02.074
79. Liu, J., Zhou, B., Zhang, H., Ma, J., Mu, B., & Zhang, W. (2019). A novel biochar modified by chitosan-FE/S for tetracycline adsorption and studies on site energy distribution. *Bioresource Technology, 294*, 122152. https://doi.org/10.1016/j.biortech.2019.122152
80. Ageev, E. P., Matushkina, N. N., Gerasimov, V. I., Zezin, S. B., Vikhoreva, G. A., Zotkin, M. A., et al. (2004). Structure and transport behavior of heat-treated chitosan films. *Polymer Science Series A, 46*(12), 1245–1250.
81. Zotkin, M. A., Vikhoreva, G. A., Smotrina, T. V., & Derbenev, M. A. (2004). Thermal modification and study of the structure of Chitosan films. *Fibre Chemistry, 36*(1), 16–20. https://doi.org/10.1023/b:fich.0000025532.80007.49
82. Hoeve, C. A. (1980). The structure of water in polymers. *ACS Symposium Series*, 135–146. https://doi.org/10.1021/bk-1980-0127.ch007
83. Miretzky, P., & Cirelli, A. F. (2011). Fluoride removal from water by chitosan derivatives and composites: A review. *Journal of Fluorine Chemistry, 132*(4), 231–240. https://doi.org/10.1016/j.jfluchem.2011.02.001
84. Frindy, S., Primo, A., Ennajih, H., El Kacem Qaiss, A., Bouhfid, R., Lahcini, M., Essassi, E. M., Garcia, H., & El Kadib, A. (2017). Chitosan–graphene oxide films and CO_2-dried porous aerogel microspheres: Interfacial Interplay and stability. *Carbohydrate Polymers, 167*, 297–305. https://doi.org/10.1016/j.carbpol.2017.03.034
85. Ahmed, M. J., & Hameed, B. H. (2019). Insights into the isotherm and kinetic models for the coadsorption of pharmaceuticals in the absence and presence of metal ions: A review. *Journal of Environmental Management, 252*, 109617. https://doi.org/10.1016/j.jenvman.2019.109617
86. Wong, S., Ngadi, N., Inuwa, I. M., & Hassan, O. (2018). Recent advances in applications of activated carbon from biowaste for wastewater treatment: A short review. *Journal of Cleaner Production, 175*, 361–375. https://doi.org/10.1016/j.jclepro.2017.12.059
87. Foo, K. Y., & Hameed, B. H. (2012). Potential of jackfruit peel as precursor for activated carbon prepared by microwave induced NaOH activation. *Bioresource Technology, 112*, 143–150. https://doi.org/10.1016/j.biortech.2012.01.178
88. Islam, M. A., Ahmed, M. J., Khanday, W. A., Asif, M., & Hameed, B. H. (2017). Mesoporous activated coconut shell-derived hydrochar prepared via hydrothermal carbonization-NaOH activation for methylene blue adsorption. *Journal of Environmental Management, 203*, 237–244.
89. Islam, M. A., Ahmed, M. J., Khanday, W. A., Asif, M., & Hameed, B. H. (2017). Mesoporous activated carbon prepared from NaOH activation of rattan (*Lacosperma secundiflorum*) hydrochar for methylene blue removal. *Ecotoxicology and Environmental Safety, 138*, 279–285.
90. Islam, M. A., Tan, I. A. W., Benhouria, A., Asif, M., & Hameed, B. H. (2015). Mesoporous and adsorptive properties of palm date seed activated carbon prepared via sequential hydrothermal carbonization and sodium hydroxide activation. *Chemical Engineering Journal, 270*, 187–195.
91. Ahmed, M. J. (2017). Adsorption of non-steroidal anti-inflammatory drugs from aqueous solution using activated carbons: Review. *Journal of Environmental Management, 190*, 274–282.
92. Hydari, S., Sharififard, H., Nabavinia, M., & Parvizi, M. R. (2012). A comparative investigation on removal performances of commercial activated carbon, chitosan biosorbent and chitosan/activated carbon composite for cadmium. *Chemical Engineering Journal, 193–194*, 276–282.

93. Wu, Y., Xia, C., Cai, L., & Shi, S. Q. (2018). Controlling pore size of activated carbon through self-activation process for removing contaminants of different molecular sizes. *Journal of Colloid and Interface Science, 518*, 41–47.
94. Yadaei, H., Beyki, M. H., Shemirani, F., & Nouroozi, S. (2018). Ferrofluid mediated chitosan@mesoporous carbon nanohybrid for green adsorption/preconcentration of toxic Cd(II): Modeling, kinetic and isotherm study. *Reactive & Functional Polymers, 122*, 85–97.
95. Han, H., Rafiq, M. K., Zhou, T., Xu, R., Mašek, O., & Li, X. (2019). A critical review of claybased composites with enhanced adsorption performance for metal and organic pollutants. *Journal of Hazardous Materials, 369*, 780–796.
96. Zhang, Z., Zhu, Z., Shen, B., & Liu, L. (2019). Insights into biochar and hydrochar production and applications: A review. *Energy, 171*, 581–598.
97. Afzal, M. Z., Sun, X.-F., Liu, J., Song, C., Wang, S.-G., & Javed, A. (2018). Enhancement of ciprofloxacin sorption on chitosan/biochar hydrogel beads. *The Science of the Total Environment, 639*, 560–569.
98. Nitayaphat, W., & Jintakosol, T. (2015). Removal of silver(I) from aqueous solutions by chitosan/bamboo charcoal composite beads. *Journal of Cleaner Production, 87*, 850–855.
99. Xiao, F., Cheng, J., Cao, W., Yang, C., Chen, J., & Luo, Z. (2019). Removal of heavy metals from aqueous solution using chitosan-combined magnetic biochars. *Journal of Colloid and Interface Science, 540*, 579–584.
100. Zhang, L., Tang, S., He, F., Liu, Y., Mao, W., & Guan, Y. (2019). Highly efficient and selective capture of heavy metals by poly(acrylic acid) grafted chitosan and biochar composite for wastewater treatment. *Chemical Engineering Journal, 378*.
101. Iijima, S. (1991). Helical microtubules of graphitic carbon. *Nature, 354*, 56–58.
102. Sarkar, B., Mandal, S., Tsang, Y. F., Kumar, P., Kim, K.-H., & Ok, Y. S. (2018). Designer carbon nanotubes for contaminant removal in water and wastewater: A critical review. *The Science of the Total Environment, 612*, 561–581.
103. Fiyadh, S. S., AlSaadi, M. A., Jaafar, W. Z., AlOmar, M. K., Fayaed, S. S., Mohd, N. S., et al. (2019). Review on heavy metal adsorption processes by carbon nanotubes. *Journal of Cleaner Production, 230*, 783–793.
104. Dou, J., Gan, D., Huang, Q., Liu, M., Chen, J., Deng, F., et al. (2019). Functionalization of carbon nanotubes with chitosan based on MALI multicomponent reaction for Cu^{2+} removal. *International Journal of Biological Macromolecules, 136*, 476–485.
105. Parlayıcı, Ş, & Pehlivan, E. (2019). Removal of chromium (VI) from aqueous solution using chitosan doped with carbon nanotubes. *Materials Today Proceedings, 18*, 1978–1985.
106. Huang, Y., Lee, X., Macazo, F. C., Grattieri, M., Cai, R., & Minteer, S. D. (2018). Fast and efficient removal of chromium (VI) anionic species by a reusable chitosan-modified multi-walled carbon nanotube composite. *Chemical Engineering Journal, 339*, 259–267.
107. Zhu, H. Y., Jiang, R., Xiao, L., & Zeng, G. M. (2010). Preparation, characterization, adsorption kinetics and thermodynamics of novel magnetic chitosan enwrapping nanosized γ-Fe_2O_3 and multi-walled carbon nanotubes with enhanced adsorption properties for methyl orange. *Bioresource Technology, 101*, 5063–5069.
108. Zhang, Y., Yan, T., Yan, L., Guo, X., Cui, L., Wei, Q., et al. (2014). Preparation of novel cobalt ferrite/chitosan grafted with graphene composite as effective adsorbents for mercury ions. *Journal of Molecular Liquids, 198*, 381–387.
109. Li, Z., Ramay, H. R., Hauch, K. D., Xiao, D., & Zhang, M. (2005). Chitosan–alginate hybrid scaffolds for bone tissue engineering. *Biomaterials, 26*(18), 3919–3928.
110. Peng, W., Li, H., Liu, Y., & Song, S. (2017). A review on heavy metal ions adsorption from water by graphene oxide and its composites. *Journal of Molecular Liquids, 230*, 496–504.
111. Sherlala, A. I. A., Raman, A. A. A., Bello, M. M., & Asghar, A. (2018). A review of the applications of organo-functionalized magnetic graphene oxide nanocomposites for heavy metal adsorption. *Chemosphere, 193*, 1004–1017.
112. Ma, Z., Liu, D., Zhu, Y., Li, Z., Li, Z., Tian, H., et al. (2016). Graphene oxide/chitin nanofibril composite foams as column adsorbents for aqueous pollutants. *Carbohydrate Polymers, 144*, 230–237.

113. Salzano de Luna, M., Ascione, C., Santillo, C., Verdolotti, L., Lavorgna, M., Buonocore, G. G., et al. (2019). Optimization of dye adsorption capacity and mechanical strength of chitosan aerogels through crosslinking strategy and graphene oxide addition. *Carbohydrate Polymers, 211*, 195–203.
114. Liu, L., Li, C., Bao, C., Jia, Q., Xiao, P., Liu, X., et al. (2012). Preparation and characterization of chitosan/graphene oxide composites for the adsorption of Au(III) and Pd(II). *Talanta, 93*, 350–357.
115. Yudin, V. E., Dobrovolskaya, I. P., Neelov, I. M., Dresvyanina, E. N., Popryadukhin, P. V., & Ivan'kova, E. M., et al. (2014). Wet spinning of fibers made of chitosan and chitin nanofibrils. *Carbohydrate Polymers, 108*, 176–182.
116. Sriupayo, J., Supaphol, P., Blackwell, J., & Rujiravanit, R. (2005). Preparation and characterization of α-chitin whisker-reinforced chitosan nanocomposite films with or without heat treatment. *Carbohydrate Polymers, 62*(2), 130–136.
117. Dresvyanina, E. N., Grebennikov, S. F., Elokhovskii, V. Y., Dobrovolskaya, I. P., Ivan'kova, E. M., Yudin, V. E., Heppe, K., & Morganti, P. (2020). Thermodynamics of interaction between water and the composite films based on Chitosan and chitin nanofibrils. *Carbohydrate Polymers, 245*, 116552. https://doi.org/10.1016/j.carbpol.2020.116552
118. Dresvyanina, E. N., Grebennikov, S. F., Dobrovol'skaya, I. P., Maslennikova, T. P., Ivan'kova, E. M., & Yudin, V. E. (2020). Effect of chitin nanofibrils on the sorption behavior of chitosan-based composite films. *Polymer Science, Series A, 62*(3), 205–212. https://doi.org/10.1134/s0965545x20030050
119. Dresvyanina, E., Yudenko, A., Yevlampieva, N., Maevskaya, E., Yudin, V., Gubarev, A., et al. (2018). The molecular mass effect on mechanical properties of chitosan. *Vlakna a Textil, 25*(2), 27–31.
120. Dobrovol'skaya, I. P., Kasatkin, I. A., Yudin, V. E., Ivan'kova, E. M., & Elokhovskii, V. Y. (2015). Supramolecular structure of chitin nanofibrils. *Polymer Science Series A, 57*(1), 52–57. https://doi.org/10.1134/s0965545x15010022
121. Wang, C., & Esker, A. R. (2014). Nanocrystalline chitin thin films. *Carbohydrate Polymers, 102*, 151–158. https://doi.org/10.1016/j.carbpol.2013.10.103
122. Grebennikov, S. F., & Chulkova, Y. U. S. (2009). The thermal coefficient of sorption of vapors by polymeric materials. *Russian Journal of Physical Chemistry, 83*(4), 660–662.
123. Grebennikov, S. F., Kynin, A. T., Zaitseva, E. I., & Chulkova, Y. U. S. (2007). Molecular structure of polymeric materials and sorption of water vapor. *Russian Journal of Applied Chemistry, 80*(12), 2113–2118.
124. Grebennikov, S. F., Chulkova, Y. U. S., & Zaitseva, E. I. (2009). Sorption processes in the nanostructures of mesomorphous polymers. *Protection of Metals and Physical Chemistry of Surfaces, 45*(2), 177–180.
125. Smotrina, T. V., Dresvyanina, E. N., Grebennikov, S. F., Kazakov, M. O., Maslennikova, T. P., Dobrovolskaya, I. P., et al. (2020). Interaction between water and the composite materials based on chitosan and chitin nanofibrils. *Polymer, 189*, 122166.
126. Shibaev, V. (2016). *Liquid crystalline polymers. Saleem Hashmi (editor-in-chief), reference module in materials science and materials engineering* (pp. 1–46). Elsevier.
127. Zhizhenkov, V. V., Kvachadze, N. G., Grebennikov, S. F., Busygin, K. N., Smotrina, T. V., & Smotrin, V. A. (2017). Formation of liquid-crystalline-type structures in poly(para-amides). *Polymer Science Series A, 59*(2), 198–205.
128. Sirajudheen, P., Poovathumkuzhi, N. C., Vigneshwaran, S., Chelaveettil, B. M., & Meenakshi, S. (2021). Applications of chitin and chitosan based biomaterials for the adsorptive removal of textile dyes from water—a comprehensive review. *Carbohydrate Polymers, 273*, 118604. https://doi.org/10.1016/j.carbpol.2021.118604
129. Amalraj, A., Jude, S., & Gopi, S. (2020). *Chapter 1. Polymer blends, composites and nanocomposites from chitin and chitosan; manufacturing, characterization and applications*. INC.
130. Sirajudheen, P., & Meenakshi, S. (2020). Lanthanum (III) incorporated chitosanmontmorillonite composite as flexible material for adsorptive removal of azo dyes from water. *Materials Today Proceedings, 27*, 318–326. https://doi.org/10.1016/j.matpr.2019.11.040

131. Huang, R., Liu, Q., Zhang, L., & Yang, B. (2015). Utilization of cross-linked chitosan/bentonite composite in the removal of methyl orange from aqueous solution. *Water Science and Technology, 71*, 174–182. https://doi.org/10.2166/wst.2014.478
132. Sirajudheen, P., & Meenakshi, S. (2020b). Encapsulation of Zn–Fe layered double hydroxide on activated carbon and its litheness in tuning anionic and Rhoda dyes through adsorption mechanism. *Asia-Pacific Journal of Chemical Engineering*, 1–13. https://doi.org/10.1002/apj.2479
133. Nithya, A., Jothivenkatachalam, K., Prabhu, S., & Jeganathan, K. (2014). Chitosan based nanocomposite materials as photocatalyst—a review. *Materials Science Forum, 781*, 79–94. https://doi.org/10.4028/www.scientific.net/MSF.781.79
134. Patel, H. (2021). Review on solvent desorption study from exhausted adsorbent. *Journal of Saudi Chemical Society, 25*, 101302. https://doi.org/10.1016/j.jscs.2021.101302
135. Lai, K. C., Lee, L. Y., Hiew, B. Y. Z., Thangalazhy-Gopakumar, S., & Gan, S. (2019). Environmental application of three-dimensional graphene materials as adsorbents for dyes and heavy metals: Review on ice-templating method and adsorption mechanisms. *Journal of Environmental Science, 79*, 174–199.
136. Garba, Z. N., Zhou, W., Lawan, I., Xiao, W., Zhang, M., Wang, L., et al. (2019). An overview of chlorophenols as contaminants and their removal from wastewater by adsorption: A review. *Journal of Environmental Management, 241*, 59–75.
137. Leudjo, A., Klink, M. J., Yangkou, X., & Bobby, E. (2021). Chitosan nanocomposites for water treatment by fixed-bed continuous flow column adsorption: A review. *Carbohydrate Polymers, 255*, 117398. https://doi.org/10.1016/j.carbpol.2020.117398
138. Khanday, W. A., Ahmed, M. J., Okoye, P. U., Hummadi, E. H., & Hameed, B. H. (2019). Single-step pyrolysis of phosphoric acid-activated chitin for efficient adsorption of cephalexin antibiotic. *Bioresource Technology, 280*, 255–259.
139. Vakili, M., Deng, S., Cagnetta, G., Wang, W., Meng, P., Liu, D., et al. (2019). Regeneration of chitosan-based adsorbents used in heavy metal adsorption: A review. *Separation and Purification Technology, 224*, 373–387.
140. Yagub, M. T., Sen, T. K., Afroze, S., & Ang, H. M. (2014). Dye and its removal from aqueous solution by adsorption: A review. *Advances in Colloid and Interface Science, 209*, 172–184.
141. Daud, M., Hai, A., Banat, F., Wazir, M. B., Habib, M., & Bharath, G., et al. (2019). A review on the recent advances, challenges and future aspect of layered double hydroxides (LDH)—containing hybrids as promising adsorbents for dyes removal. *Journal of Molecular Liquids, 288*.
142. Ahmed, M. J., & Hameed, B. H. (2018). Removal of emerging pharmaceutical contaminants by adsorption in a fixed bed column: A review. *Ecotoxicology and Environmental Safety, 149*, 257–266.
143. Wu, M., Zhao, S., Jing, R., Shao, Y., Liu, X., & Lv, F., et al. (2019). Competitive adsorption of antibiotic tetracycline and ciprofloxacin on montmorillonite. *Applied Clay Science, 180*.
144. Hejazi, F., Ghoreyshi, A. A., & Rahimnejad, M. (2019). Simultaneous phenol removal and electricity generation using a hybrid granular activated carbon adsorption-biodegradation process in a batch recycled tubular microbial fuel cell. *Biomass & Bioenergy, 129*.
145. Shah, J., Jan, M. R., & Tasmia. (2018). Magnetic chitosan graphene oxide composite for solid phase extraction of phenylurea herbicides. *Carbohydrate Polymers, 199*, 461–472.
146. Vasile, C. (2018). Polymeric nanocomposites and nanocoatings for food packaging: A review. *Materials, 11*(10), 1834. https://doi.org/10.3390/ma11101834
147. Souza, V. G., Fernando, A. L., Pires, J. R., Rodrigues, P. F., Lopes, A. A. S., & Fernandes, F. M. (2017). Physical properties of chitosan films incorporated with natural antioxidants. *Industrial Crops and Products, 107*, 565–572. https://doi.org/10.1016/j.indcrop.2017.04.056
148. Kerch, G. (2015). Chitosan films and coatings prevent losses of fresh fruit nutritional quality: A review. *Trends in Food Science & Technology, 46*(2), 159–166. https://doi.org/10.1016/j.tifs.2015.10.010
149. Dehghani, S., Hosseini, S. V., & Regenstein, J. M. (2018). Edible films and coatings in seafood preservation: A review. *Food Chemistry, 240*, 505–513. https://doi.org/10.1016/j.foodchem.2017.07.034

150. Yuan, G., Chen, X., & Li, D. (2016). Chitosan films and coatings containing essential oils: The antioxidant and antimicrobial activity, and application in food systems. *Food Research International, 89*, 117–128. https://doi.org/10.1016/j.foodres.2016.10.004
151. Ganiari, S., Choulitoudi, E., & Oreopoulou, V. (2017). Edible and active films and coatings as carriers of natural antioxidants for lipid food. *Trends in Food Science & Technology, 68*, 70–82. https://doi.org/10.1016/j.tifs.2017.08.009
152. Sorrentino, A., Gorrasi, G., & Vittoria, V. (2007). Potential perspectives of bio-nanocomposites for food packaging applications. *Trends in Food Science & Technology, 18*(2), 84–95. https://doi.org/10.1016/j.tifs.2006.09.004
153. Dutta, P. K., Tripathi, S., Mehrotra, G. K., & Dutta, J. (2009). Perspectives for chitosan based antimicrobial films in food applications. *Food Chemistry, 114*(4), 1173–1182. https://doi.org/10.1016/j.foodchem.2008.11.047
154. Coma, V. (2008). Bioactive packaging technologies for extended shelf life of meat-based products. *Meat Science, 78*(1-2), 90–103. https://doi.org/10.1016/j.meatsci.2007.07.035
155. Soares, N., Pires, A. C., Camilloto, G., Santiago-Silva, P., Espitia, P. J., & Silva, W. (2009). Recent patents on active packaging for food application. *Recent Patents on Food, Nutrition & Agriculture, 1*(2), 171–178. https://doi.org/10.2174/2212798410901020171
156. Ahmed, I., Lin, H., Zou, L., Brody, A. L., Li, Z., Qazi, I. M., Pavase, T. R., & Lv, L. (2017). A comprehensive review on the application of active packaging technologies to muscle foods. *Food Control, 82*, 163–178. https://doi.org/10.1016/j.foodcont.2017.06.009
157. Souza, V. G., Pires, J. R. A., Vieira, É. T., Coelhoso, I. M., Duarte, M. P., & Fernando, A. L. (2019). Activity of chitosan-montmorillonite bionanocomposites incorporated with rosemary essential oil: From in vitro assays to application in fresh poultry meat. *Food Hydrocolloids, 89*, 241–252. https://doi.org/10.1016/j.foodhyd.2018.10.049
158. Pires, J. R., de Souza, V. G., & Fernando, A. L. (2018). Chitosan/Montmorillonite bionanocomposites incorporated with rosemary and ginger essential oil as packaging for fresh poultry meat. *Food Packaging and Shelf Life, 17*, 142–149. https://doi.org/10.1016/j.fpsl.2018.06.011
159. Müller, K., Bugnicourt, E., Latorre, M., Jorda, M., Echegoyen Sanz, Y., Lagaron, J., Miesbauer, O., Bianchin, A., Hankin, S., Bölz, U., Pérez, G., Jesdinszki, M., Lindner, M., Scheuerer, Z., Castelló, S., & Schmid, M. (2017). Review on the processing and properties of polymer nanocomposites and nanocoatings and their applications in the packaging, automotive and solar energy fields. *Nanomaterials, 7*(4), 74. https://doi.org/10.3390/nano7040074
160. Goddard, J. M., & Hotchkiss, J. H. (2007). Polymer surface modification for the attachment of bioactive compounds. *Progress in Polymer Science, 32*(7), 698–725. https://doi.org/10.1016/j.progpolymsci.2007.04.002
161. Munteanu, B. S., Aytac, Z., Pricope, G. M., Uyar, T., & Vasile, C. (2014). Polylactic acid (PLA)/silver-NP/vitamine bionanocomposite electrospun nanofibers with antibacterial and antioxidant activity. *Journal of Nanoparticle Research, 16*(10). https://doi.org/10.1007/s11051-014-2643-4
162. Stoleru, E., Dumitriu, R. P., Munteanu, B. S., Zaharescu, T., Tănase, E. E., Mitelut, A., Ailiesei, G.-L., & Vasile, C. (2016). Novel procedure to enhance PLA surface properties by chitosan irreversible immobilization. *Applied Surface Science, 367*, 407–417. https://doi.org/10.1016/j.apsusc.2016.01.200
163. Stoleru, E., Munteanu, S. B., Dumitriu, R. P., Coroaba, A., Drobotă, M., Zemljic, L. F., Pricope, G. M., & Vasile, C. (2016). Polyethylene materials with multifunctional surface properties by electrospraying chitosan/vitamin E formulation destined to biomedical and food packaging applications. *Iranian Polymer Journal, 25*(4), 295–307. https://doi.org/10.1007/s13726-016-0421-0
164. Pâslaru, E., Fras Zemljic, L., Bračič, M., Vesel, A., Petrinić, I., & Vasile, C. (2013). Stability of a chitosan layer deposited onto a polyethylene surface. *Journal of Applied Polymer Science, 130*(4), 2444–2457. https://doi.org/10.1002/app.39329
165. Xu, Y. X., Kim, K. M., Hanna, M. A., & Nag, D. (2005). Chitosan–starch composite film: Preparation and characterization. *Industrial Crops and Products, 21*(2), 185–192. https://doi.org/10.1016/j.indcrop.2004.03.002

166. Jindal, M., Kumar, V., Rana, V., & Tiwary, A. K. (2013). An insight into the properties of Aegle Marmelos pectin–Chitosan cross-linked films. *International Journal of Biological Macromolecules, 52*, 77–84. https://doi.org/10.1016/j.ijbiomac.2012.10.020
167. Luo, Y., & Wang, Q. (2014). Recent development of chitosan-based polyelectrolyte complexes with natural polysaccharides for drug delivery. *International Journal of Biological Macromolecules, 64*, 353–367. https://doi.org/10.1016/j.ijbiomac.2013.12.017
168. Ferreira, A. R. V., Torres, C. A. V., Freitas, F., Sevrin, C., Grandfils, C., Reis, M. A. M., Alves, V. D., & Coelhoso, I. M. (2016). Development and characterization of bilayer films of Fucopol and Chitosan. *Carbohydrate Polymers, 147*, 8–15. https://doi.org/10.1016/j.carbpol.2016.03.089
169. Freitas, F., Alves, V. D., Reis, M. A., Crespo, J. G., & Coelhoso, I. M. (2013). Microbial polysaccharide-based membranes: Current and future applications. *Journal of Applied Polymer Science, 131*(6). https://doi.org/10.1002/app.40047
170. Rivero, S., García, M. A., & Pinotti, A. (2009). Composite and bi-layer films based on gelatin and Chitosan. *Journal of Food Engineering, 90*(4), 531–539. https://doi.org/10.1016/j.jfoodeng.2008.07.021
171. Ferreira, C. O., Nunes, C. A., Delgadillo, I., & Lopes-da-Silva, J. A. (2009). Characterization of chitosan–whey protein films at ACID PH. *Food Research International, 42*(7), 807–813. https://doi.org/10.1016/j.foodres.2009.03.005
172. Kurek, M., Galus, S., & Debeaufort, F. (2014). Surface, mechanical and barrier properties of bio-based composite films based on Chitosan and whey protein. *Food Packaging and Shelf Life, 1*(1), 56–67. https://doi.org/10.1016/j.fpsl.2014.01.001
173. Souza, V. G., & Fernando, A. L. (2016). Nanoparticles in food packaging: Biodegradability and potential migration to food—a review. *Food Packaging and Shelf Life, 8*, 63–70. https://doi.org/10.1016/j.fpsl.2016.04.001
174. Souza, V. G., Pires, J. R. A., Rodrigues, P. F., Lopes, A. A. S., Fernandes, F. M. B., Duarte, M. P., Coelhoso, I. M., & Fernando, A. L. (2018). Bionanocomposites of Chitosan/Montmorillonite incorporated with rosmarinus officinalis essential oil: Development and physical characterization. *Food Packaging and Shelf Life, 16*, 148–156. https://doi.org/10.1016/j.fpsl.2018.03.009
175. Beigzadeh Ghelejlu, S., Esmaiili, M., & Almasi, H. (2016). Characterization of chitosan–nanoclay bionanocomposite active films containing milk thistle extract. *International Journal of Biological Macromolecules, 86*, 613–621. https://doi.org/10.1016/j.ijbiomac.2016.02.012
176. Vlacha, M., Giannakas, A., Katapodis, P., Stamatis, H., Ladavos, A., & Barkoula, N.-M. (2016). On the efficiency of oleic acid as plasticizer of chitosan/clay nanocomposites and its role on thermo-mechanical, barrier and antimicrobial properties—comparison with glycerol. *Food Hydrocolloids, 57*, 10–19. https://doi.org/10.1016/j.foodhyd.2016.01.003
177. Qin, Y.-Y., Zhang, Z.-H., Li, L., Yuan, M.-L., Fan, J., & Zhao, T.-R. (2013). Physio-mechanical properties of an active chitosan film incorporated with montmorillonite and natural antioxidants extracted from Pomegranate Rind. *Journal of Food Science and Technology, 52*(3), 1471–1479. https://doi.org/10.1007/s13197-013-1137-1
178. Nouri, A., Yaraki, M. T., Ghorbanpour, M., Agarwal, S., & Gupta, V. K. (2018). Enhanced antibacterial effect of chitosan film using Montmorillonite/Cuo nanocomposite. *International Journal of Biological Macromolecules, 109*, 1219–1231. https://doi.org/10.1016/j.ijbiomac.2017.11.119
179. Souza, V. G., Pires, J. R., Rodrigues, C., Rodrigues, P. F., Lopes, A., Silva, R. J., Caldeira, J., Duarte, M. P., Fernandes, F. B., Coelhoso, I. M., & Fernando, A. L. (2019). Physical and morphological characterization of Chitosan/Montmorillonite films incorporated with ginger essential oil. *Coatings, 9*(11), 700. https://doi.org/10.3390/coatings9110700
180. Giannakas, A., Vlacha, M., Salmas, C., Leontiou, A., Katapodis, P., Stamatis, H., Barkoula, N.-M., & Ladavos, A. (2016). Preparation, characterization, mechanical, barrier and antimicrobial properties of chitosan/PVOH/clay nanocomposites. *Carbohydrate Polymers, 140*, 408–415. https://doi.org/10.1016/j.carbpol.2015.12.072

181. Souza, V., Pires, J., Vieira, É., Coelhoso, I., Duarte, M., & Fernando, A. (2018). Shelf life assessment of fresh poultry meat packaged in novel Bionanocomposite of chitosan/montmorillonite incorporated with Ginger Essential Oil. *Coatings, 8*(5), 177. https://doi.org/10.3390/coatings8050177
182. Celebi, H., & Kurt, A. (2015). Effects of processing on the properties of chitosan/cellulose nanocrystal films. *Carbohydrate Polymers, 133*, 284–293. https://doi.org/10.1016/j.carbpol.2015.07.007
183. Mao, H., Wei, C., Gong, Y., Wang, S., & Ding, W. (2019). Mechanical and water-resistant properties of eco-friendly chitosan membrane reinforced with cellulose nanocrystals. *Polymers, 11*(1), 166. https://doi.org/10.3390/polym11010166
184. Abdul Khalil, H. P. S., Saurabh, C. K., Adnan, A. S., Nurul Fazita, M. R., Syakir, M. I., Davoudpour, Y., Rafatullah, M., Abdullah, C. K., Haafiz, M. K., & Dungani, R. (2016). A review on chitosan-cellulose blends and nanocellulose reinforced chitosan biocomposites: Properties and their applications. *Carbohydrate Polymers, 150*, 216–226. https://doi.org/10.1016/j.carbpol.2016.05.028
185. Pires, J. R., de Souza, V. G., & Fernando, A. L. (2018). Production of nanocellulose from lignocellulosic biomass wastes: Prospects and limitations. *Innovation, Engineering and Entrepreneurship*, 719–725. https://doi.org/10.1007/978-3-319-91334-6_98
186. Tang, Y., Zhang, X., Zhao, R., Guo, D., & Zhang, J. (2018). Preparation and properties of chitosan/guar gum/nanocrystalline cellulose nanocomposite films. *Carbohydrate Polymers, 197*, 128–136. https://doi.org/10.1016/j.carbpol.2018.05.073
187. Noorbakhsh-Soltani, S. M., Zerafat, M. M., & Sabbaghi, S. (2018). A comparative study of gelatin and starch-based nano-composite films modified by nano-cellulose and chitosan for food packaging applications. *Carbohydrate Polymers, 189*, 48–55. https://doi.org/10.1016/j.carbpol.2018.02.012
188. Youssef, A. M., Abou-Yousef, H., El-Sayed, S. M., & Kamel, S. (2015). Mechanical and antibacterial properties of novel high performance chitosan/nanocomposite films. *International Journal of Biological Macromolecules, 76*, 25–32. https://doi.org/10.1016/j.ijbiomac.2015.02.016
189. Al-Naamani, L., Dobretsov, S., & Dutta, J. (2016). Chitosan-zinc oxide nanoparticle composite coating for active food packaging applications. *Innovative Food Science & Emerging Technologies, 38*, 231–237. https://doi.org/10.1016/j.ifset.2016.10.010
190. Zhang, X., Xiao, G., Wang, Y., Zhao, Y., Su, H., & Tan, T. (2017). Preparation of chitosan-TIO_2 composite film with efficient antimicrobial activities under visible light for food packaging applications. *Carbohydrate Polymers, 169*, 101–107. https://doi.org/10.1016/j.carbpol.2017.03.073
191. Mallakpour, S., & Madani, M. (2015). Effect of functionalized TiO_2 on mechanical, thermal and swelling properties of chitosan-based nanocomposite films. *Polymer-Plastics Technology and Engineering, 54*(10), 1035–1042. https://doi.org/10.1080/03602559.2014.974194
192. Yun, Y.-H., Yun, J.-W., Yoon, S.-D., & Byun, H.-S. (2016). Physical properties and photocatalytic activity of chitosan-based nanocomposites added titanium oxide nanoparticles. *Macromolecular Research, 24*(1), 51–59. https://doi.org/10.1007/s13233-016-4008-6
193. De Silva, R. T., Mantilaka, M. M. M. G. P. G., Ratnayake, S. P., Amaratunga, G. A. J., & de Silva, K. M. N. (2017). Nano-MGO reinforced chitosan nanocomposites for high performance packaging applications with improved mechanical, thermal and barrier properties. *Carbohydrate Polymers, 157*, 739–747. https://doi.org/10.1016/j.carbpol.2016.10.038
194. Jahed, E., Khaledabad, M. A., Almasi, H., & Hasanzadeh, R. (2017). Physicochemical properties of Carum Copticum essential oil loaded chitosan films containing organic nanoreinforcements. *Carbohydrate Polymers, 164*, 325–338. https://doi.org/10.1016/j.carbpol.2017.02.022
195. Hafsa, J., Smach, M. A., Ben Khedher, M. R., Charfeddine, B., Limem, K., Majdoub, H., & Rouatbi, S. (2016). Physical, antioxidant and antimicrobial properties of chitosan films containing eucalyptus globulus essential oil. *LWT—Food Science and Technology, 68*, 356–364. https://doi.org/10.1016/j.lwt.2015.12.050

196. Perdones, Á., Escriche, I., Chiralt, A., & Vargas, M. (2016). Effect of chitosan–lemon essential oil coatings on volatile profile of strawberries during storage. *Food Chemistry, 197*, 979–986. https://doi.org/10.1016/j.foodchem.2015.11.054
197. Yuan, G., Lv, H., Yang, B., Chen, X., & Sun, H. (2015). Physical properties, antioxidant and antimicrobial activity of chitosan films containing carvacrol and pomegranate peel extract. *Molecules, 20*(6), 11034–11045. https://doi.org/10.3390/molecules200611034
198. Kalaycıoğlu, Z., Torlak, E., Akın-Evingür, G., Özen, İ, & Erim, F. B. (2017). Antimicrobial and physical properties of chitosan films incorporated with turmeric extract. *International Journal of Biological Macromolecules, 101*, 882–888. https://doi.org/10.1016/j.ijbiomac.2017.03.174
199. Lekjing, S. (2016). A chitosan-based coating with or without clove oil extends the shelf life of cooked pork sausages in refrigerated storage. *Meat Science, 111*, 192–197. https://doi.org/10.1016/j.meatsci.2015.10.003
200. Galus, S., & Kadzińska, J. (2015). Food applications of emulsion-based edible films and coatings. *Trends in Food Science & Technology, 45*(2), 273–283. https://doi.org/10.1016/j.tifs.2015.07.011
201. Hromiš, N. M., Lazić, V. L., Markov, S. L., Vaštag, ŽG., Popović, S. Z., Šuput, D. Z., Džinić, N. R., Velićanski, A. S., & Popović, L. M. (2015). Optimization of chitosan biofilm properties by addition of caraway essential oil and beeswax. *Journal of Food Engineering, 158*, 86–93. https://doi.org/10.1016/j.jfoodeng.2015.01.001
202. Vargas, M., Perdones, Á., Chiralt, A., Cháfer, M., & González-Martínez, C. (2011). Effect of homogenization conditions on physicochemical properties of chitosan-based film-forming dispersions and films. *Food Hydrocolloids, 25*(5), 1158–1164. https://doi.org/10.1016/j.foodhyd.2010.11.002
203. Sanuja, S., Agalya, A., & Umapathy, M. J. (2015). Synthesis and characterization of zinc oxide–neem oil–Chitosan bionanocomposite for food packaging application. *International Journal of Biological Macromolecules, 74*, 76–84. https://doi.org/10.1016/j.ijbiomac.2014.11.036
204. Ojagh, S. M., Rezaei, M., Razavi, S. H., & Hosseini, S. M. (2010). Development and evaluation of a novel biodegradable film made from chitosan and cinnamon essential oil with low affinity toward water. *Food Chemistry, 122*(1), 161–166. https://doi.org/10.1016/j.foodchem.2010.02.033
205. Antony, R., Arun, T., & Manickam, S. T. D. (2019). A review on applications of Chitosan-based Schiff bases. *International Journal of Biological Macromolecules, 129*, 615–633. https://doi.org/10.1016/j.ijbiomac.2019.02.047
206. Epure, V., Griffon, M., Pollet, E., & Avérous, L. (2011). Structure and properties of glycerol-plasticized chitosan obtained by mechanical kneading. *Carbohydrate Polymers, 83*(2), 947–952. https://doi.org/10.1016/j.carbpol.2010.09.003
207. Galvis-Sánchez, A. C., Castro, M. C., Biernacki, K., Gonçalves, M. P., & Souza, H. K. S. (2018). Natural deep eutectic solvents as green plasticizers for chitosan thermoplastic production with controlled/desired mechanical and barrier properties. *Food Hydrocolloids, 82*, 478–489. https://doi.org/10.1016/j.foodhyd.2018.04.026
208. Matet, M., Heuzey, M.-C., Pollet, E., Ajji, A., & Avérous, L. (2013). Innovative thermoplastic Chitosan obtained by thermo-mechanical mixing with polyol plasticizers. *Carbohydrate Polymers, 95*(1), 241–251. https://doi.org/10.1016/j.carbpol.2013.02.052
209. Matet, M., Heuzey, M.-C., Ajji, A., & Sarazin, P. (2015). Plasticized chitosan/polyolefin films produced by extrusion. *Carbohydrate Polymers, 117*, 177–184. https://doi.org/10.1016/j.carbpol.2014.09.058
210. Dang, K. M., & Yoksan, R. (2015). Development of thermoplastic starch blown film by incorporating plasticized chitosan. *Carbohydrate Polymers, 115*, 575–581. https://doi.org/10.1016/j.carbpol.2014.09.005
211. Dang, K. M., & Yoksan, R. (2016). Morphological characteristics and barrier properties of thermoplastic starch/chitosan blown film. *Carbohydrate Polymers, 150*, 40–47. https://doi.org/10.1016/j.carbpol.2016.04.113

212. Mendes, J. F., Paschoalin, R. T., Carmona, V. B., Sena Neto, A. R., Marques, A. C. P., Marconcini, J. M., Mattoso, L. H. C., Medeiros, E. S., & Oliveira, J. E. (2016). Biodegradable polymer blends based on corn starch and thermoplastic chitosan processed by extrusion. *Carbohydrate Polymers, 137*, 452–458. https://doi.org/10.1016/j.carbpol.2015.10.093
213. Galvis-Sánchez, A. C., Sousa, A. M., Hilliou, L., Gonçalves, M. P., & Souza, H. K. (2016). Thermo-compression molding of chitosan with a deep eutectic mixture for biofilms development. *Green Chemistry, 18*(6), 1571–1580. https://doi.org/10.1039/c5gc02231b
214. Kim, S.-K. (2014). *Chitin and chitosan derivatives: Advances in drug discovery and developments*. CRC Press.
215. Venkatesan, J., & Kim, S.-K. (2010). Chitosan composites for bone tissue engineering—an overview. *Marine Drugs, 8*(8), 2252–2266. https://doi.org/10.3390/md8082252
216. Huang, R. Y. M., Pal, R., & Moon, G. Y. (1999). Crosslinked chitosan composite membrane for the pervaporation dehydration of alcohol mixtures and enhancement of structural stability of chitosan/polysulfone composite membranes. *Journal of Membrane Science, 160*(1), 17–30.
217. Kong, L., Gao, Y., Cao, W., Gong, Y., Zhao, N., & Zhang, X. (2005). Preparation and characterization of nano-hydroxyapatite/chitosan composite scaffolds. *Journal of Biomedical Materials Research Part A, 75*(2), 275–282.
218. Tan, W., Krishnaraj, R., & Desai, T. A. (2001). Evaluation of nanostructured composite collagen–chitosan matrices for tissue engineering. *Tissue Engineering, 7*(2), 203–210.
219. Zhang, Y., & Zhang, M. (2002). Calcium phosphate/chitosan composite scaffolds for controlled in vitro antibiotic drug release. *Journal of Biomedical Materials Research, 62*(3), 378–386.
220. Hein, S., Wang, K., Stevens, W. F., & Kjems, J. (2008). Chitosan Composites for biomedical applications: Status, challenges and perspectives. *Materials Science and Technology, 24*(9), 1053–1061. https://doi.org/10.1179/174328408x341744
221. Muzzarelli, C., & Muzzarelli, R. A. A. (2002). Natural and artificial chitosan–inorganic composites. *Journal of Inorganic Biochemistry, 92*(2), 89–94. https://doi.org/10.1016/s0162-0134(02)00486-5
222. Kenny, S. M., & Buggy, M. (2003). *Journal of Materials Science: Materials in Medicine, 14*(11), 923–938. https://doi.org/10.1023/a:1026394530192
223. Ohashi, E., & Koriyama, T. (1992). Simple and mild preparation of an enzyme-immobilized membrane for a biosensor using β-type crystalline chitin. *Analytica Chimica Acta, 262*(1), 19–25. https://doi.org/10.1016/0003-2670(92)80003-p
224. Magalhães, J. (1998). Urea potentiometric biosensor based on urease immobilized on chitosan membranes. *Talanta, 47*(1), 183–191. https://doi.org/10.1016/s0039-9140(98)00066-6
225. Miao, Y., & Tan, S. N. (2000). Amperometric hydrogen peroxide biosensor based on immobilization of peroxidase in chitosan matrix crosslinked with glutaraldehyde. *The Analyst, 125*(9), 1591–1594. https://doi.org/10.1039/b003483p
226. Luo, X.-L., Xu, J.-J., Du, Y., & Chen, H.-Y. (2004). A glucose biosensor based on chitosan–glucose oxidase–gold nanoparticles biocomposite formed by one-step electrodeposition. *Analytical Biochemistry, 334*(2), 284–289. https://doi.org/10.1016/j.ab.2004.07.005
227. Kerman, K., Saito, M., Morita, Y., Takamura, Y., Ozsoz, M., & Tamiya, E. (2004). Electrochemical coding of single-nucleotide polymorphisms by monobase-modified gold nanoparticles. *Analytical Chemistry, 76*(7), 1877–1884. https://doi.org/10.1021/ac0351872
228. Lu, Y., Yang, M., Qu, F., Shen, G., & Yu, R. (2007). Amperometric biosensors based on platinum nanowires. *Analytical Letters, 40*(5), 875–886. https://doi.org/10.1080/000327107 01242071
229. Qu, F., Yang, M., Shen, G., & Yu, R. (2007). Electrochemical biosensing utilizing synergic action of carbon nanotubes and platinum nanowires prepared by template synthesis. *Biosensors and Bioelectronics, 22*(8), 1749–1755. https://doi.org/10.1016/j.bios.2006.08.003
230. Liu, Y., Yang, Y., Yang, H., Liu, Z., Shen, G., & Yu, R. (2005). Nanosized flower-like zno synthesized by a simple hydrothermal method and applied as matrix for horseradish peroxidase immobilization for electro-biosensing. *Journal of Inorganic Biochemistry, 99*(10), 2046–2053. https://doi.org/10.1016/j.jinorgbio.2005.07.001

231. Wang, Z., Yang, Y., Li, J., Gong, J., Shen, G., & Yu, R. (2006). Organic–inorganic matrix for electrochemical immunoassay: Detection of human IGG based on ZnO/chitosan composite. *Talanta, 69*(3), 686–690. https://doi.org/10.1016/j.talanta.2005.11.004
232. Yang, Y., Wang, Z., Yang, M., Li, J., Zheng, F., Shen, G., & Yu, R. (2007). Electrical detection of deoxyribonucleic acid hybridization based on carbon-nanotubes/nano zirconium dioxide/chitosan-modified electrodes. *Analytica Chimica Acta, 584*(2), 268–274. https://doi.org/10.1016/j.aca.2006.11.055
233. Yang Y., Yang, H., Yang, M., Liu, Y., Shen, G., & Yu, R. (2004). Amperometric glucose biosensor based on a surface-treated nanoporous ZRO_2/chitosan composite film as immobilization matrix. *Analytica Chimica Acta, 525*(2), 213–220. https://doi.org/10.1016/j.aca.2004.07.071
234. Luo, X., Killard, A. J., & Smyth, M. R. (2006). Reagentless glucose biosensor based on the direct electrochemistry of glucose oxidase on carbon nanotube-modified electrodes. *Electroanalysis, 18*(11), 1131–1134. https://doi.org/10.1002/elan.200603513
235. Ding, L., Hao, C., Xue, Y., & Ju, H. (2007). A bio-inspired support of gold nanoparticles−chitosan nanocomposites gel for immobilization and electrochemical study of K562 leukemia cells. *Biomacromolecules, 8*(4), 1341–1346. https://doi.org/10.1021/bm061224y
236. Chen, T. W., Chang, S. J., Niu, G.C.-C., Hsu, Y. T., & Kuo, S. M. (2006). Alginate-coated chitosan membrane for guided tissue regeneration. *Journal of Applied Polymer Science, 102*(5), 4528–4534. https://doi.org/10.1002/app.24945
237. López-Pérez, P. M., Marques, A. P., Silva, R. M., Pashkuleva, I., & Reis, R. L. (2007). Effect of chitosan membrane surface modification via plasma induced polymerization on the adhesion of osteoblast-like cells. *Journal of Materials Chemistry, 17*(38), 4064. https://doi.org/10.1039/b707326g
238. Mi, F.-L., Shyu, S.-S., Wu, Y.-B., Lee, S.-T., Shyong, J.-Y., & Huang, R.-N. (2001). Fabrication and characterization of a sponge-like asymmetric chitosan membrane as a wound dressing. *Biomaterials, 22*(2), 165–173. https://doi.org/10.1016/s0142-9612(00)00167-8
239. Tuzlakoglu, K., & Reis, R. L. (2007). Formation of bone-like apatite layer on chitosan fiber mesh scaffolds by a biomimetic spraying process. *Journal of Materials Science: Materials in Medicine, 18*(7), 1279–1286. https://doi.org/10.1007/s10856-006-0063-4
240. Bruyere, O., & Reginster, J.-Y. (2007). Glucosamine and chondroitin sulfate as therapeutic agents for knee and hip osteoarthritis. *Drugs & Aging, 24*(7), 573–580. https://doi.org/10.2165/00002512-200724070-00005
241. Sinha, V. R., & Kumria, R. (2001). Polysaccharides in colon-specific drug delivery. *International Journal of Pharmaceutics, 224*(1–2), 19–38. https://doi.org/10.1016/s0378-5173(01)00720-7
242. Chen, W.-B., Wang, L.-F., Chen, J.-S., & Fan, S.-Y. (2005). Characterization of polyelectrolyte complexes between chondroitin sulfate and chitosan in the Solid State. *Journal of Biomedical Materials Research Part A, 75A*(1), 128–137. https://doi.org/10.1002/jbm.a.30393
243. Ganza-González, A., Anguiano-Igea, S., Otero-Espinar, F. J., & Blanco, M. J. (1999). Chitosan and chondroitin microspheres for oral-administration controlled release of metoclopramide. *European Journal of Pharmaceutics and Biopharmaceutics, 48*(2), 149–155. https://doi.org/10.1016/s0939-6411(99)00040-5
244. Liu, H., Yin, Y., & Yao, K. (2006). Construction of chitosan—gelatin—hyaluronic acid artificial skin in vitro. *Journal of Biomaterials Applications, 21*(4), 413–430. https://doi.org/10.1177/0885328206065241
245. Srinoulprasert, Y., Kongtawelert, P., & Chaiyaroj, S. C. (2006). Chondroitin sulfate B and heparin mediate adhesion of *Penicillium marneffei* conidia to host extracellular matrices. *Microbial Pathogenesis, 40*(3), 126–132. https://doi.org/10.1016/j.micpath.2005.12.001
246. Barańska-Rybak, W., Sonesson, A., Nowicki, R., & Schmidtchen, A. (2005). Glycosaminoglycans inhibit the antibacterial activity of LL-37 in biological fluids. *Journal of Antimicrobial Chemotherapy, 57*(2), 260–265. https://doi.org/10.1093/jac/dki460

247. Hamdine, M., Heuzey, M.-C., & Bégin, A. (2005). Effect of organic and inorganic acids on concentrated chitosan solutions and Gels. *International Journal of Biological Macromolecules, 37*(3), 134–142. https://doi.org/10.1016/j.ijbiomac.2005.09.009
248. Yang, R., Li, H., Huang, M., Yang, H., & Li, A. (2016). A review on chitosan-based flocculants and their applications in water treatment. *Water Research, 95*, 59–89. https://doi.org/10.1016/j.watres.2016.02.068
249. Wu, M., Long, Z., Xiao, H., & Dong, C. (2016). Recent research progress on preparation and application of *N*, *N*, *N*-trimethyl chitosan. *Carbohydrate Research, 434*, 27–32. https://doi.org/10.1016/j.carres.2016.08.002
250. Malini, M., Thirumavalavan, M., Yang, W.-Y., Lee, J.-F., & Annadurai, G. (2015). A versatile chitosan/zno nanocomposite with enhanced antimicrobial properties. *International Journal of Biological Macromolecules, 80*, 121–129. https://doi.org/10.1016/j.ijbiomac.2015.06.036
251. Rai, M., Yadav, A., & Gade, A. (2009). Silver nanoparticles as a new generation of antimicrobials. *Biotechnology Advances, 27*(1), 76–83. https://doi.org/10.1016/j.biotechadv.2008.09.002
252. Madhumathi, K., Sudheesh Kumar, P. T., Abilash, S., Sreeja, V., Tamura, H., Manzoor, K., et al. (2010). Development of novel chitin/nanosilver composite scaffolds for wound dressing applications. *Journal of Material Science: Materials in Medicine, 21*, 807–813.
253. Lim, S.-H., & Hudson, S. M. (2004). Synthesis and antimicrobial activity of a water-soluble chitosan derivative with a fiber-reactive group. *Carbohydrate Research, 339*(2), 313–319. https://doi.org/10.1016/j.carres.2003.10.024
254. Liu, N., Chen, X.-G., Park, H.-J., Liu, C.-G., Liu, C.-S., Meng, X.-H., & Yu, L.-J. (2006). Effect of MW and concentration of chitosan on antibacterial activity of *Escherichia coli*. *Carbohydrate Polymers, 64*(1), 60–65. https://doi.org/10.1016/j.carbpol.2005.10.028
255. Reesha, K. V., Panda, S. K., Bindu, J., & Varghese, T. O. (2015). Development and characterization of an LDPE/chitosan composite antimicrobial film for chilled fish storage. *International Journal of Biological Macromolecules, 79*, 934–942. https://doi.org/10.1016/j.ijbiomac.2015.06.016
256. Kong, M., Chen, X. G., Xing, K., & Park, H. J. (2010). Antimicrobial properties of Chitosan and mode of action: A state of the art review. *International Journal of Food Microbiology, 144*(1), 51–63. https://doi.org/10.1016/j.ijfoodmicro.2010.09.012
257. Tsai, G., & Su, W. (1999). Antibacterial activity of shrimp chitosan against *Escherichia coli*. *Journal of Food Protection, 62*(3), 239–243. https://doi.org/10.4315/0362-028x-62.3.239
258. Taha, S. M. A., & Swailam, H. M. H. (2002). Antibacterial activity of chitosan against *Aeromonas hydrophila*. *Nahrung/Food, 46*(5), 337–340. https://doi.org/10.1002/1521-3803(20020901)46:5%3C337::AID-FOOD337%3E3.0.CO;2-3
259. Chen, C.-Y., & Chung, Y.-C. (2012). Antibacterial effect of water-soluble chitosan on representative dental pathogens *Streptococcus mutans* and *Lactobacilli brevis*. *Journal of Applied Oral Science, 20*(6), 620–627. https://doi.org/10.1590/s1678-77572012000600006
260. Jung, E. J., Youn, D. K., Lee, S. H., No, H. K., Ha, J. G., & Prinyawiwatkul, W. (2010). Antibacterial activity of chitosans with different degrees of deacetylation and Viscosities. *International Journal of Food Science & Technology, 45*(4), 676–682. https://doi.org/10.1111/j.1365-2621.2010.02186.x
261. Fei Liu, X., Lin Guan, Y., Zhi Yang, D., Li, Z., & De Yao, K. (2000). Antibacterial action of chitosan and carboxymethylated chitosan. *Journal of Applied Polymer Science, 79*(7), 1324–1335. https://doi.org/10.1002/1097-4628(20010214)79:7%3c1324::aid-app210%3e3.0.co;2-1
262. Chung, Y.-C., & Chen, C.-Y. (2008). Antibacterial characteristics and activity of acid-soluble chitosan. *Bioresource Technology, 99*(8), 2806–2814. https://doi.org/10.1016/j.biortech.2007.06.044
263. Anitha, A., Sowmya, S., Kumar, P. T. S., Deepthi, S., Chennazhi, K. P., Ehrlich, H., Tsurkan, M., & Jayakumar, R. (2014). Chitin and Chitosan in selected biomedical applications. *Progress in Polymer Science, 39*(9), 1644–1667. https://doi.org/10.1016/j.progpolymsci.2014.02.008

264. Venkatesan J., Vinodhini, P. A., Sudha, P. N., & Kim, S.-K. (2014). Chitin and chitosan composites for bone tissue regeneration. *Marine Carbohydrates: Fundamentals and Applications, Part B*, 59–81. https://doi.org/10.1016/b978-0-12-800268-1.00005-6
265. Giri T. K., Thakur, A., Alexander, A., Ajazuddin, Badwaik, H., & Tripathi, D. K. (2012). Modified Chitosan Hydrogels as drug delivery and tissue engineering systems: Present status and applications. *Acta Pharmaceutica Sinica B, 2*(5), 439–449. https://doi.org/10.1016/j.apsb.2012.07.004
266. Wan, A. C. A., & Tai, B. C. U. (2013). Chitin—a promising biomaterial for tissue engineering and stem cell technologies. *Biotechnology Advances, 31*(8), 1776–1785. https://doi.org/10.1016/j.biotechadv.2013.09.007
267. Wang, J., Liu, C., & Chi, P. (2008). In situ preparation of glycoconjugate hollow microspheres mimics the extracellular matrix via interfacial polymerization. *International Journal of Biological Macromolecules, 42*(5), 450–454. https://doi.org/10.1016/j.ijbiomac.2008.02.007
268. Jayakumar, R., Menon, D., Manzoor, K., Nair, S. V., & Tamura, H. (2010). Biomedical applications of chitin and chitosan based nanomaterials—a short review. *Carbohydrate Polymers, 82*(2), 227–232. https://doi.org/10.1016/j.carbpol.2010.04.074
269. Valerio, P., Pereira, M. M., Goes, A. M., & Leite, M. F. (2004). The effect of ionic products from bioactive glass dissolution on osteoblast proliferation and collagen production. *Biomaterials, 25*(15), 2941–2948. https://doi.org/10.1016/j.biomaterials.2003.09.086
270. Peter, M., Binulal, N. S., Nair, S. V., Selvamurugan, N., Tamura, H., & Jayakumar, R. (2010). Novel biodegradable chitosan–gelatin/nano-bioactive glass ceramic composite scaffolds for alveolar bone tissue engineering. *Chemical Engineering Journal, 158*(2), 353–361. https://doi.org/10.1016/j.cej.2010.02.003
271. Kishore, V., Bullock, W., Sun, X., Van Dyke, W. S., & Akkus, O. (2012). Tenogenic differentiation of human mscs induced by the topography of electrochemically aligned collagen threads. *Biomaterials, 33*(7), 2137–2144. https://doi.org/10.1016/j.biomaterials.2011.11.066
272. Kinneberg, K. R., Galloway, M. T., Butler, D. L., & Shearn, J. T. (2011). Effect of implanting a soft tissue autograft in a central-third patellar tendon defect: Biomechanical and histological comparisons. *Journal of Biomechanical Engineering, 133*(9). https://doi.org/10.1115/1.4004948
273. Funakoshi, T., Majima, T., Iwasaki, N., Suenaga, N., Sawaguchi, N., Shimode, K., Minami, A., Harada, K., & Nishimura, S.-I. (2005). Application of tissue engineering techniques for rotator cuff regeneration using a chitosan-based hyaluronan hybrid fiber scaffold. *The American Journal of Sports Medicine, 33*(8), 1193–1201. https://doi.org/10.1177/0363546504272689
274. Funakoshi, T., Majima, T., Suenaga, N., Iwasaki, N., Yamane, S., & Minami, A. (2006). Rotator cuff regeneration using chitin fabric as an acellular matrix. *Journal of Shoulder and Elbow Surgery, 15*(1), 112–118. https://doi.org/10.1016/j.jse.2005.05.012
275. Sato, M., Maeda, M., Kurosawa, H., Inoue, Y., Yamauchi, Y., & Iwase, H. (2000). Reconstruction of rabbit achilles tendon with three bioabsorbable materials: Histological and biomechanical studies. *Journal of Orthopaedic Science, 5*(3), 256–267. https://doi.org/10.1007/s007760050161
276. Shao, H.-J., Chen, C. S., Lee, Y.-T., Wang, J.-H., & Young, T.-H. (2009). The phenotypic responses of human anterior cruciate ligament cells cultured on poly(ϵ-caprolactone) and Chitosan. *Journal of Biomedical Materials Research Part A*, 9999A. https://doi.org/10.1002/jbm.a.32629
277. Shao, H.-J., Lee, Y.-T., Chen, C.-S., Wang, J.-H., & Young, T.-H. (2010). Modulation of gene expression and collagen production of anterior cruciate ligament cells through cell shape changes on polycaprolactone/Chitosan blends. *Biomaterials, 31*(17), 4695–4705. https://doi.org/10.1016/j.biomaterials.2010.02.037
278. Sarukawa, J., Takahashi, M., Abe, M., Suzuki, D., Tokura, S., Furuike, T., & Tamura, H. (2011). Effects of chitosan-coated fibers as a scaffold for three-dimensional cultures of rabbit fibroblasts for ligament tissue engineering. *Journal of Biomaterials Science, Polymer Edition, 22*(4–6), 717–732. https://doi.org/10.1163/092050610x491067

279. Hayami, J. W., Surrao, D. C., Waldman, S. D., & Amsden, B. G. (2009). Design and characterization of a biodegradable composite scaffold for ligament tissue engineering. *Journal of Biomedical Materials Research Part A*, 9999A. https://doi.org/10.1002/jbm.a.32472
280. Nettles, D. L., Elder, S. H., & Gilbert, J. A. (2002). Potential use of chitosan as a cell scaffold material for cartilage tissue engineering. *Tissue Engineering, 8*(6), 1009–1016. https://doi.org/10.1089/107632702320934100
281. Park, H., Choi, B., Hu, J., & Lee, M. (2013). Injectable chitosan hyaluronic acid hydrogels for cartilage tissue engineering. *Acta Biomaterialia, 9*(1), 4779–4786. https://doi.org/10.1016/j.actbio.2012.08.033
282. Xi, L. J., Prudhommeaux, F., Meunier, A., Sedel, L., & Guillemin, G. (1999). Effects of chitosan on Rat knee cartilages. *Biomaterials, 20*(20), 1937–1944. https://doi.org/10.1016/s0142-9612(99)00097-6
283. Jin, R., Moreira Teixeira, L. S., Dijkstra, P. J., Karperien, M., van Blitterswijk, C. A., Zhong, Z. Y., & Feijen, J. (2009). Injectable chitosan-based hydrogels for cartilage tissue engineering. *Biomaterials, 30*(13), 2544–2551. https://doi.org/10.1016/j.biomaterials.2009.01.020
284. Tan, H., Chu, C. R., Payne, K. A., & Marra, K. G. (2009). Injectable in situ forming biodegradable chitosan–hyaluronic acid based hydrogels for cartilage tissue engineering. *Biomaterials, 30*(13), 2499–2506. https://doi.org/10.1016/j.biomaterials.2008.12.080
285. Ragetly, G. R., Slavik, G. J., Cunningham, B. T., Schaeffer, D. J., & Griffon, D. J. (2009). Cartilage tissue engineering on fibrous chitosan scaffolds produced by a replica molding technique. *Journal of Biomedical Materials Research Part A*, 9999A. https://doi.org/10.1002/jbm.a.32514

Agar Based Composite as a New Alternative Biopolymer

Ridhwan Jumaidin

1 Introduction

Since few decades ago, plastic has been used as main product in many sectors especially in packaging application whether can be reuse again or only for single-use [1]. In global, the plastic production had continued rising for more than five decades. In 2013, the plastic produced are 299 million tons and it had been recorded to increase about 3.9% in a year [2]. The demand for plastic production had increased from time to time due to the low price and the convenient use. However, the plastics produced are non-biodegradable which is cannot be recycle and cannot be decomposed as it is fully made by chemical compositions such as propylene and ethylene [3]. These had caused the accumulation of the plastics and the present of plastic waste in environment had gave a huge impact towards environment and humans [1]. This situation has provided a major threat to the flora and fauna, especially to the aquatic life. Endangered aquatic life such as turtle, sea-lion, dolphin, fish, etc. are among the most affected species from the pollution. Hence, this life-threatening situation has attracted the attention of scientist around the world on developing alternative bio-based plastic which is more environmentally friendly then the synthetic plastic.

Biopolymer is one of the most promising alternatives which could provide a similar performance to the synthetic plastics, while at the same time preserving the nature from long-term pollution. There are many types of biopolymers which can be categorized into plant based, animal based, and synthetic based. To date, many research has been carried out on development of biodegradable materials using natural resources such as pandanus amaryfllifolius [4], dioscorea hispida [5], cassava starch [6–8], sugar palm [9, 10], palm wax [11], seaweed [12–15], lemongrass [16], etc.

R. Jumaidin (✉)
Fakulti Teknologi Kejuruteraan Mekanikal dan Pembuatan, Universiti Teknikal Malaysia Melaka, 76100 Durian Tunggal, Melaka, Malaysia
e-mail: ridhwan@utem.edu.my

Sapuan S. M. and I. Ahmad (eds.), *Composites from the Aquatic Environment*, Composites Science and Technology, https://doi.org/10.1007/978-981-19-5327-9_3

Seaweed is a macro-alga which can be found on most seawater around the world. In general, seaweed can be divided into three category which are red seaweed, green seaweed, and brown seaweed. Among them, red seaweed is one of the most common species being agriculture for its hydrocolloid namely agar.

Agar is one type of biopolymer derived from red seaweed. It is a gelatinous substance which commonly used in food preparation, cosmetics, and pharmaceutical industry. This biopolymer has ability to form hard gels even at very low concentration i.e. 0.04% [17]. In recent, this biopolymer has been tested for various applications, one of the most significant application is as packaging materials. This is due to the interesting behaviour of agar which possess good thermoplasticity, biocompatibility, biodegradability, and moderate water resistance [17].

The most important attribute of agar is its ability to form hard gels at very low concentrations (0.04%) and it has been broadly utilized as a gelling agent in processed foods, pharmaceutical products and cosmetics, besides applications in biotechnology and medicine. Due to its thermoplasticity, biocompatibility, biodegradation and moderate water resistance, agar has been tested as an alternative source for the petroleum plastic packaging materials [7]. In comparison to other polymers, agar is more stable at low pH and high temperature. The films obtained from agar are clear, transparent, strong and flexible at low moisture contents. Unfortunately, as for other biopolymers, their application has been limited for food packaging due to their hydrophilic characteristic, weak barrier and mechanical properties of the films. Hence, in this review, various studies carried out to improve the properties of agar biopolymer will be discussed to provide a new insight to the versatility of this bio material.

2 Biopolymer

Biopolymers or also known as natural polymers are polymers that is formed during all organism's growth cycles, under natural conditions. These are formed by complex metabolic processes within cells [18]. Bio-based and biodegradable products have been in high demand this few years, as it had contributed a lot to the sustainable development with less impact to environment. About 10–20% of enhancement in the market of this product materials each year, and this product is manufactured from biodegradable polymer or known as biopolymers. Biopolymers have been an interesting way to deal with the decreasing of petrochemical resources in the future. Green agricultural resources can take place the gas and fossil fuel and can reduce the emission of carbon dioxide, CO_2. Biopolymers can be applied in different course such as food, medicine and petroleum industries [19]. In general, there are three major groups that exist in biopolymers which is protein, polysaccharides and polynucleotics [20]. Table 1 shows the classification of each biopolymers while Fig. 1 shows the application of biopolymers.

Table 1 Classification of polymers [19, 21]

Classification	Origin	Biopolymers
Polysaccharides	Plant Algal Animal Bacterial Fungal Lipids/surfactants	Starch, Cellulose Agar, Alginate, Pectin, Various gums Chitin/chitosan, Hyaluronic acid Xanthan, Dextran, Gellan, Levan, Curdlan, Cellulose (bacterial) Pullulan, Elsinan, Yeast glucans Acetoglycerides, waxes, surfactants, Emulsan
Proteins		Silks, Collagen/gelatina, Elastin, Resilin, Adhesives, Polyamino acids, Soy, Serum albumin
Polyesters		Polyhydroxyalkanoates, Polylactic acid
Specialty polymers		Shellac, Natural rubber, Synthetic polymers from natural fats and oils, Nylon from castor oil, Poly-gamma-glutamic acid

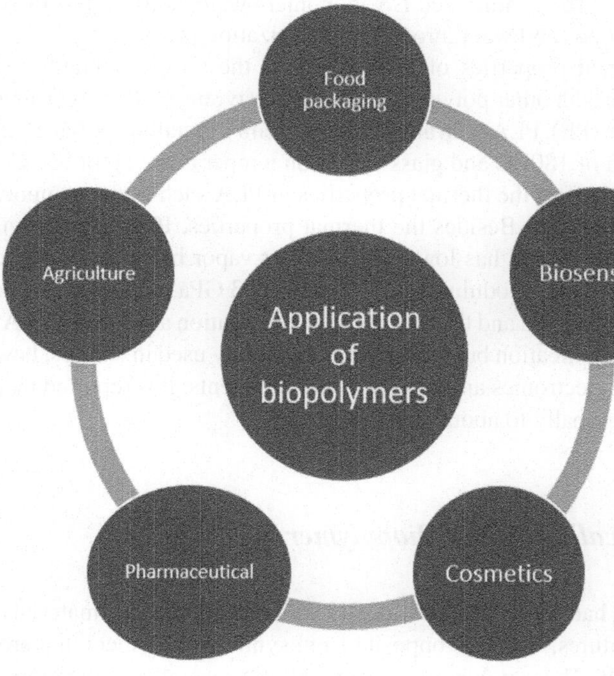

Fig. 1 Applications of biopolymers

2.1 Synthetic Derived Biopolymer

Recently, the production of synthetic biopolymers has rising due to the development of technology. These includes poly-lactic acid (PLA), poly-caprolactone (PCL), poly-glycolic acid (PGA) and polyvinyl alcohol (PVOH). Poly-lactic acid (PLA) which is one of the synthetic biopolymers, is an aliphatic polyester that is made from lactic acid (2-hydroxypropionic acid) through polymerization is one of the polymers being used in plastic applications. PLA is a polymer with a helical conformation that contains asymmetric carbon atoms [22]. It is used to decrease the negative impact towards the environment which is plastic waste accumulation. Hamad et al. (2015) stated that in 1932, a low-molecular weight of PLA was created by Carothers and continued by DuPont in 1954 to create a higher-molecular weight of PLA. To produce lactic acid (LA), which is the basic block of PLA, the method used are synthetization of chemical based on feedstock of petrochemicals and fermentation of carbohydrate. Mainly, the method used to produce LA is natural material's fermentation that contains carbohydrates (>90%). The synthesized LA monomers were converted to PLA by using the last two methods and went through polymerization processes.

The mechanical properties of PLA which is the tensile strength and flexural strength is better than other polymers such as polystyrene (PS), polypropylene (PP), and polyethylene (PE). PLA is an amorphous or semi-crystalline polymer has melting temperature, Tm of 180 °C and glass transition temperature, Tg of 55 °C. Different parameters could affect the thermal properties of PLA such as the compositions and molecular weights [23]. Besides the thermal properties, PLA usually will softs at temperature of 60 °C and has low gas and water vapor barriers compared to other polymers. The Young's modulus of PLA is around 3 GPa, while the tensile strength varies from 50 to 70 MPa and the percentage of elongation at 4% [24]. PLA is mainly used in medical application but now it has been widely used in textiles, beverage and food packaging, electronics and automotive components. It is reported that PLA has been produced globally to about 800 kt/year [25].

2.2 Biologically Derived Biopolymer

Zhang et al. [26] had stated that biologically derived polymers are materials produced from living creatures, which is opposite from synthetic polymers that are made by the humans itself. These polymers are derived into several classifications as shown in Fig. 2.

There are various type of biologically derived polymers, chitin is one of them and categorized under natural polysaccharides. It is the second essential natural biopolymers after cellulose, produced from two main sources which is shrimp and crabs [27]. Chitin or poly (β-(1 → 4)-*N*-acetyl-D-glucosamine) can be found in two form of allomorphs, α and β forms. Chitin has the same structure with cellulose but has an acetamide group (-$NHCOCH_3$) at the C2 position. It has biomaterial characteristics

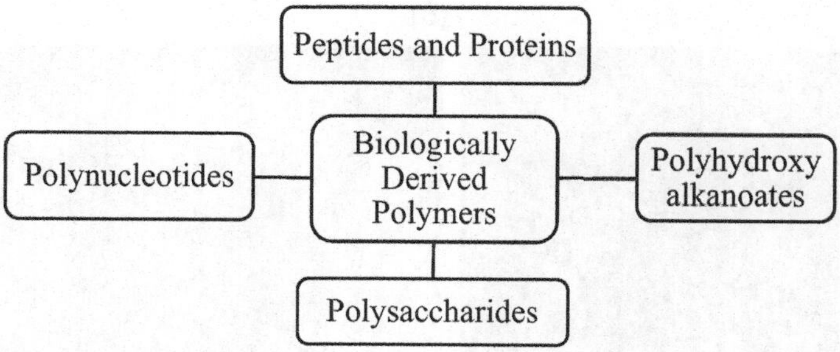

Fig. 2 Classification of biologically derived polymers

such as biocompatible, environmentally friendly, can be additive in water treatment and so on. Unfortunately, chitin has lower solubility in diluted aqueous solvents and common organic solvents due to its hydrophobic properties. Due to this advantage, chitosan is derived from chitin and has high solubility in dilute acidic solutions (pH ≤ 6.0). According to Rajeswari et al. [21], chitosan is composed of β-(1 → 4)-linked D-glucosamine and N-acetyl D-glucosamine. When the pH is low, the amines of chitosan will be protonated and charged positively, so the solubility towards water will increase. These solubility transition will happen when the pKa value is between 6 and 6.5 [28].

3 Origin of Agar

Agar is mainly produced from red marine seaweeds of Rhodophyta species. In terms of structure analogy, agar has similar function as the cellulose role in green plants. However, the function slightly differs since seaweed needs more elastic and flexible properties in order to suit the high current and wave motion in the sea. Agar is extracted from the cell matrix of seaweeds of the Gelidiaceae and Gracilariaceae families. Figure 3 shows the picture of the two red seaweed species. According to Freiler et al. [29], agar is a hydrophilic colloid consisting of polysaccharides that have the ability to form reversible gels simply by cooling a hot aqueous solution. It is composed of alternating 1,3-linked-D-galactose and 1,4-linked 3,6-anhydro-L-galactose units. This disaccharide can be substituted by sulfate esters and methoxyl, and may also carry pyruvic acid residues. The type, amount and location of these substitutes strongly affect the physical properties of the gel and, therefore, its functionality [29].

Gimenez et al. [30] states that agar is a fibrous polysaccharide extracted from marine algae such as Gelidium sp. and Gracilaria sp., consisting in a mixture of agarose and agaropectin, which is slightly branched and sulphated. Meanwhile, according to Madera-Santana et al. [31], agars are synthesized by species of red seaweeds belonging to the genus Gracilaria, Gelidium, and Pterocladia, and constitute a complex mixture of molecules containing several extremes in their structure.

Fig. 3 Agar-bearing red seaweeds; **a** *Gelidium amansii*; **b** *Gracilaria* [32]

In terms of the processing, agar gel will melt on heating and resets on cooling, this cycle can be repeated for an indefinite number of times without compromising gel mechanical properties [29].

3.1 Application of Agar

Owing to its ability to form very hard gels at very low concentrations, agar has been used extensively as a gelling agent in the food industry, cosmetic, pharmaceutical, and in other applications such as microbiology and molecular biology techniques. More recent uses of agar include dental moulds, casting of archaeological pieces and sculpture moulds [29]. Due to its combination of renewability and biodegradability, its enormous gelling power, and the simplicity of the extraction process, agar has been singled out as a promising candidate for future use in plastic materials [29].

4 Agar as Biopolymer

Agar has received much attention in biopolymer development due to its ability to form film that possess good characteristics as alternative packaging material [33–35]. Distinctive characteristics of agar as gelling and thickening agent has brought to comprehensive investigation on the potential of this polysachharide as alternative material for non-degradable plastic and other applications. Almost all studies conducted on agar film utilizing solution casting method that produced thin and flexible films.

Agar films are reported to have a high retraction ratio which is mainly due to the syneresis of agar gel while drying. They are also transparent, heat-sealable, and biodegradable. Moreover, agar films are biologically inert makes them possible to interact easily with different bioactive substances and coat the surface of food products [36]. However, in comparison with plastic-based packaging materials, the pure agar film is relatively brittle. It also has low elasticity, poor thermal stability, high water sensitivity, and high water vapor permeability (WVP). All these drawbacks limit the application of pure agar film. In order to improve the properties of agar film, various kind of material were incorporated into agar to produce composite films with enhanced properties.

5 Agar Composite Film

In this subchapter, the modification of agar film with the addition of filler, reinforcement, or other biopolymer will be discussed. To overcome the limitation of agar film, most of the studies reported were utilizing agar film as a polymer matrix while the effects of filler addition were evaluated.

A study reported by Sousa et al. [37] were using locust bean gum as additive for agar film where various mass ratios of locust bean gum were added i.e. 25, 50, and 75%. The preparation of this composite film were adapting solution casting method to produce all samples. Overall, the study shows that LBG addition made agar films easier to process. This is shown by increase in the viscosity and decrease in gelling character of the film-forming solutions. The best films were obtained at 50/50 and 25/75 agar/LBG ratios. In this study, they also investigating two different type of agar namely native agar and alkali modified agar. It was concluded that sing a cheaper native agar and LBG amounts as high as 50–75% could significantly reduce the cost-production and improve the properties of agar films.

Apart from using natural additive, investigation on the blending performance of agar with synthetic polymer were also evaluated. This study gives a new insight to the synthetic polymer which could initiate the biodegradable characteristic of the polymer blend, hence, reducing the negative impact of this material to the environment. Robledo et al. [38] investigated the effects of LDPE blended with agar. In this study, LDPE-agar biocomposites were mixed at various agar proportions i.e. 0, 10, 20, 30 and 40%wt. Resulting blends were hot pressed and characterized with regard to their torque-rheological, mechanical, dynamic-mechanical, thermal, and morphological properties. The torque rhfeological properties were determined using classical power law model. Tensile properties of LDPE-agar biocomposites showed that agar improves the tensile modulus but compromise the tensile strength and elongation at break. Viscoelastic behavior of the matrix is clearly influenced by the presence of agar biofiller as shown by the dynamic mechanical analysis (DMA). Morphological observations by scanning electron microscopy (SEM) show the ductile to brittle fracture of LDPE-agar biocomposites subjected to tensile test.

Polyurethane is a widely use polymer due to the excellent mechanical properties, elongation, flexibility, and good abrasion resistance. Oprea [39] investigates the performance of PU/agar polymer blend. Films of the composite were prepared by blending polyurethane resin with agar (0–12.5 wt%) and were characterized through analysis of their mechanical, dynamic-mechanical, thermal, and morphological properties. The agar filler decreases breaking strains from 450 to 250% and improves tensile strengths from 1.8 to 2.7 MPa depending on the agar content. The intermolecular interactions between polyurethane and agar in composites have been studied by Fourier transform infrared spectroscopy and by mechanical measurements. The water contact angle of the polyurethane–agar composite surface indicated that the presence of agar improves the hydrophilicity of the composite. The morphological observations obtained by optic micrograph show a random dispersion of the agar filler in the polyurethane matrix. Thus, blends of agar with castor oil–polyurethane elastomers can be used to produce composite materials with increased natural raw materials and improved hydrophilicity.

Food packaging film is one of the most famous potential application for agar film. Hence, many study were conducted to evaluate the potential of this film for this application. Rhim et al. [33], investigated the properties of agar film modified by several biopolymer and filler namely k-carrageenan, konjac glucomannan powder, and nanoclay (Cloisite® 30B). The composite film were evaluated for their mechanical and water barrier properties such as water vapor permeability (WVP), water contact angle (CA), water solubility (WS), water uptake ratio (WUR), water vapor uptake ratio (WVUR). Mechanical, water vapor barrier, and water resistance properties of the ternary blend film exhibited middle range of individual component films, however, they increased significantly after formation of nanocomposite with the clay. Especially, the water holding capacity of the ternary blend biopolymer films increased tremendously, from 800 to 1681% of WUR for agar and k-carrageenan films up to 5118% and 5488% of WUR for the ternary blend and ternary blend nanocomposite films, respectively. Water vapor adsorption behavior of films was also tested by water vapor adsorption kinetics and water vapor adsorption isotherms test. Preliminary test result for fresh spinach packaging revealed that the ternary blend biohydrogel films had a high potential for the use as an antifogging film for packaging highly respiring agricultural produce. In addition, the ternary blend nanocomposite film showed an antimicrobial activity against Gram-positive bacteria, Listeria monocytogenes.

The potential of agar as food packaging film were also reported by Roy et al. [40]. However, this study were using different approach where Copper sulfide nanoparticles (CuS NP) were used as additived for the film. The films were also incorporated with cornstarch as the stabilizing agent and ammonia as a hydrolyzing agent. Th bioactive agar/CuS NP nanocomposite film. Prior to the fabrication of the composite film, CuS NP was characterized using UV–vis spectroscopy, X-ray diffraction pattern, scanning electron microscopy (SEM) and transmission electron microscopy (TEM). CuS NP was roughly spherical and irregular with a size in the range of 2–10 nm with an average of 4.6 ± 1.3 nm. In terms of the composite film, agar-based films with different concentrations (0.25, 0.50, 1.0 and 2.0 wt%) of CuS NP were prepared. It was then characterized using SEM and Fourier transforms infrared spectroscopy (FTIR), and the film properties such as UV-barrier, mechanical, water vapor barrier, swelling ratio, water solubility, hydrophobicity, thermal stability, antibacterial properties and cyto-toxicity. The morphological investigation shows that the CuS NP was well dispersed in the polymer matrix to form compatible nano-composite films. The swelling ratio and moisture content of the composite films decreased while the water solubility increased slightly after the addition of CuS NP. The composite films showed significantly increased UV-barrier without much sacrifice of transparency, and they also showed increased mechanical strength and water vapor barrier properties. Also, in vitro analysis showed excellent biocompatibility of CuS NP and nanocomposite films on skin fibroblast L929 cell lines with cell viability above 90%. Also, they exhibited distinctive antibacterial activity against food-borne pathogenic bacteria, E. coli and some activity against L. monocytogenes.

Another study by Roy et al. [41] evaluated the potential of Melanin nanoparticles (MNP) as additive for agar packaging film. MNP were isolated from the sepia ink using a centrifugation method and used as a functional filler for the preparation of agar-based functional films. The MNP were spherical with an average diameter of 95.6 ± 21.2 nm. Field emission scanning electron microscopy (FE-SEM) and Fourier transform infrared spectroscopy (FTIR) test results indicated that the MNP were well dispersed in agar polymer to form free-standing composite films. The addition of MNP enhanced the UV-blocking, hydrophobicity, mechanical, and water vapor barrier properties of the agar film. Also, the agar/MNP composite films showed a high antioxidant activity comparable to ascorbic acid. The MNP separated from sepia ink can be used as a functional filler to develop antioxidant biopolymer films for food packaging and biomedical applications.

Extended study by Roy et al. [42] developed food packaging film by using modified agar/pectin film as the polymer matrix. Instead of using pure agar film as matrix, the author claim that pectin/agar blend have better functional characteristics than the pure agar film. Then, a hybridized filler consists of melanin nanoparticles (MNP) and grapefruit seed extract (GSE) were incorporated onto agar film. It was found that the physical properties of composite films made by mixing pectin and agar have been improved. The MNP and GSE were well dispersed in the pectin/agar blend film. The addition of MNP and GSE significantly increased the composite film's UV-blocking property, with some sacrificing transparency. The addition of MNP increased the mechanical strength of the agar/pectin film over GSE, but the addition of MNP and GSE together synergistically increased the film's strength. The film's thermal stability was not affected by the addition of MNP and GSE, but the water vapor barrier property was significantly increased. The water resistance and water swelling properties were increased considerably by the addition of MNP and GSE. The pectin/agar composite film showed intense antioxidant activity and excellent antibacterial activity against foodborne pathogens by adding MNP and GSE.

Silver is a well known as an effective antimicrobial agent for bacteria and virus. Hence, the application of silver nanoparticles in agar film for food packaging application is a promising study to be carried out. Rhim et al. [43] reported a development of antimicrobial film from silver nanoparticles (AgNPs) and agar by solvent casting method. The author produce the AgNP using laser ablation method. UV–vis absorbance test and transmission electron microscopy (TEM) analysis results revealed that non-agglomerated spherical AgNPs were formed by the laser ablation method. The surface color of the resulting agar/AgNPs films exhibited the characteristic plasmonic effect of the AgNPs with the maximum absorption peaks of 400–407 nm. X-ray diffraction (XRD) test results also exhibited characteristic AgNPs crystals with diffraction peaks observed at 2? values of 38.39°, 44.49°, and 64.45°, which were corresponding to (1 1 1), (2 0 0), and (2 2 0) crystallographic planes of face-centered cubic (fcc) silver crystals, respectively. Thermogravimetric analysis (TGA) results showed that thermal stability of the agar/AgNPs composite films was increased by the inclusion of metallic silver. Water vapor barrier properties and surface hydrophobicity of the agar/AgNPs films increased slightly with the increase in AgNPs content but they were not statistically significant ($p > 0.05$), while mechanical strength and stiffness of the composite films decreased slightly ($p < 0.05$). The agar/AgNPs films exhibited distinctive antimicrobial activity against both Gram-positive (Listeria monocytogenes) and Gram-negative (Escherichia coli O157:H7) bacterial pathogens.

In addition to silver particles, it is known that the combination of two metallic materials possess better optical, interfacial, and catalytic properties. Hence, a study on using bimetallic materials of silver-copper combination into agar film were carried out by Arfat et al. [44]. Agar-based active nanocomposite films were prepared by incorporating silver-copper (Ag–Cu) alloy nanoparticles (NPs) (0.5–4 wt%) into glycerol plasticized agar solution. Thermo-mechanical, morphological, structural, and optical properties of the nanocomposite films were characterized by texture analyzer, differential scanning calorimetry (DSC), scanning electron microscope (SEM), X-ray diffraction (XRD), Fourier transforms infrared (FTIR) spectroscopy, and surface color measurement. Tensile strength and the melting temperature of the film increased linearly with NPs loading concentration. Color, transparency and UV barrier properties of agar films were influenced by the reinforcement of Ag–Cu NPs. XRD analysis confirmed the crystalline structure of the Agar/Ag–Cu nanocomposite films, whereas the smoothness and the homogeneity of film surface strongly reduced as observed through the SEM. The nanocomposite films exhibited a profound antibacterial activity against both Gram-positive (Listeria monocytogenes) and Gram-negative (Salmonella enterica sv typhimurium) bacteria. Overall, the agar nanocomposite films could be used as packaging material for food preservation by controlling foodborne pathogens and spoilage bacteria.

Banana is a well known fruit to the most places in the world. Apart from being used as food, the potential of banana powder as filler in agar composites film were studied by Orsuwan et al. [45]. Binary blend films of agar and banana powder and Agar/Banana composite films reinforced with silver nanoparticles (Agar/Banana/AgNPs) were prepared using a solution casting method and their properties were characterized. The SEM micrographs and FT-IR results confirmed the formation of physical interactions between polymer matrices and nanofillers. Apparent surface color and transmittance of the composite film were greatly influenced not only by the mixing of banana powder with agar but also by the incorporation of AgNPs. The UV light absorption, water vapor barrier properties, and antioxidant activity of Agar/Banana blend films increased with the increase in the concentration of the banana powder, while the mechanical properties decreased.

The Agar/Banana/AgNPs composite film exhibited distinctive antimicrobial activity against food-borne pathogenic bacteria, Escherichia coli and Listeria monocytogenes with stronger antibacterial activity against Gram-negative bacteria than Gram-positive bacteria. The binary blend of Agar/ Banan films are expected to be used for the edible film or coating of foods and their nanocomposite films with antimicrobial activity have a potential to be used as food packaging material for maintaining the safety and extending the shelf life of packaged food.

More recent study by Xiao et al. [46] evaluate a multi-component films namely emulsified film based on konjac glucomannan (KGM)/agar/gum Arabic (GA) incorporated with virgin coconut oil (VCO). The effects of VCO on the physical, structural, and water barrier properties of the film were investigated. The values of the mechanical and water barrier properties were different with statistical significance ($p < 0.05$), with VCO contents ranging from 0.1 to 0.6% of the film solid weights. The addition of VCO decreased the tensile strength but effectively increased the elongation at the break of the films. Increased VCO concentrations resulted in decreased water vapor permeability, reduced water swelling, solubility, and adsorption, and increased water contact angle. Compared with cucumber without packaging, packed cucumber with emulsified film showed significantly lower weight loss and firmness reduction during storage up to 12 days at 7 ± 1 °C. Fourier transform infrared spectroscopy results indicated intermolecular hydrogen bonds between KGM, agar, and GA occurred, and X-ray diffraction results suggested that all the films were in the amorphous status. Combining all the above results, the mechanism of the water barrier property improvement was proposed. This study offers an alternative emulsion polysaccharide-based edible film with high potential to be used in cucumber packaging.

Agar film were also explored for other unique potential such as flame retardant film. Hou et al. [47] reported the development of novel flame retardant film of agar/sodium alginate/boric acid (AG/SA/BA). This composite film was prepared in presence of various concentration of BA (2.5, 5, 10, 15 wt%) through solution casting method. The result showed that C-O-B bonds were formed between BA and matrix. The addition of BA enhanced the limiting oxygen index (LOI), which might be related to the interaction formed between BA and matrix and excessive BA existing in molecular skeleton. The thermal stability of crosslinking film containing BA concentration above 5 wt% was improved when compared to that of the film without BA. Overall, the author concluded that the addition of BA enhanced flame retardancy and thermal stability at appropriate BA concentration.

Another interesting potential application of agar film were discussed in a study by Huang et al. [48]. In their research, the performance of agar as novel indicator film were evaluated. The function of the film is for monitoring fish freshness based on agar incorporated with natural dye extracted from Arnebia euchroma root (AEREs). The Fourier transform infrared spectroscopy results reflected that some new interactions have occurred between polymer matrix and natural dye. X-ray diffraction and scanning electron microscopy indicated that AEREs were well dispersed in the agar base film. The tensile strength, stiffness, water vapor permeability, and water contact angle has improved following the addition of AERE. This finding was accompanied with decrease in water solubility, swelling ratio and elongation at break of the colorimetric film. The prepared indicator films were used as freshness labels in an application trial, which was conducted to monitor the freshness of Wuchang bream (Megalobrama amblycephala) under refrigeration (4 °C) and at room temperature (25 °C). The total volatile basic nitrogen (TVB-N) and total viable count (TVC) of fish sample were determined periodically, and the color change of freshness labels was recorded simultaneously. The results showed that the indicator film with lower content of AEREs demonstrated more conspicuous color change during fish spoilage. And the color response of freshness label was consistent with the spoilage threshold of TVC and TVB-N content in fish sample. Thus, these colorimetric indicator films could indicate the fish spoilage by visible color change. Overall, the developed colorimetric indicator film show promising potential to provide a convenient, non-destructive and visual method to estimate fish freshness during storage.

Different perspective of agar in polymer electrolyte application were explored by Raphael et al. [49]. New types of polymer electrolytes based on agar have been prepared and characterized by impedance spectroscopy, X-ray diffraction measurements, UV–vis spectroscopy and scanning electronic microscopy (SEM). The best ionic conductivity has been obtained for the samples containing a concentration of 50 wt.% of acetic acid. As a function of the temperature the ionic conductivity exhibits an Arrhenius behavior increasing from 1.1×10^{-4} S/cm at room temperature to 9.6×10^{-4} S/cm at 80 °C. All the samples showed more than 70% of transparency in the visible region of the electromagnetic spectrum, a very homogeneous surface and a predominantly amorphous structure. All these characteristics imply that these polymer electrolytes can be applied in electrochromic devices.

Composites of magnetic particles in a polymeric matrix have received increasing interest due to their capacity to respond to external magnetic or electromagnetic fields. In biomedicine, these hybrid compounds with micro- or nano-particles, can be used as auxiliary elements for treatment and diagnosis of diverse diseases. Diaz-Bleis et al. [50] reported a study on performance of carbonyl iron particle (CIP) onto agar film. The amount of CIP were varied at 0 to 30% w/w. The mixture was mechanically agitated during 3 s using a blade stirrer and poured in a Petri dish and left to settle down for 15 min to allow the gel formation. When the film was formed, it was separated from the dish plate using a spatula. The film was set in a plastic container, and placed in an oven to obtain a thin biofilm of 30–35 μm. The samples were analyzed using the photothermal radiometry (PTR) technique in the back-propagation emission configuration performing a modulation frequency scan. The amplitude and phase of the PTR experimental data were fitted simultaneously using a one-layer thermal-wave model considering homogeneous optical and thermal properties. The results indicate a systematic increase of the thermal diffusivity and optical absorption coefficient when the magnetic particle content increases. Scanning electron microscopy surface morphology of the agar/carbonyl iron composite indicates that a homogeneous distribution of particles can be obtained with the reported procedure and also provides evidence of agglomeration at high concentrations. The author concluded that the samples prepared with less than 20% of CIP is the optimum ratio due to the optimal particle dispersion in the agar matrix and lack of agglomerates formation.

Starch is among the most applicable biopolymer available in nature. However, the biopolymer derived from starch, namely thermoplastic starch have several weakness such as poor thermal and mechanical properties. Hence, few study has been reported on utilizing agar as the polymer blend to improve the properties of thermoplastic starch [51–54]. It was found that agar has excellent compatibility with thermoplastic starch due to the similar hydrophilic nature of the materials. This has led to significant improvement in the mechanical properties of the material i.e. tensile and flexural properties. However, this finding were accompanied with more hydrophilic behavior of the material which was attributed to the fact that agar is a sulfated polysaccharide and the presence of charged groups resulted in more extended chains with a higher hydrophilicity.

6 Conclusion

The performance of agar film in various potential application were reviewed. Overall agar shows a promising characteristic in food packaging application and able to show good performance especially in terms of physical and food safety criteria. This were indicated by the ability to show antimicrobial activities towards several types of bacteria, mainly due to the addition of antimicrobial agent into the film. Agar film also shows potential in other application such as indicator film and polymer electrolyte. Despite various research that has been carried out on agar film, lack of study reported on the potential of agar as a rigid material. Hence, this research can be an interesting path for developing a new alternative biopolymer from a versatile seaweed-based polysaccharide. In conclusion, agar is a versatile biomaterial which possess high potential as an alternative material for countless number of applications. More research is ultimately necessary to further explore the potential of this material in other beneficial application.

Acknowledgements The authors would like to express sincere gratitude to Universiti Teknikal Malaysia Melaka and Ministry of Higher Education Malaysia for the financial support provided through research grant RACER/2019/FTKMP-CARE/F00413, the article proof reading was supported by publication incentive grant JURNAL/FTK/2018/Q00004.

References

1. Geyer, R., Jambeck, J. R., & Law, K. L. (2017). Production, use, and fate of all plastics ever made. *Science Advances, 3*, e1700782. https://doi.org/10.1126/sciadv.1700782
2. Gourmelon, G., Mármol, Z., Páez, G., Rincón, M., Araujo, K., Aiello, C., Global Plastic Production Rises, Recycling Lags, & Rev. Tcnocientifica URU. (2015). 1–7. 2244-775X.

3. Rosso, M. (2006). Ceramic and metal matrix composites: Routes and properties. *Journal of Materials Processing Technology, 175*, 364–375. https://doi.org/10.1016/j.jmatprotec.2005.04.038
4. Diyana, Z. N., Jumaidin, R., Selamat, M. Z., Alamjuri, R. H., & Yusof, F. A. M. (2021). Extraction and characterization of natural cellulosic fiber from pandanus amaryllifolius leaves. *Polymers (Basel)., 13*, 4171.
5. Hazrati, K. Z., Sapuan, S. M., Zuhri, M. Y. M., & Jumaidin, R. (2021). Preparation and characterization of starch-based biocomposite films reinforced by Dioscorea hispida fibers. *Journal of Materials Research and Technology, 15*, 1342–1355. https://doi.org/10.1016/j.jmrt.2021.09.003
6. Zakuan, N. D., Jumaidin, R., & Selamat, M. Z. (2020). Biopolymer from starch. *Proceedings of Mechanical Engineering Research Day, 2020*, 323–324.
7. Diyana, Z. N., Jumaidin, R., Selamat, M. Z., Ghazali, I., Julmohammad, N., Huda, N., & Ilyas, R. A. (2021). Physical properties of thermoplastic starch derived from natural resources and its blends: A review. *Polymers (Basel), 13*, 5–20. https://doi.org/10.3390/polym13091396
8. Kamaruddin, Z. H., Jumaidin, R., & Selamat, M. Z. (2020). Thermoplastic starch as biodegradable materials. *Proceedings of Mechanical Engineering Research Day*, 104–105.
9. Mahamud, A., Ilyas, R. A., Sapuan, S. M., & Ishak, M. R. (2019). Degradation and physical properties of sugar palm starch/sugar palm nanofibrillated cellulose bionanocomposite. *Polimery, 64*, 680–689. https://doi.org/10.14314/polimery.2019.10.5
10. Ilyas, R. A., Sapuan, S. M., Ibrahim, R., Abral, H., Ishak, M. R., Zainudin, E. S., Atiqah, A., Atikah, M. S. N., Syafri, E., Asrofi, M., & Jumaidin, R. (2020). Thermal, biodegradability and water barrier properties of bio-nanocomposites based on plasticised sugar palm starch and nanofibrillated celluloses from sugar palm fibres. *Journal of Biobased Materials and Bioenergy, 14*, 234–248. https://doi.org/10.1166/jbmb.2020.1951
11. Hafila, K. Z., Jumaidin, R., Ilyas, R. A., Selamat, M. Z., & Asyadi, F. (2022). Effect of palm wax on the mechanical, thermal, and moisture absorption properties of thermoplastic cassava starch composites. *International Journal of Biological Macromolecules, 194*, 851–860. https://doi.org/10.1016/j.ijbiomac.2021.11.139
12. Jumaidin, R., Sapuan, S. M., Jawaid, M., Ishak, M. R., & Sahari, J. (2017). Effect of seaweed on mechanical, thermal, and biodegradation properties of thermoplastic sugar palm starch/agar composites. *International Journal of Biological Macromolecules, 99*, 265–273. https://doi.org/10.1016/j.ijbiomac.2017.02.092
13. Jumaidin, R., Sapuan, S. M., Jawaid, M., Ishak, M. R., & Sahari, J. (2016). Effect of seaweed on physical properties of thermoplastic sugar palm starch/agar composites. *Journal of Mechanical Engineering Science, 10*, 2214–2225. https://doi.org/10.15282/jmes.10.3.2016.1.0207
14. Jumaidin, R., Sapuan Salit, M., Jawaid, M., Ridzwan Ishak, M., Sahari, J. (2017). Seaweeds as renewable sources for biopolymers and its composites: A review. *Current Analytical Chemistry, 13*, 249–267. https://doi.org/10.2174/1573411013666171009164355
15. Jumaidin, R., Sapuan, S. M., Jawaid, M., Ishak, M. R., & Sahari, J. (2017). Characteristics of *Eucheuma cottonii* waste from East Malaysia: Physical, thermal and chemical composition. *European Journal of Phycology, 52*, 200–207. https://doi.org/10.1080/09670262.2016.1248498
16. Hafila, Z., & Jumaidin, R. (2021). Characteristics and properties of lemongrass (Cymbopogan Citratus): A comprehensive review. *Journal of Natural Fibers, 00*, 1–18. https://doi.org/10.1080/15440478.2021.1958439
17. Bao, X., Hayashi, K., Li, Y., Teramoto, A., & Abe, K. (2010). Novel agarose and agar fibers: Fabrication and characterization. *Materials Letters, 64*, 2435–2437. https://doi.org/10.1016/j.matlet.2010.08.008
18. Rao, M. G., Bharathi, P., Akila, R. M. (2014). A comprehensive review on biopolymers. *4*, 61–68.
19. Avérous, L., & Pollet, E. (2012). *Biodegradable Polymers*. https://doi.org/10.1007/978-1-4471-4108-2

20. Rebelo, R., Fernandes, M., & Fangueiro, R. (2017). Biopolymers in medical implants: A brief review. *Procedia Engineering, 200*, 236–243. https://doi.org/10.1016/j.proeng.2017.07.034
21. Rajeswari, S. (2017). Natural polymers: A recent review. *World Journal of Pharmaceutical Sciences, 6*, 472–494. https://doi.org/10.20959/wjpps20178-9762
22. Lasprilla, A. J. R., Martinez, G. A. R., Lunelli, B. H., Jardini, A. L., & Maciel, R. (2012). Polylactic acid synthesis for application in biomedical devices—A review. *Biotechnology Advances, 30*, 321–328. https://doi.org/10.1016/j.biotechadv.2011.06.019
23. Hamad, K., Kaseem, M., Yang, H. W., Deri, F., Ko, Y. G. (2015). Properties and medical applications of polylactic acid : A review. *9*, 435–455. https://doi.org/10.3144/expresspolymlett.2015.42
24. Motaung, T. E., Linganiso, L. Z., Mohomane, S. M. (2017). Agricultural waste fibers and biopolymer matrices used in biocomposites. *Biocomposites: Properties, performance and applications.*
25. Murariu, M., & Dubois, P. (2016). PLA composites: From production to properties. *Advanced Drug Delivery Reviews, 107*, 17–46. https://doi.org/10.1016/j.addr.2016.04.003
26. Zhang, Z., Ortiz, O., Goyal, R., & Kohn, J. (2014). *Biodegradable polymers*. Elsevier Inc. https://doi.org/10.1016/B978-0-323-22805-3.00013-X
27. Rinaudo, M. (2006). Chitin and chitosan: Properties and applications. *Progress in Polymer Science, 31*, 603–632. https://doi.org/10.1016/j.progpolymsci.2006.06.001
28. Zargar, V., Asghari, M., & Dashti, A. (2015). A review on chitin and chitosan polymers: Structure, chemistry, solubility, derivatives, and applications. *ChemBioEng Reviews, 2*, 204–226. https://doi.org/10.1002/cben.201400025
29. Freile-Pelegrín, Y., Madera-Santana, T., Robledo, D., Veleva, L., Quintana, P., Azamar, J. A. (2007). Degradation of agar films in a humid tropical climate: Thermal, mechanical, morphological and structural changes. *Polymer Degradation and Stability, 92*, 244–252. https://doi.org/10.1016/j.polymdegradstab.2006.11.005
30. Giménez, B., López de Lacey, A., Pérez-Santín, E., López-Caballero, M. E., & Montero, P. (2013). Release of active compounds from agar and agar–gelatin films with green tea extract. *Food Hydrocolloids, 30*, 264–271. https://doi.org/10.1016/j.foodhyd.2012.05.014
31. Madera-Santana, T. J., Robledo, D., & Freile-Pelegrín, Y. (2011). Physicochemical properties of biodegradable polyvinyl alcohol-agar films from the red algae Hydropuntia cornea. *Marine Biotechnology (New York, N.Y.), 13*, 793–800. https://doi.org/10.1007/s10126-010-9341-8
32. Qin, Y. (2018). Seaweed bioresources. In Y. Qin (Ed.), *Bioactive seaweeds for food applications* (pp. 3–24). Academic Press.
33. Rhim, J.-W., & Wang, L.-F. (2013). Mechanical and water barrier properties of agar/κ-carrageenan/konjac glucomannan ternary blend biohydrogel films. *Carbohydrate Polymers, 96*, 71–81. https://doi.org/10.1016/j.carbpol.2013.03.083
34. Phan, D., Debeaufort, F., & Luu, D. (2005). Functional properties of edible agar-based and starch-based films for food quality preservation. *Journal of Agriculture and Food Chemistry, 53*, 973–981.
35. Sousa, A. M. M., Sereno, A. M., Hilliou, L., & Gonçalves, M. P. (2010). Biodegradable agar extracted from gracilaria vermiculophylla: Film properties and application to edible coating. *Materials Science Forum, 636–637*, 739–744. https://doi.org/10.4028/www.scientific.net/MSF.636-637.739
36. Sadat, F., & Zaeim, D. (2020). Agar-based edible films for food packaging applications—A review. *International Journal of Biological Macromolecules, 159*, 1165–1176. https://doi.org/10.1016/j.ijbiomac.2020.05.123
37. Sousa, A. M. M., & Gonçalves, M. P. (2015). Strategies to improve the mechanical strength and water resistance of agar films for food packaging applications. *Carbohydrate Polymers, 132*, 196–204. https://doi.org/10.1016/j.carbpol.2015.06.022
38. Robledo, D., Azamar, J. A., & Rı, C. R. (2010). Preparation and characterization of low density polyethylene-agar biocomposites: Torque-rheological, mechanical, thermal and morphological properties. *Polymer Engineering & Science.* https://doi.org/10.1002/pen

39. Oprea, S. (2011). Preparation and characterization of the agar/polyurethane composites. *Journal of Composite Materials, 45*, 2039–2045. https://doi.org/10.1177/0021998310392415
40. Roy, S., Rhim, J., & Jaiswal, L. (2019). Food hydrocolloids bioactive agar-based functional composite film incorporated with copper sulfide nanoparticles. *Food Hydrocolloids, 93*, 156–166. https://doi.org/10.1016/j.foodhyd.2019.02.034
41. Roy, S., & Rhim, J. (2019). Food hydrocolloids agar-based antioxidant composite films incorporated with melanin nanoparticles. *Food Hydrocolloids, 94*, 391–398. https://doi.org/10.1016/j.foodhyd.2019.03.038
42. Roy, S., & Rhim, J. (2021). Food hydrocolloids fabrication of pectin/agar blended functional film: Effect of reinforcement of melanin nanoparticles and grapefruit seed extract. *Food Hydrocolloids, 118*, 106823. https://doi.org/10.1016/j.foodhyd.2021.106823
43. Rhim, J.-W., Wang, L.-F., Lee, Y., & Hong, S.-I. (2014). Preparation and characterization of bio-nanocomposite films of agar and silver nanoparticles: Laser ablation method. *Carbohydrate Polymers, 103*, 456–465. https://doi.org/10.1016/j.carbpol.2013.12.075
44. Arfat, Y. A., Ahmed, J., & Jacob, H. (2017). Preparation and characterization of agar-based nanocomposite films reinforced with bimetallic (Ag-Cu) alloy nanoparticles. *Carbohydrate Polymers, 155*, 382–390. https://doi.org/10.1016/j.carbpol.2016.08.097
45. Orsuwan, A., Shankar, S., Wang, L. F., Sothornvit, R., & Rhim, J. W. (2016). Preparation of antimicrobial agar/banana powder blend films reinforced with silver nanoparticles. *Food Hydrocolloids, 60*, 476–485. https://doi.org/10.1016/j.foodhyd.2016.04.017
46. Xiao, M., Luo, L., Tang, B., Qin, J., Wu, K., & Jiang, F. (2022). Physical, structural, and water barrier properties of emulsified blend film based on konjac glucomannan/agar/gum Arabic incorporating virgin coconut oil. *Lebensmittel-Wissenschaft & Technologie, 154*, 112683. https://doi.org/10.1016/j.lwt.2021.112683
47. Hou, X., Xue, Z., & Xia, Y. (2018). Preparation of a novel agar/sodium alginate fire-retardancy film. *Materials Letters, 233*, 274–277. https://doi.org/10.1016/j.matlet.2018.09.026
48. Huang, S., Xiong, Y., Zou, Y., Dong, Q., Ding, F., & Liu, X. (2019). Food Hydrocolloids A novel colorimetric indicator based on agar incorporated with Arnebia euchroma root extracts for monitoring fish freshness. *Food Hydrocolloids, 90*, 198–205. https://doi.org/10.1016/j.foodhyd.2018.12.009
49. Raphael, E., Avellaneda, C. O., Manzolli, B., & Pawlicka, A. (2010). Agar-based films for application as polymer electrolytes. *Electrochimica Acta, 55*, 1455–1459. https://doi.org/10.1016/j.electacta.2009.06.010
50. Diaz-Bleis, D., Freile-Pelegrín, Y., Vales-Pinzón, C., Martínez-Torres, P., & Alvarado-Gil, J. J. (2012). Preparation and characterization of algal polysaccharides/magnetite microparticles composite films. *International Journal of Thermophysics, 33*, 2125–2131. https://doi.org/10.1007/s10765-012-1252-7
51. Jumaidin, R., Sapuan, S. M., Jawaid, M., Ishak, M. R. (2017). Effect of agar on physical properties of thermoplastic starch derived from sugar palm tree. *Pertanika Journal of Science & Technology, 25*, 1235–1248.
52. Jumaidin, R., Sapuan, S. M., Jawaid, M., Ishak, M. R., & Sahari, J. (2016). Characteristics of thermoplastic sugar palm starch/agar blend: Thermal, tensile, and physical properties. *International Journal of Biological Macromolecules, 89*, 575–581. https://doi.org/10.1016/j.ijbiomac.2016.05.028
53. Jumaidin, R., Sapuan, S. M., Jawaid, M., Ishak, M. R., & Sahari, J. (2017). Effect of agar on flexural, impact, and thermogravimetric properties of thermo-plastic sugar palm starch. *Current Organic Synthesis, 14*, 200–205. https://doi.org/10.1386/ctl.9.1.5
54. Jumaidin, R., Salit, M. S., Firdaus, M. S., Ghani, A. F. A., Yaakob, M. Y., Zakaria, N. H., Munir, F. A., Zakaria, A. A., & Jenal, N. (2018). Effect of agar on dynamic mechanical properties of thermoplastic sugar palm starch: Thermal behavior. *Journal of Advanced Research in Fluid Mechanics and Thermal Sciences, 47*, 89–96.

Aquatic Hydroxyapatite (HAp) Sources as Fillers in Polymer Composites for Bio-Medical Applications

C. N. Aiza Jaafar and I. Zainol

1 Introduction

Nowadays, a large number of researchers are focusing on developing of new biomaterials for various applications. Biomaterials can be natural or synthetic material used in fabrication of implant to replace the damaged or diseased biological human organ or structure. This field has turn into an electrifying area because these materials has improved the quality and longevity of human life.

Biomaterial have been used since a few decades in medicine and dentistry with a purpose to replace and repair a body feature, tissues or organ. The most importance property of biomaterial is non-toxicity especially when implanted in human body and they should exhibit antibacterial properties [1, 2]. The biomaterial used in fabrication of different parts of the human body are fabricated from polymer, ceramic, metal and composite materials. The applications of biomaterial are expected to increase by year due to ageing population all over the world [3].

Polymer is a material that widely used in engineering and medical applications. It defines as a type of material that is made of long chains or networks with these networks or chains created by repeating units of molecules. They can be easily fabricated into many forms such as fibre, textiles, films, rods and viscous liquids. These have close resemblance to natural tissues component such as collagen. Currently polymer is chosen to be a best material for biomaterial due to low cost, good biocompatibility and versatile applications.

C. N. Aiza Jaafar (✉)
Department of Mechanical and Manufacturing Engineering, Faculty of Engineering, Universiti Putra Malaysia, Selangor, Malaysia
e-mail: cnaiza@upm.edu.my

I. Zainol
Department of Chemistry, Faculty of Science and Mathematics, Universiti Pendidikan Sultan Idris, Tanjong Malim, Perak, Malaysia

Sapuan S. M. and I. Ahmad (eds.), *Composites from the Aquatic Environment*, Composites Science and Technology, https://doi.org/10.1007/978-981-19-5327-9_4

Wide range of polymer was used in biomaterials such as natural and synthetic polymers. Natural polymer includes collagen, gelatine, chitosan, alginate, and hyaluronic acid [4] whereas synthetic polymers includes biodegradable polymers such as poly (lactic acid) (PLA), poly (glycolic acid) (PGA), poly (lactic acid-co-glycolic acid) (PLGA) [5] and non-biodegradable polymer such as polypropylene, acrylic, polyethylene, nylon, silicone, polyurethane and ultra-high molecular weight polyethylene (UHMWPE), with excellent properties for load bearing applications [6].

Research in bone replacement have found promising biomaterial known as hydroxyapatite (HAp) and its derivative. HAp is ceramic materials and this material is of interest due to their excellent in mechanical and biological properties as their chemical structure are very similar to bone chemical composition. Application of HAp material as bone substitute was started since 1980s [7] and beyond has been mainly used as bone defects filler. HAp materials is highly stable in body fluids and they can interact with bone tissues upon implantation and enhance bone cell proliferation. Natural HAp materials have been found to replace synthetic HAp in biomedical application due to their low production cost. Recently, natural HAp is extracted from fish scale and reported to be biocompatible as its chemical structure is similar to synthetic HAp [8].

HAp has been used as fillers in polymeric materials since 1981 as reported by Bonfield et al. [9]. In their research HAp was used as fillers high density polyethylene (HDPE). Wans et al. has developed high density polyethylene (HDPE)/hydroxyapatite (HAp) composites known as HAPEX™ for medical applications [10]. They found that composites with smaller hydroxyapatite particles had higher torsional modulus, tensile modulus and tensile strength, but lower strain to failure. Other researchers also studied this composite systems [11, 12]. However, the HAp fillers used mostly focusing on synthetic HAp. There are limited research has been reported about biocomposite materials based on HDPE matrix filled with biogenic HAp from fish scale [13]. From previous studies, HDPE/HAp composites is a promising candidate for biomaterial implant in medium strength applications [14].

Ultra-high molecular weight polyethylene (UHMWPE) with outstanding impact resistance, low friction coefficient and high wear resistance was used as matrix in HAp polymer composite [15]. Beside excellent mechanical properties, UHMWPE is also biocompatible and this make them suitable material for bone replacement [16].

2 Hydroxyapatite (HAp)

In recent years, HAp has been an important inorganic material which has attracted the attention of researchers related to biomaterial application [17, 18]. HAp has formula molecule of Ca_{10} $(PO_4)_6(OH)_2$ has crystal structure as shown in Fig. 1. HAp as ceramic materials can be produced from chemical synthesis or extracted from natural sources such as fish scale, animal bone and coral [19]. It is well reported that

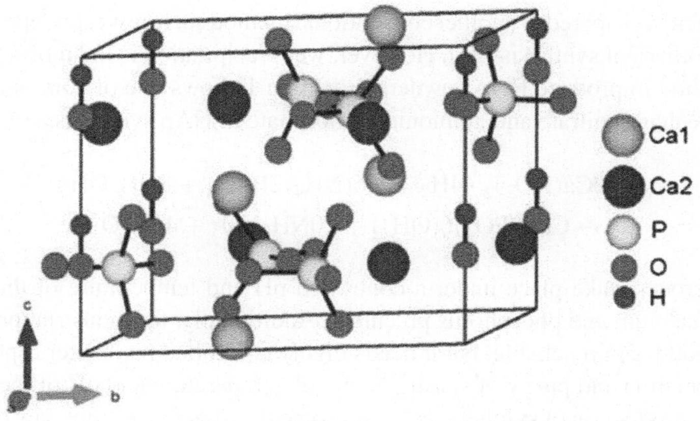

Fig. 1 Crystal structure of hydroxyapatite (HAp) [21]

Table 1 Composition of biological apatite and HAp materials [22]

Major element	Natural apatite		HAp (wt%)
	In enamel (wt%)	In bone (wt%)	
Ca	36.00	24.50	39.60
P	17.70	11.50	18.50
Na	0.50	0.70	–
K	0.08	0.03	–
Mg	0.44	0.55	–
F	0.01	0.02	–
Cl	0.30	0.10	–
$CO^{2-}{}_3$	3.20	5.80	–
Ca: P (Molar ratio)	1.62	1.65	1.67

HAp has been used as implant for biomedical application due to their structure close or similarity in composition to mineral component of human bone and teeth as well as they have excellent in biocompatibility, bioactivity, biodegrable and osteoconductive material [20]. Table 1 shows the composition of biological apatite in human enamel and HAp bone.

2.1 *Synthesis of Hydroxyapatite*

Sol- gel synthesis of HAp is one of the methods that used to synthesis HAp. This method offers mixing of calcium and phosphorus precursors and enhance chemical homogeneity compared to another conventional method such as wet precipitation [23] and hydrothermal synthesis [24]. However, wet precipitation is the most commonly used method to produce HAp powders. Equation 1 shows one of common reaction between calcium nitrate and ammonium phosphate in HAp synthesis.

$$
\begin{aligned}
&10Ca(NO_3)_2 \cdot 4H_2O + 6(NH_4)2HPO_4 + 8NH_4OH \\
&\longrightarrow Ca_{10}(PO_4)6(OH)_2 + 20NH_4NO_3 + 46H_2O
\end{aligned}
\tag{1}
$$

This process take place under a controlled pH and temperature of the solution between calcium and phosphorus precursor. Additionally, for industrial production, this method is approachable, but it needs highly controlled parameter especially on the composition and purity of starting material, temperature and pH of the solution. However, production of synthetic HAp has some drawback where the raw material is expensive as well as time consuming process. Therefore, an alternative was found to extract HAp from natural resources which is cheaper and eco-friendly than chemical synthesis approached. Synthetic HA obtained high degree of crystallinity thus exhibit high structural stability [25]. Some synthesis methods however lead to formation of some toxic chemicals that may have adverse effects in the medical applications [26].

2.2 *Natural Hydroxyapatite*

Alternative production of HAp have been focussed on extraction from natural resources such as bovine and pig bone. Goren et al. (2004) have found that the morphology and chemical structure of animal bone was similar to human bone and synthetic HAp [27]. Over the past few years, extraction of HAp from other natural resources such as seashell, fish scale was reported [19]. Bano et al. (2017) also reported that natural HAp which is similar chemical structure to synthetic HAp by hydrothermal method and calcination at 1100° C [28]. Natural HAp is good alternative to synthetic HAp due to their low manufacturing cost.

HAp from natural resources is biologically safe as no chemical are used and is more economical due to cheaper raw material [29]. Kusrini et al. (2013) in their study stated that strong chemical bond with bone tissues can be formed by using HAp material that extracted from natural resources [30]. Biowaste such as bovine bone is widely available and can very useful in biomedical applications as it economical and environment friendly. It is found that, bovine bone consists of 93% HAp and 7% of tricalcium phosphate composition that has similar properties to human bone mineral and teeth. The common techniques of producing natural HA from the various natural sources are summarized in Fig. 2.

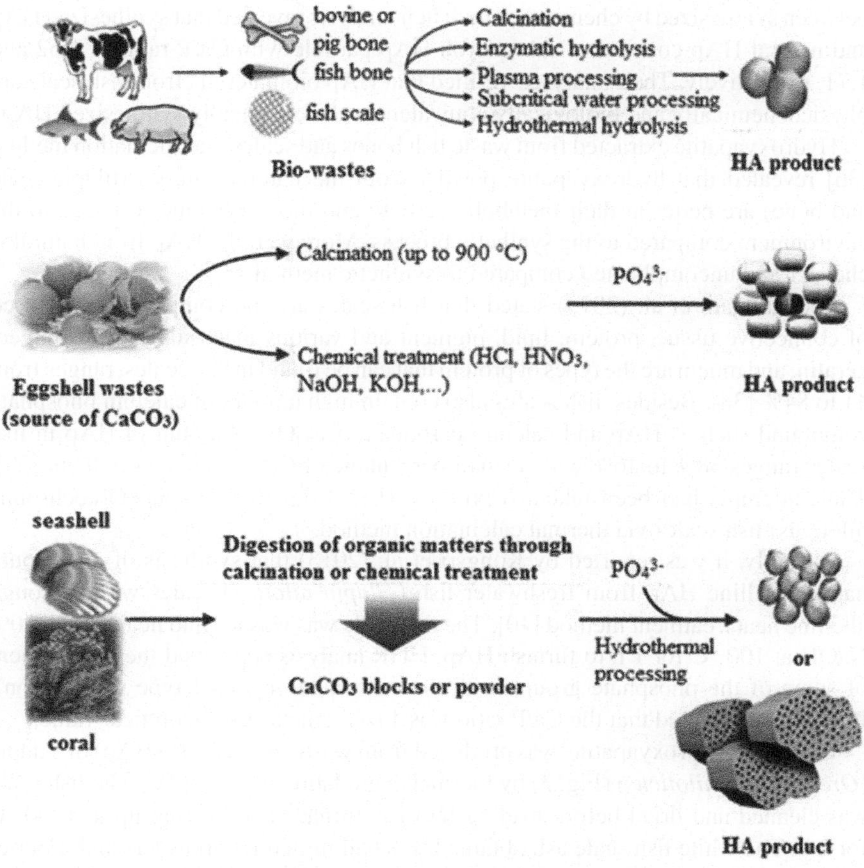

Fig. 2 Summary of preparation of HAp from natural sources [31]

2.3 *Natural HAp Extracted from Fish Scale (FsHAp)*

In the last years, HAp from fish bone and scales has emerged as an alternative to substitute synthetic and bovine HAp because similar chemical properties have been achieved by simple and inexpensive methods [32]. It has been demonstrated that fish sources are safe and present low risks of disease transmission [33]. Additionally, fishes are abundant in the environment, and the application of this by products is suitable for biomedical application.

Attempts have been taken to isolate fish scale derived HAp and use them as an alternate for synthetic HAp [34]. Conventional methods in extraction HAp from fish scales are subcritical water process, calcination, enzymatic hydrolysis and alkaline heat treatment. HAp powder from freshwater fish scales such as *Labeo rohita* and *Catla catla* have been synthesized using calcination method at 800 °C for 1 h for tissue engineering scaffolds [35]. The fish scales HAp were analysed for comparison between synthesized by chemical route. Their analysis revealed that synthesized HAp and natural HAp consists of sub-micron HAp particle with Ca/P ratio of 1.62 and 1.71, respectively. Their analysed verified that HAp biomaterials from fish scale are physicochemically and biologically equivalent to the chemically synthesized HAp.

Hydroxyapatite extracted from waste fish bones and scales via calcination method [36] revealed that hydroxyapatite powder from the natural sources (tilapia scale and bone) are better in their metabolic activity and more dynamic response to the environment compared to the synthetic process. Moreover, the HAp from natural is cheaper and uncomplicated compared to synthetic method.

Sockalingam et al. (2015) stated that fish scales are bio-composites composed of connective tissue, protein, lipid, pigment and various materials [37]. Collagen, keratin, and mucin are the types of protein that can be found in fish scales, ranges from 41 to 84% [38]. Besides, fish scales also contain high amount of calcium phosphate compound such as HAp and calcium carbonate ($CaCO_3$). Amount of HAp in fish scales ranges 38% to 46% with a small percentage of $CaCO_3$ content in them [39]. Thus, attempts had been taken to produce HAp from Red Tilapia (Oreochromis niloticus) fish scales via thermal calcination method.

Recently, it was reported by Kongsri et al. (2013) that synthesis of phase-pure nanocrystalline HAp from freshwater fish (*Tilapia nilotica*) scales waste through alkaline heat treatment method [40]. The fish scale was washed and heated with 50% NaOH at 100 °C for 1 h to furnish HAp. FTIR analysis confirmed the replacement of some of the phosphate groups with the carbonate group (B-type substitution). ICP-OES confirmed that the Ca/P ratio was 1.67, same as the theoretical value.

Currently, hydroxyapatite was produced from waste fish scale (FsHAp) of Tilapia (*Oreochromis niloticus*) (Fig. 3) by thermal degradation method [41]. The fish scale was cleaned and dried before loading into gas furnace and heating up to 1200 °C for 2 h. The white fish scale ash obtained was ball milled into powders and Fourier transform infrared (FTIR) spectroscopy analysis verified that the materials produced was HAp.

Figure 4 shows FTIR spectrum of fish scale ash powder. The sharp peak appeared at 3569 cm^{-1} which correspond to OH group from HAp. It can be seen that the peaks in the regions of 472, 569, 601, 632, 1046 and 1091 cm^{-1} were corresponded to phosphate groups. The results of FTIR have shown the typical spectrum of highly crystalline HAp.

Fig. 3 The nile tilapia (Oreochromis niloticus) [42]

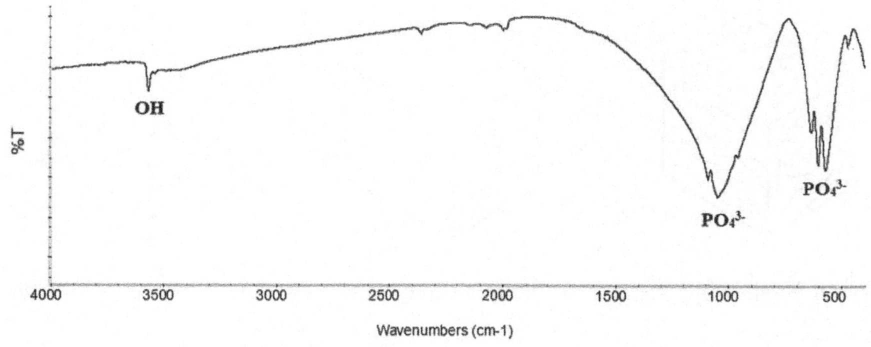

Fig. 4 FTIR spectrum of fish scale ash powder [43]

3 Hydroxyapatite as Fillers in Polymer Composite

Importance components for composite materials are matrix and filler materials. Application of hydroxyapatite as fillers in many type of polymers has been reported by many reseachers [5, 44]. Polymer/HAp composite have been widely used as bone tissue replacement. Most of polymers are in active to the human tissue, thus addition of hydroxyapatite to the polymer matrix does not only improved the mechanical strength but also enhanced the biological properties of the composite. Being biocompatible, the present of HAp in the polymer scaffold will significantly improve the osteoconductivity and bone bonding ability while polymer component offer design flexibility and strength [45].

Every combination of HAp as fillers in the polymeric matrix will result in various configuration of property of composites. The property of a polymer composite will strongly depend on the way the fibres are arranged in the composite [46]. Figure 5 shows the type of fillers in composite materials. HAp can be generally categories as particulate and flakes fillers depending on it source and technique to synthesise.

HAp have been used as filler in fabrication of HAp/sodium alginate/chitosan composite for microspheres in drug delivery and bone tissue engineering [48]. Sodium alginate is the natural polysaccharide that obtained from brown seaweed and it is one of the polymers that have good toughness and flexibility. The combination of HAp and sodium alginate composite improved the strength, fracture toughness and stiffness of the composite.

4 Polymer Used in Biomaterial Composite

Polymer composites as biomaterial played an importance role in biomedical applications. They are widely used in medical devices such as artificial organs, prostheses, implant, dentistry and other medical applications. Stanisławska, 2014 reported that

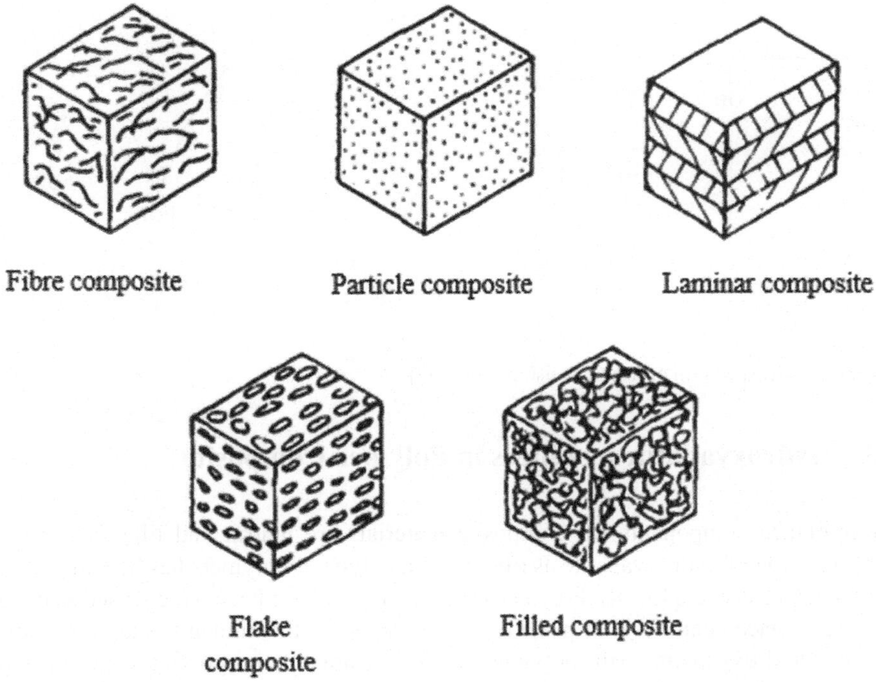

Fig. 5 Types of filler in composites [47]

polymer composite to be the most popular in biomedical tissue engineering due to their similarity structural characteristics of tissues and good mechanical properties [49]. The most common polymer used in composites materials are thermoplastic polyesters such as poly lactic acid (PLA), poly glycolic acid (PGA), poly caprolactone (PCL) and copolymers.

Synthetic biodegradable polymer such as poly (L-lactic acid) (PLLA) has attract intention in fabrication implant medical devices since they are biodegradable materials which have advantages over metal or non-degradable materials. As stated by Middleton et al. (2000), PLLA and bio biodegradable polymers are very suitable materials used for maxillofacial repair and orthopaedic fixation devices such as pins and rods for bone fracture fixation and screw and plates [50].

In dental composite, the most commonly used resin is bisphenol A-glycidyl methacrylate (Bis-GMA) and other dimethacrylate monomer (TEGMA, UDMA,) and filler materials such as silica. Among other polymers used are epoxy resin, polycarbonate, polyethylene and etc. The material used is dependent on its ability to resemble physical, mechanical and aesthetic properties of natural tooth structure. Other example of material composite that can reduce the cost and time consuming is carbon fibre/poly(methyl-methacrylate) (CF/PMMA), ultra-high molecular weight polyethylene/poly (methyl-methacrylate) (UHMWPE/PMMA) composite, composite bridges and dentures.

4.1 High Density Polyethylene (HDPE)/HAp Composites

Bonfield et al. (1981) and Wang et al. (1994) reported the application of particulates HAp reinforced filler in high density polyethylene (HDPE) composites for bone substitution [9, 51]. This composite has a modulus value approaching that of cortical bone. Their finding have solved the problem of bone resorption occurring to implant of conventional materials which much higher modulus value than cortical bone.

Besides, Wang et al. (1998), have investigated the hydroxyapatite-polyethylene (HAp-PE) composites for bone substitution known as HAPEXTM [10]. In their research, they reported that the higher filler loadings of HAp led to increase tensile strength and modulus of composite but simultaneous reduction in strain to failure. They also reported the HAp filler particle size and morphology play a crucial role that have significant effects on the mechanical properties of HAPEXTM. It was revealed that the composite with smaller particle size of HAp increased the tensile strength and modulus but lowered strain to failure [52].

Zhang et al. (2007) studied the impact behaviour HAp reinforced HDPE composite [52]. They found that the fracture toughness of HDPE/HAp composites increased with HDPE molecular weight but decreased with increasing HAp filler loadings.

Recently Liu et al. (2019) was applied self-made loop oscillatory push–pull molding (LOPPM) equipment to produce HDPE/HA composites with high tensile strength, modulus and toughness up to 95.1 MPa, 4.2 GPa, 58.4 kJ/m^2, respectively [53].

4.2 Mechanical Properties of HDPE/HAp Composites

There are several mechanical properties can be investigated from mechanical test such as tensile, flexural and impact properties. All has been explained as follows:

4.2.1 Tensile Properties

Tensile test can be referred to the ability of composite materials to withstand the forces that tend to pull it apart and to determine to what extent the materials stretches before failure [54]. From tensile test, the most commons behaviours or properties of composite materials can be determined such as yield strength, tensile strength or ultimate tensile strength, Young's modulus and percent elongation.

According to Cheang et al. (2003) have studied different morphology of HAp fillers on the tensile properties of the HDPE/HAp composites [55]. They have found that the tensile strength of HDPE/HAp composites with rough surface of HAp filler higher than that of using smooth surface of HAp fillers. The rough surface promoted mechanical interlocking, thus restrain and stiffening of the composites. On the other hand, smooth surface HAp detach from the polymer matrix, thus lowered tensile strength of composites.

Husin et al. (2011) reported that tensile strength properties and Young's modulus of HDPE/HAp composite increased while elongation at break decreased along with the HAp content from 10 to 50 phr [56]. This is because addition of HAp to this polymer enhances rigidity and brittleness of these polymer composites. Besides that, the increment in tensile strength is believed due to the uniform dispersion of HAp in the composite.

Balakrishnan et al. (2013), in their works revealed that the tensile strength and tensile modulus of HDPE/HAp composite was also improved with increasing of HAp content [57]. This finding can be related to the improved of HAp dispersion and distribution in the composite. Salmoria et al. (2013) in his paper discovered that the tensile strength and elongation values for the HDPE/HAp composite decreased as increasing HAp filler loading. This is probably affected by low chemical affinity between the polymeric and the ceramic phases [58].

Balakrishnan et al. (2013) reported that the Young's modulus and tensile strength of HDPE/HAp composite increased when HAp was pre-treated with triethanolamine coupling agent [57]. No significant change was found in the value of elongation at break despite better dispersion of the fillers. This can be related by the absence of strong chemical interaction between the phases.

4.2.2 Flexural Properties

Flexural strength is also known as modulus of rupture, bend strength or fracture strength. It can be defined as the stress in a material just before it yields in a flexural test. Flexural modulus is measure of the stiffness during the first part of the initial part of the bending process. It is important to produce the composites materials which have good flexural.

Lim et al. (2006) in their research found that flexural strength of HDPE/HAp gradually decreasing starting from 10 vol% HAp content [59]. The lower flexural strength indicated that the higher rigidity of composite leads to brittle characteristics in which failure occurs before the sample able to reach it real strength. Brittle failure occurs when applied stress is unable to be fairly distributed, causing local stress concentration that leads to crack formation especially near defect frozen stress area, particle matrix interface and particle–particle interface.

Balakrishnan et al. (2013) have observed that the flexural strength and modulus of HDPE/HAp increased with increasing of HAp filler loading and this indicates the stiffening effect of HAp fillers [57]. This finding can be related to the improve HAp particles dispersion where they were uniformly distribution in the matrix of composite.

Based on previous researchers, the flexural modulus value for the HDPE/HAp composite decreased as increasing HAp content filler [58]. This is probably affected by low chemical affinity between the polymeric and the ceramic phases.

4.2.3 Impact Properties

Impact strength is ability of a part to absorb energy. It can be explained by its ability to develop an internal force multiplied by the deformation of the part without failure. The effect of filler loading on impact strength of HDPE/HAp composites were studied [56]. They found that at 10 phr of HAp there a significantly reduction and remain stable with increasing HAp content. This is probability because of low compatibility between HAp and HDPE matrix disrupts continuity of the matrix and restricted the capability of the matrix to dissipate impact energy applied. Therefore, lowering the energy absorbed and reduced the impact strength values.

Lim et al. (2006) in their experiment work had found that impact strength obvious decreased in value after the initial loading of HAp filler [59]. This happen because the introduction of filler into polymer matrix disrupt the continuity of the matrix. HAp has relatively low compatibility with PE. This is turn lowered the capability of the matrix to distribute the impact strength applied.

The same pattern was found by Balakrishnan et al. [57]. They mentioned the value of impact strength of HDPE/HAp composite decreased with increasing HAp fillers content. This indicates embrittlement of the composite was due to particles act as stress concentration in the matrix while giving initiating crack propagation leading to a brittle failure of the composite.

5 Potential Application of HDPE/HAp Composites

Biomaterial commonly used to make component to replace damage or diseased human organ in safe, economic, reliable and physiologically acceptable manner. The commonly used materials to design biomaterials are polymers, ceramic, metal and composites. They have been used in human or animal body due to their excellent biocompatibility and mechanical properties for replacement of a body part which has lost function due to diseases or trauma. This type of material also applied to assist healing, to enhance organ functionality as well as to correct abnormalities. Besides that, biomaterial can also use to repair a body feature, tissues or organ [60].

One of the most material used in biomedical applications is polymeric material. There are two types of polymer which are synthetic and natural polymer. The example of synthetic polymeric normally used as biomaterials includes polyesters, polyethylene, polyurethane and polyamides. Based on processability, the synthetic polymer is easy to be used to fabricate tissues and implantation medical devices engineering, joint prosthesis and dentistry. Table 2 shows the examples of successful used of polymer in total knee replacement (TKR). TKR component has complicated geometry and biomechanics of movement. Common TKR components consist of femora

Table 2 Application of synthetic polymer biomaterial [61]

Synthetic polymer	Application in biomedical	Example of application
• Ultra-high-molecular-weight polyethylene (UHMWPE)	Joint prothesis • *Total knee replacement (TKR)* • *Total hip replacement*	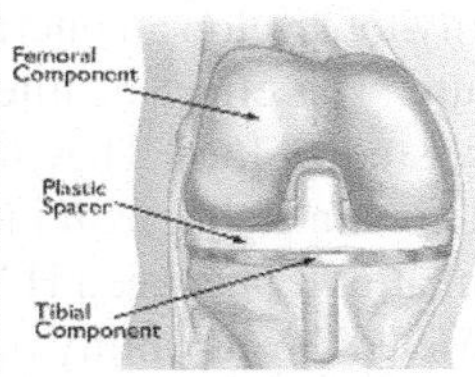 Total knee replacement (TKR)
• Ultra-high-molecular-weight polyethylene (UHMPE) • Poly (methyl methacrylate) (PMMA) • Polytetrafluoroethylene (PTFE)	Dentistry • *Dental implant* • *Prosthetic*	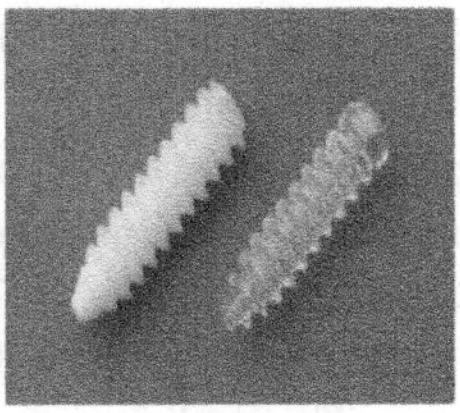 Prosthetic

and tibia which tibia part is made up from ultra-high molecular weight polyethylene (UHMWPE).

In recent years, dental treatment is found to be among popular medical treatment performed upon human being. Among materials used in dental treatment are pit lining, prosthetic, pit filling, endodontic, connect teeth and so on. They are made from wide range of polymer including UHMWPE. The used of this type of polymer has been developed in order to avoid reoperation which may result in reduction in cost and psychological benefit.

Ceramics materials are also widely used in biomaterial applications. This material was chosen due to their inertness towards living tissues, excellent wear characteristics, high compressive strength and their convenient to design variety of shape and different porosities [62]. Although ceramic has been used for structure biomaterials, but their weaknesses due to brittleness and poor elasticity have made them less favoured as compared to other materials such as metals or polymers. Currently, ceramic materials such hydroxyapatite (HAp) has been found increasing in demand as a filler in polymeric materials [63]. They are used as a filler in biomaterial like nasal septal bone, middle ear, bio-eye HAp orbital implant [64] and HAp block ceramic [65]. Table 3 shows HAp materials used for bio medical application.

Table 3 HAp material used for biomedical application [66]

Materials	Application in biomedical
• Hydroxyapatite (HAp)	Nasal septal nose 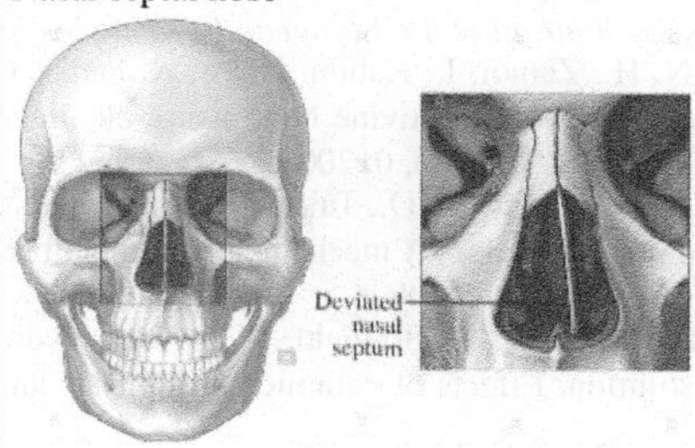

6 Conclusions

Hydroxyapatite (HAp) widely used as fillers in polymeric composite especially high density (HDPE)/HAp composite. HAPEX is one of example of HDPE/HAp composite commercially used in middle ear implant. The synthetic HAp used was chemically synthesised and expensive due to high cost of chemicals and laborious procedures. Biogenic HAp from natural resources such as animal bone has been used as an implant for biomedical application due to its close similarity in composition to mineral component of human bone. However, due to safety concern of animal diseases such as bovine spongiform encephalopathy (BSE) and foot and mouth diseases (FMD), biogenic HAp from animals was not used. HAp from aquatic sources such as fish scales is the best alternatives to replace animal source. The properties of fish scales HAp are biocompatible, bioactivity and osteoconductive materials which make it qualify as reinforcement in the biomaterials composite. Additionally, the presence of HAp filler in the composite increases the biological and mechanical properties of the HDPE/HAp composites, thus, this give good alternative to produce composite for bio-medical applications.

References

1. Shivani, P., Jashandeep, S., & Sing, K. (2021). Ceramic biomaterials: Properties, state of the art and future prospectives. *Ceramics International, 47*, 28059–28074.
2. Nath, S., Kalmodia, S., & Basu, B. (2010). Densification, phase stability and in vitro biocompatibility property of hydroxyapatite—10 wt% silver composites. *Journal of Materials Science: Materials in Medicine., 21*(4), 1273–1287.
3. Narayan, R. J. (2010). The next generation of biomaterial development. *Philosophical Transactions of the Royal Society A: Mathematical, Physical and Engineering Sciences.*
4. Ma, P. X., Eyster, T. W., & Doleyres, Y. (2018). Tissue engineering biomaterials. In: Mark, H. F. (Ed.), *Encyclopedia of polymer science and technology* (4th edn.). John Wiley & Sons, New York.
5. Rahasudha, K., Guobao, W., Ma, P. X. (2022). Synthetic polymeric biomaterials for tissue engineering. In *Chapter 2: Tissue engineering using ceramics and polymers* (p. 41). Elsevier Ltd.

6. Anahita R. S., Alireza, N., Cuie, W. (2021). Structural polymer biomaterials. In *Chapter 12: Structural biomaterials: Properties, characteristics and selection, woodhead publishing series in biomaterials* (p. 395).
7. Chevalier, J., & Gremillard, L. (2009). Ceramics for medical applications: A picture for the next 20 years. *Journal of the European Ceramic Society, 29*(7), 1245–1255.
8. Adenan, N. H., Zainol, I., Rahim, N. A., & Jaafar, C. N. A. (2018). Extraction of nanohydroxyapatite from waste bovine bone using alkaline digestion method. *Journal of Physics: Conference Series, 1082*(1), 012005.
9. Bonfield, W., Grynpas, M. D., Tully, A. E., Bowman, J., & Abram, J. (1981). Hydroxyapatite reinforced polyethylene—A mechanically compatible implant material for bone replacement. *Biomaterials, 2*(13), 185–186.
10. Wang, M., Joseph, R., & Bonfield, W. (1998). Hydroxyapatite-polyethylene composites for bone substitution: Effects of ceramic particle size and morphology. *Biomaterials, 19*, 2357–2366.
11. Oladele, I. O., Agbabiaka, O. G., Adediran, A. A., Akinwekomi, A. D., & Balogun, A. O. (2019). Structural performance of poultry eggshell derived hydroxyapatite based high density polyethylene bio-composites. *Heliyon, 5*(10), e02552.
12. Kribaa, O. K., Latif, S., Saifi, F., & Chahbaoui, N. (2022). Elaboration and chemical characterization of a composite material based on hydroxyapatite/polyethylene. *Materials Today: Proceedings, 49*, 1017–1022.
13. Aiza Jaafar, C. N., Zainol, I., Izyan Khairani, M. I., & Dele-Afolabi, T. T. (2022). Physical and mechanical properties of tilapia scale hydroxyapatite-filled high-density polyethylene composites. *Polymers, 14*(2), 251.
14. Navarro, M., Michiardi, A., Castano, O., & Planell, J. A. (2008). Biomaterials in orthopaedics. *Journal of the Royal Society Interface., 5*(27), 1137–1158.
15. Kurtz, S. M. (2015). UHMWPE biomaterials handbook: ultra-high molecular weight polyethylene in total joint replacement and medical devices.
16. Mirsalehi, S. A., Sattari, M., Khavandi, A., Mirdamadi, S., & Naimi-Jamal, M. R. (2016). Tensile and biocompatibility properties of synthesized nano-hydroxyapatite reinforced ultrahigh molecular weight polyethylene nanocomposite. *Journal of Composite Materials, 50*, 1725–1737.
17. Sathiyavimal, S., Vasantharaj, S., LewisOscar, F., Selvaraj, R., Brindhadevi, K., & Pugazhendhi, A. (2020). Natural organic and inorganic–hydroxyapatite biopolymer composite for biomedical applications. *Progress in Organic Coatings, 147*, 105858.
18. Zhu, J., Xiong, R., Zhao, F., Peng, T., Hu, J., Xie, L., Xie, H., Wang, K., & Jiang, C. (2019). Lightweight, high-strength, and anisotropic structure composite aerogel based on hydroxyapatite nanocrystal and chitosan with thermal insulation and flame retardant properties. *ACS Sustainable Chemistry & Engineering., 8*(1), 71–83.
19. Akram, M., Ahmed, R., Shakir, I., Ibrahim, W. A., & Hussain, R. (2014). Extracting hydroxyapatite and its precursors from natural resources. *Journal of Materials Science, 49*(4), 1461–1475.
20. Canillas, M., Rivero, R., García-Carrodeguas, R., Barba, F., & Rodríguez, M. A. (2017). Processing of hydroxyapatite obtained by combustion synthesis. *Boletín de la Sociedad Española de Cerámica y Vidrio., 56*(5), 237–242.
21. Yousefi, A. M., Oudadesse, H., Akbarzadeh, R., Wers, E., & Lucas-Girot, A. (2014). Physical and biological characteristics of nanohydroxyapatite and bioactive glasses used for bone tissue engineering. *Nanotechnology Reviews., 3*(6), 527–552.
22. Wong, J. Y., & Bronzino, J. D. (2007). *Biomaterials*. CRC press.
23. Fathi, M. H., & Hanifi, A. (2007). Evaluation and characterization of nanostructure hydroxyapatite powder prepared by simple sol-gel method. *Materials Letters, 61*, 3978–3983.
24. Manafi, S. A., & Joughehdoust, S. (2002). Synthesis of hydroxyapatite nanostructure by hydrothermal condition for biomedical application. *Iranian Journal of Pharmaceutical Science, 15*, 385–393.
25. Kavitha, M., Subramanian, R., Somasundara Vinoth, K., Narayanan, R., Venkatesh, G., & Esakkiraja, N. (2015). Optimization of process parameters for solution combustion synthesis of strontium substituted hydroxyapatite nanocrystals using design of experiments approach. In *Powder technology handbook* (Vol. 271, pp. 167–181).

26. Choudhur, P., & Agrawal, D. C. (2012). *CSJM university*. India Woodhead Publishing Limited.
27. Gören, Ş., Gökbayrak, H., Altıntaş, S. (2004). Production of hydroxylapatite from animal bone. In *Key engineering materials* (vol. 264, pp. 1949–1952).
28. Bano, N., Jikan, S. S., Basri, H., Bakar, S. A., & Nuhu, A. H. (2017). Natural hydroxyapatite extracted from bovine bone. *Journal of Science and Technology, 9*(2).
29. Fara, A. N., & Abdullah, H. Z. (2015). Characterization of derived natural hydroxyapatite (HAp) obtained from different types of tilapia fish bones and scales. In *AIP conference proceedings* (Vol. 1669, p. 02007).
30. Kusrini, E., Sofyan, N., Nurjaya, D. M., Santoso, S., & Tristantini, D. (2013). Removal of heavy metals from aqueous solution by hydroxyapatite/chitosan composite. In *Advanced materials research* (Vol. 789, pp. 176–179).
31. Sadat-Shojai, M., Khorasani, M. T., Dinpanah-Khoshdargi, E., & Jamshidi, A. (2013). Synthesis methods for nanosized hydroxyapatite with diverse structures. *Acta Biomaterialia, 9*(8), 7591–7621.
32. Venkatesan, J., Pallela, R., Bhatnagar, I., & Kim, S. K. (2012). Chitosan–amylopectin/hydroxyapatite and chitosan–chondroitin sulphate/hydroxyapatite composite scaffolds for bone tissue engineering. *International Journal of Biological Macromolecules, 51*(5), 1033–1042.
33. Hoyer, B., Bernhardt, A., Heinemann, S., Stachel, I., Meyer, M., & Gelinsky, M. (2012). Biomimetically mineralized salmon collagen scaffolds for application in bone tissue engineering. *Biomacromolecules, 13*(4), 1059–1066.
34. Mondal, S., Mahata, S., Kundu, S., & Mondal, B. (2010). Processing of natural resourced hydroxyapatite ceramics from fish scale. *Advances in Applied Ceramics., 109*(4), 234–239.
35. Panda, N. N., Pramanik, K., & Sukla, L. B. (2014). Extraction and characterization of biocompatible hydroxyapatite from fresh water fish scales for tissue engineering scaffold. *Bioprocess and Biosystems Engineering, 37*(3), 433–440.
36. Mustafa, N., Ibrahim, M. H., Asmawi, R., & Amin, A. M. (2015). Hydroxyapatite extracted from waste fish bones and scales via calcination method. In *Applied mechanics and materials* (Vol. 773, pp. 287–290).
37. Sockalingam, K. A., Yahya, M. A., & Abdullah, H. Z. (2015). Preparations of hydroxyapatite from Tilapia scales. In *Advanced materials research* (Vol. 1087, pp. 30–34).
38. Mori, A. S., Shiono, T., Koide, D., Kitagawa, R., Ota, A. T., & Mizumachi, E. (2013). Community assembly processes shape an altitudinal gradient of forest biodiversity. *Global Ecology and Biogeography., 22*(7), 878–888.
39. Wang, X., Kong, R., Pan, X., Xu, H., Xia, D., Shan, H., & Lu, J. R. (2009). Role of ovalbumin in the stabilization of metastable vaterite in calcium carbonate biomineralization. *The Journal of Physical Chemistry B, 113*(26), 8975–8982.
40. Kongsri, S., Janpradit, K., Buapa, K., Techawongstien, S., & Chanthai, S. (2013). Nanocrystalline hydroxyapatite from fish scale waste: Preparation, characterization and application for selenium adsorption in aqueous solution. *Chemical Engineering Journal, 215*, 522–532.
41. Pielichowska, K. (2016). The influence of polyoxymethylene molar mass on the oxidative thermal degradation of its nanocomposites with hydroxyapatite. *Journal of Thermal Analysis and Calorimetry, 124*(2), 751–765.
42. Valenzuela-Rojo, D. R., López-Cervantes, J., & Sánchez-Machado, D. I. (2018). Tilapia (Oreochromis aureus) collagen for medical biomaterials. *Seaweed Biomaterials, 3*(10), 55–77.
43. Zainol, I., Adenan, N. H., Rahim, N. A., & Jaafar, C. A. (2019). Extraction of natural hydroxyapatite from tilapia fish scales using alkaline treatment. *Materials Today: Proceedings, 16*, 1942–1948.
44. Wei, G. B., & Ma, P. X. (2004). Structure and properties of nano-hydroxyapatite/polymer composite scaffolds for bone tissue engineering. *Biomaterials, 25*, 4749–4757.
45. Legeros, R. Z. (2002). Properties of osteoconductive biomaterials: Calcium phosphates. *Clinical Orthopaedics and Related Research, 395*, 81–98.
46. Pickering, K. L., Efendy, M. A., & Le, T. M. (2016). A review of recent developments in natural fibre composites and their mechanical performance. *Composites Part A: Applied Science and Manufacturing, 83*, 98–112.

47. Bunsell, A. R. (2005). *Fundamentals of fibre reinforced composite materials, France*. CRC Press.
48. Yong-guang, B., Zi-ting, L., & Shi-ting, D. (2019). *Materials Science and Engineering: C, 100*, 576–583.
49. Stanisławska, A. (2014). Biomaterials and implants in cardiac and vascular surgery-review. *Advances in Materials Science, 14*(3), 5–17.
50. Middleton, J. C., & Tipton, A. J. (2000). Synthetic biodegradable polymers as orthopedic devices. *Biomaterials, 21*, 2334–2346.
51. Wang, M., Porter, D., & Bonfield, W. (1994). Processing, characterisation, and evaluation of hydroxyapatite reinforced polyethylene. *British Ceramic Transaction and Journal, 93*, 91–95.
52. Zhang, B., Zhu, W., Huang, S.-P., & Zhou, K. C. (2007). Mechanical properties and microstructure of nano HAP/HDPE extrusion bio-composite modified by silane-coupling agents. Gongneng Cailiao. *Journal of Functional Materials., 38*, 1912–1915.
53. Liu, T., Huang, K., Li, L., Gu, Z., Liu, X., Peng, X., & Kuang, T. (2019). High performance high-density polyethylene/hydroxyapatite nanocomposites for load-bearing bone substitute: Fabrication, in vitro and in vivo biocompatibility evaluation. *Composites Science and Technology, 175*, 100–110.
54. Fairuz, A. M., Sapuan, S. M., Zainudin, E. S., & Jaafar, C. N. A. (2016). Effect of filler loading on mechanical properties of pultruded kenaf fibre reinforced vinyl ester composites. *Journal of Mechanical Engineering and Sciences, 10*(1), 1931–1942.
55. Cheang, P., & Khor, K. (2003). Effect of particulate morphology on the tensile behaviour of polymer–hydroxyapatite composites. *Materials Science and Engineering, 345*(1–2), 47–54.
56. Husin, M. R., Wahit, M. U., Kadir, M. R. A., & Rahman, W. A. W. A. (2011). Effect of hydroxyapatite reinforced high density polyethylene composites on mechanical and bioactivity properties. *Key Engineering Materials, 471*, 303–308.
57. Balakrishnan, H., Husin, M. R., Wahit, M. U., & Abdul Kadir, M. R. (2013). Maleated high density polyethylene compatibilized high density polyethylene/hydroxyapatite composites for biomedical applications: Properties and characterization. *Polymer-Plastics Technology and Engineering, 52*(8), 774–782.
58. Salmoria, G. V., Fancello, E. A., Roesler, C. R., & Dabbas, F. (2013). Functional graded scaffold of HDPE/HA prepared by selective laser sintering: Microstructure and mechanical properties. *The International Journal of Advanced Manufacturing Technology, 65*(9–12), 1529–1534.
59. Lim, K. L. K., Ishak, Z. M., Ishiaku, U. S., Fuad, A. M. Y., Yusof, A. H., Czigany, T., Pukanzsky, B., & Ogunniyi, D. S. (2006). High density polyethylene/ultrahigh molecular weight polyethylene blend. II. Effect of hydroxyapatite on processing, thermal, and mechanical properties. *Journal of Applied Polymer Science, 100*(5), 3931–3942.
60. Hench, L. L., & Erthridge, E. C. (1982). *Biomaterials: An interfacial approach*. Academic Press.
61. Blunn, G., Del Preva, E. B., Costa, L., Fisher, J., & Freeman, M. A. R. (2002). Ultra-High Molecular-Weight Polyethylene (UHMWPE) in total knee replacement: Fabrication, sterilisation and wear. *The Journal of Bone and Joint Surgery, 84*(7), 946–949.
62. Ramakrisha, S., Majer, J., Wintermantel, E., & Leong, K. W. (2001). Biomedical application of polymer composite material. *A review. Composites Science and Technology, 61*(9), 189–1224.
63. Alizadeh-Osgouei, M., Li, Y., & Wen, C. (2019). A comprehensive review of biodegradable synthetic polymer-ceramic composites and their manufacture for biomedical applications. *Bioactive Materials, 4*(1), 22–36.
64. Jordan, D. R., Gilberg, S., & Bawazeer, A. (2004). Coralline Hydroxyapatite Orbital Implant (bio-eye): Experience with 158 patients. *Ophthalmic Plastic & Reconstructive Surgery, 20*(1), 69–74.
65. Sousa, R. A., Reis, R. L., Cunha, A. M., & Bevis, M. J. (2002). Structure development and interfacial in high density polyethylene/hydroxyapatite (HDPE/HA) composites molded with preferred orientation. *Journal of Applied Polymer Science, 86*, 2873–2886.
66. Kanasan, N., Adzila, S., Suid, M. S., & Gurubaran, P. (2016). *Proceeding from AIP Conference: Preparation and characterization of hydroxyapatite/sodium alginate biocomposites for bone implant application*. American Institute of Physics.

Biocomposites from Microalgae

Natasha Nabila Ibrahim, Imran Ahmad, Norhayati Abdullah, Iwamoto Koji, Shaza Eva Mohamad, and Fazrena Nadia Binti Md. Akhir

1 Introduction

In parallel with world growth and globalization, plastic production has become imperative in our daily life as they are very advantageous in terms of endurance, production expense, operation simplicity and convenience. Plastics are vastly utilized in industries like healthcare, automotive, construction, agriculture, packaging, and components of electronic [1]. From back in 1950, world production of plastic has expanded to 8.4% of annual growth rate and every year, global plastic manufacture has exceeded 400 million tons. It has been predicted that by 2025, generation of plastic would extend to 500 million tons.

As described by Horton [2], regardless of various application of petroleum-derived plastics and petrochemical-based polymers, the downside of them are they are not environmentally friendly because they are non-biodegradable and poses threat to earth ecosystem. Swift upscale of plastic manufacture has resulted in greenhouse gases and harmful chemicals emission alongside extensive energy usage [3]. Nearly 60% plastic waste end up in oceans, rivers, lakes, and landfills [4, 5].

Over the course of decades, scientists and researchers strived and still are seeking possible solutions that are eco-friendly and sustainable to substitute conventional plastics generation. Chia et al. [6] mentioned that biopolymer synthesis has been recognized by researchers as potential promising composite to replace petroleum-derived plastics to cater global daily needs. Bioplastics can be manufactured from microorganisms, sustainable biomass resources as well as by-products of agriculture [3].

N. N. Ibrahim · I. Ahmad (✉) · N. Abdullah (✉) · I. Koji · S. E. Mohamad · F. N. B. Md. Akhir
Malaysia-Japan International Institute of Technology, Universiti Teknologi Malaysia, Jalan Sultan Yahya Petra, 54100, Kuala Lumpur, Malaysia
e-mail: mustafwibinqamar@gmail.com

N. Abdullah
e-mail: norhayati@utm.my

Sapuan S. M. and I. Ahmad (eds.), *Composites from the Aquatic Environment*, Composites Science and Technology, https://doi.org/10.1007/978-981-19-5327-9_5

Karan et al. [7] explained that, historically, biopolymers can be categorized into three generations, i.e., feedstock from petroleum with plant monomers, bio-derived monomers with polymers and feedstock from plant biomass. Biopolymers of first generation are created from mixing of petroleum-based plastics with raw natural resources which necessitates farmable land with fresh water and nutrients. This leads to food versus fuel debate and jeopardizing food security. First and second generation of bioplastics both have similarities regarding the material blends. Due to partial mixing of petrochemical based plastics, the plastic waste would still be persistent and potentially harmful to environment since the degradation would only be achieved partially. Finally, the study of bioplastic production has moved on to the third generation, which exploits terrestrial plants as feedstock like potatoes and corn. This approach is to ensure the easiness in biodegrading ability without any microplastic footprint. Alas, this strategy too is unsustainable in the distant future. Wastes that are generated from agricultural activities might solve this issue, nevertheless, agricultural wastes are meagre and deficient to produce bioplastic [8].

Henceforth, tremendous research has been targeted towards bioplastics production from microorganisms like microalgae, cyanobacteria, and bacteria due their fast growth rate feature, with microalgae as the main spotlight. Das et al. [9] reported that in pursuance of global circular bioeconomy, microalgae are regarded as the best suited candidate for biomass feedstock for the mass generation of biopolymer. Freshwater and marine water are common habitats of photosynthetic organisms-microalgae. These eukaryotes are categorized by their size and colours. The latter characteristic includes Rhodophyta as red algae, Chlorophyta as green algae, and Phaeophyta as brown algae. According to Garbowski et al. [10], the swift-growing microorganism has various sizes that ranges from 0.02 to up to 2000 μm. Hamid et al. [11] also described the size of macroalgae that are better known as seaweed, can be identified in microscopic size to a maximum of 200 ft. This highly sought-after photosynthetic microorganism possesses basic cell system that grows well with the presence of illuminance, nutrients, water, and carbon dioxide.

Rahman and Miller [12] mentioned that the current microalgal technology utilizes direct microalgal biomass or exploited as raw material for downstream processes. This technology has been recognized as probable resource material in enhancing assorted industries, inclusive of bioplastic manufacture. Microalgae are known for their efficiency in accumulating high number of bioactive compounds namely, carbohydrates (18–46%), lipids (12–48%), proteins (18–46%) and carotenoids (10–14%) [13, 14]. These macromolecules are the most vital constituents in bioderived products, viz. biopolymers, bioplastics, and biobased polyurethane [15].

In addition, one of the advantageous characteristics of microalgae is the high accumulation of biomass. Elrayies [16] explained that microalgae are extremely capable of mitigating carbon dioxide. Captured carbon dioxide that is amounted to 0.8 kg resulted in 75% of oxygen being released into the environment at minimum [8]. Moreover, in contrast to conventional terrestrial crops, microalgal cultivation is possible without arable land and lesser amount of water is needed. They are also robust since they thrive in harsh cultivation conditions like municipal and industrial wastewaters [17]. Madadi et al. [18] discussed that production of bioplastics from microalgae would be less likely to impose food scarcity issue considering that microalgae are not human beings source for staple food.

Thus far, *Chlorella sp.* and *Spirulina sp.* are the two strains that are well-studied for microalgae-derived bioproducts and are already commercialized in the market in various industrial fields such as dietary supplements, food and beverage, nutraceuticals as well as pharmaceuticals [6]. According to research study by Beckstrom et al. [19], it has been proven that biopolymer that is produced from microalgae have demonstrated improved mechanical attributes in comparison to petroleum-based polymers. Another beneficial side of microalgae-based bioplastics is refitting is possible by augmentation with plasticizer, additives and compatibilizer. This augmentation is conducted to enhance the material durability, robustness, flexibility, and interaction forces between intermolecular compounds [20].

Koyande et al. [14] reported that amongst all plants that has existed on Earth, *Chlorella* is regarded as one of the most primeval organisms since its existence is from about 3.5 billion years ago. *Chlorella* was sprung from the *Chlorellaceae* family and is classified into the genus of Chlorophyta, which means green algae. *Chlorella* is native to fresh water with various cell sizes from 3.2 to up to 10.2 μm [21]. Koyande et al. [14] also mentioned that *Chlorella* species is usually exploited for nutraceutical line as protein is the most abundant macromolecule in *Chlorella* which accounts for up to 60% of its biomass. In contrast to artificial polymers, protein from *Chlorella* is made up of complex heteropolymer rather than long chains of monomers which makes it a viable biopolymer feedstock [6].

Besides that, Sakarika and Kornaros [22] indicated that *Chlorella* has high resistance against microbial anaerobic digestion. The reason behind this characteristic is because *Chlorella* is surrounded by a strong sturdy cell wall which is beneficial in bioplastic making as has substantial pliancy feature. This is proven in an experiment by Sabathini et al. [23], when they determined the tensile strength of *Chlorella*-distilled water mixture to be 35.1 kgf/cm^2. This reading shows that *Chlorella* is a potential and effective feedstock for bioplastic manufacture.

Another promising microalga that is broadly studied for bioplastic manufacture is *Spirulina* species. *Spirulina* is widely known as bluish green algae, a prokaryotic cyanobacteria which is originated from *Oscillatoriaceae* family that also primitively grows in fresh water sources. According to Rout et al. [24], *Spirulina* grows in spiral shape with trichomes that has cell width of 5–7 mm, diameter of helix from 20 to 30 μm and helix pitch of 20 to 30 μm. This shape feature distinguishes *Spirulina* from other microalgae species. *Spirulina* has higher macromolecular protein of nearly 60–70% from its dry biomass, if compared to *Chlorella* [25]. This quality makes them highly adaptable to even harsh environmental conditions.

Similar to *Chlorella, Spirulina* has sturdy cell wall and owing to its tiny size of cells, this microalgae species does not need extraction process thus reducing production cost [26]. In the same article, Dianursanti et al. [26] stated that *Spirulina* is widely used for bioplastic blends with other materials. The researchers discovered that *Spirulina platensis*-polyvinyl alcohol blend created an improved composite with rigid interactions and strong tensile structure. In addition, *Spirulina* is extremely resilient in surrounding conditions that are highly saline and have high pH value. This feature is advantageous for bio-derived plastic making as the bioplastic manufacturing requires vigorous processes [27].

Khalis [28] discussed that *Spirulina* exhibited different properties of bioplastics upon mixing with polyethylene (PE). This is caused from different compositions of amino acids and proteins in these two species. *Spirulina* accumulates higher protein in their cells than *Chlorella* which is a viable protein source and is highly adaptable to stress environment. Despite that, *Chlorella* showed efficient resistance against breakage or breach because of its round shape, as well as the strong cell that are made up of pectin and cellulose. Table 1 summarizes examples of microalgal strains that are studied for the production of bioplastics.

Table 1 Several strains of microalgae and cyanobacteria that are utilised for bioplastic manufacture

Microalgae species	Product type	References
Chlorella	*Chlorella* blended with polyvinyl alcohol (PVA) film	[23]
Chlorella vulgaris	*C. vulgaris* blended with polyvinyl alcohol (PVA) mixture	[28]
Chlorella sorokiniana	Starch granules	[29]
Spirulina sp.	*Spirulina sp.* blended with ultra-high molecular weight polyethylene (UHMPE)	[20]
Spirulina platensis	*S. platensis* as bio-filler	[30]
Spirulina platensis	*S. platensis* blended with polyvinyl alcohol (PVA)	[26]
Spirulina	*Spirulina* as plasticiser (bioplastic derived from *Spirulina*)	[26]
Spirulina	*Spirulina* blended with poly(butylene succinate) (PBS)	[31]
Spirulina	Polylactic acid (PLA) production	[32]
Microalgae-cyanobacteria consortium; *Scenedesmus sobliquus, Desmodesmus communi* (microalgae), *Nannochloropsis gaditana, Arthrospira platensis* (cyanobacteria)	Microalgae-cyanobacteria consortium blended with glycerol for bioplastic	[33]
Chlorogloea fritschii	Poly-3-hydroxybutyrate (P3HB) bioplastic	[34]
Neochloris oleoabundans, Calothrix scytonemicola, and *Scenedesmus almeriensis*	Bio-derived plastic film	[35]
Calothrix scytonemicola	PHA production	[35]

2 Microalgae Cultivation and Harvesting

Recent trends of microalgae cultivation portrays that it has been extensively studied and applied with various methods and technologies to bring about the optimised methodology and meet the targeted biomass accumulation. Chakraborty et al. [36] expressed that prior to culturing microalgae for commercial use, the cultivation system and the chosen species needs to be decided since microalgae is eminently correlated to its natural water ambiance. Numerous techniques are being conducted for microalgal biomass synthesis that are curated for different species of microalgae and objectives. Ahmad et al. [37] described that in microalgae culture system, the most extrusive systems are outdoor or open pond structures and closed photobioreactor (PBR) structures. There is also hybrid cultivation system for lower cost of production approach [38].

2.1 Open Pond Cultivation Systems

Historically, open pond culturing system is the most conventional and typical structure for microalgal cultivation. Examples of open cultivation design are raceway ponds, circular ponds, tanks, and lakes. These outdoor systems are broadly used and are already commercialized. According to Apel et al. [39], the commercially available raceway ponds come in various sizes with depth from 1 to up to 100 cm. The size of the ponds could be available as big as a few acres. The productivity of biomass could be increased with several configurations within the pond structure itself that are designed in different shapes like circular and raceways. A recent development of multi-layered pond system that incorporates a few open tanks installed at different heights made it more effective [40].

Basically, static pond system does not have mixing properties which means this type of system promotes simplicity and cost-effective mass production. In addition, raceway ponds system is a possible cultivation system for upscaled culture with low capital cost, but some form of agitation is necessary to facilitate the culture to move and be mixed homogenously within the pond [41]. Researchers commonly favor paddlewheels as the form of turbulence inside the raceway ponds which are more advantageous if compared to the conventional open ponds structure.

The paddle-wheel raceway pond looks like a racetrack, but it comes with paddle wheels that are used to stir and ensure proper liquid flow within the whole pond. This culture approach is broadly used for industrial scale cultivation.

When it comes to open culture system, the benefits than can be obtained from them are they are easily constructible and manageable, as well as suitable for pilot-scale installation [38]. Outdoor cultivations also do not demand great operational cost because the main energy input is from sunlight and only little operational energy is needed to mix the culture [41].

Regardless of the popularity and preference, this cultivation system do have some downsides that includes irregularity in light penetration and nutrients mixing, as well as carbon dioxide uptake. This culturing system is not recommended because it requires large and high land. The outdoor condition also made the microalgal culture to be exposed to contamination. The nutrient medium makeup is also prone to changes owing to seasons and weather inconsistencies such as precipitation, not to mention being too concentrated or dried up because of evaporation. These changes would pose variations in biomass productivity [42].

2.2 Closed Photo Bioreactor (PBR) Systems

Next up, another culturing system for microalgae is a closed systems, which are also known as photobioreactors (PBRs). There are many types of PBRs that are applied for microalgae culturing and they are usually made up of transparent polyvinyl chloride (PVC) or glass. The PBRs come with lots of shapes and sizes and can be applied for indoor and outdoor cultivations, depending on the purpose and target of the experiment. Cultivation of microalgae is typically conducted in tubular shaped PBRs. Several other examples of the different shapes are flat panels or laminar, helical, hanging plastic sleeves, bubble column, fermenter-like tank reactors and airlift PBRs [43].

In parallel with improvements and globalizations, some other types of PBRs system are also being developed and improved. Despite that, vertical or horizontal installation of tubular PBRs is chosen in most studies. The reason behind the prevalence of PBRs over to raceway pond culturing, is owing to better operational control in terms of growth rate and lower risk of contamination from algae, grazers, and bacteria [44].

Nonetheless, these PBRs do have some drawbacks. Firstly, the expensive cost of designing and manufacturing of PBR system as well as their maintenance limits the full utilization of this system for pilot studies as we speak [45]. Furthermore, growth medium production and retaining the consistency of culture mixture is also not cost-effective. Researchers had determined to use air lift pump within the reactors to prevent microalgal cells from settling at the bottom, which added up to expenditure [46]. The expenditure offsets other benefits of PBRs culturing system [47]. However, Kothari et al. [48] stated that the costly expense regarding PBRs construction could be undertaken with economical PBRs materials, as well as employing industrial waste products and wastewater as microalgal growth medium. PBRs can also be operated with energy-saving pumps to the system.

Other than that, closed PBRs systems also face some hindrance that are less precarious for open ponds systems, namely, gaseous exchange and cooling requirement. Microalgae requires ample carbon dioxide to ensure photosynthesis running smoothly as well as controlling the pH. When carbon dioxide is dissolved in water, the water becomes slightly acidic and carbon capture by microalgae will increse the pH value. This is particularly problematic in saturated culture because pH level may increase to risky level, leading to growth inhibition with and improper gaseous exchange.

Furthermore, microalgal growth may be hindered and eventually enter cell death due to swift build-up of oxygen in dense culture. This means that oxygen removal needs to be efficient. Besides that, extended sun exposure will increase temperature of the culture, especially in closed models where the process would happen so quickly. Temperature that exceeds 30 °C will consequently force the cultures to enter cell cessation. Hence, appropriate cooling is essential and there are several means to gauge this issue such as installing water cooling section that is interconnected to the reactor model. Spraying and sparging the cultivation models with cooling water is an another technique [41].

Albeit the abovementioned issues regarding PBRs models, several researchers still prefer closed systems and figuring out on solutions to decrease the problems. The money-extensive manufacturing cost can possibly be lessened with appropriate decision on shape, materials, size, and design of the closed system [49]. Arcigni et al. [50] reported that to curb any incidents associated with technical flaws of system construction, the system design must be highly efficient and practical. Be that as it may, Vo et al. [51] stated that the efficient design of the system must be in parallel with growth conditions that are species-specific and location of the PBRs system.

Additionally, Chakraborty et al. [36] discovered critical findings through their study on growth rates of various microalgae species cultivated in PBR systems that were designated with different environmental conditions. *Chlorella vulgaris, Euglena granulata* and *Scenedesmus quadricauda* were experimented in different growth conditions and the resulting observation concluded that different growth conditions give different outcome to these microalgae [52]. Consequently, there is the latest recommendation by researchers on hybrid design which in particular, is a cultivating system that merges flat panel and tubular together in a single PBRs system [38]. Efficient hybrid PBR models altogether with improved insight and comprehension on growth factors will be significantly advantageous in curbing challenges of microalgal biomass accumulation. The system would be highly productive, profitable and eco-friendly [52].

In conclusion, from an operational viewpoint, capital cost and system of production are the two major disparity amongst both systems. In contrast to open cultivation design, closed PBR system comprised of costly components and units that requires higher energy consumption. Yet, on the other hand, closed PBRs usually accumulated higher productivity compared to open ponds and this may compensate the additional costs. Table 2 represents a tabulated summary on pros and cons of various microalgal cultivation systems that is adapted from Cheng et al. [46].

Table 2 Summary on advantages and disadvantages of various microalgal cultivation systems

Cultivation system	Advantages	Disadvantages
Flat panel/laminar PBRs 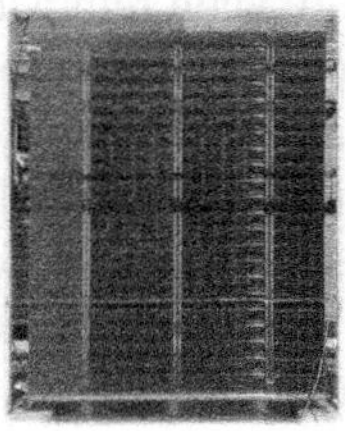	• Less prone to contamination • Low hydrodynamic cell damage • Non-space intensive if they are compactly designed • Non species specific • High biomass and metabolites yield • Spacious area to volume ratio for illumination • Great light penetration	• High risk of photo inactivity • Laborious maintenance and cleaning • High operational and principal cost • Prone to thermal cell cessation • Complex upscale operation
Airlift PBRs 	• Low operational cost • Low energy utilization • Small biofilm growth • Better mixing via aeration and draft tubes • Great mixing and mass transfer • Low hydrodynamic cell damage • High biomass and metabolites yield • Non species specific • Less prone to photo inactivity • Less prone to contamination • Non-space intensive if they are compactly designed	• Laborious maintenance and cleaning • High principal cost • Upscale operation would incur extra cost for extra units • Limited area to volume ratio for illumination • Poor light penetration for large columns
Bubble column PBRs 	• Moderate energy utilization • Simple mixing via aeration • Good mixing and mass transfer • Adequate mixing and mass transfer • High biomass and metabolites yield • Less prone to contamination • Small biofilm growth • Low hydrodynamic cell damage • Less prone to photo inactivity • Non-space intensive if they are compactly designed • Non species specific	• Laborious maintenance and cleaning • High operational and principal cost • Poor light penetration for large columns • Limited area to volume ratio for illumination • Upscale operation would incur extra cost for extra units
Tubular PBRs 	• Spacious area to volume ratio for illumination • Good mixing and mass transfer • Less prone to contamination • High biomass and metabolites yield • Non species specific • Great light penetration • Non-space intensive if they are compactly designed	• High operational and principal cost • Energy-intensive mixing via pump • Large biofilm growth • High energy utilization • Laborious maintenance and cleaning • High hydrodynamic cell damage • High risk of photo inactivity • Complex upscale operation

(continued)

Table 2 (continued)

Cultivation system	Advantages	Disadvantages
Open raceway pond 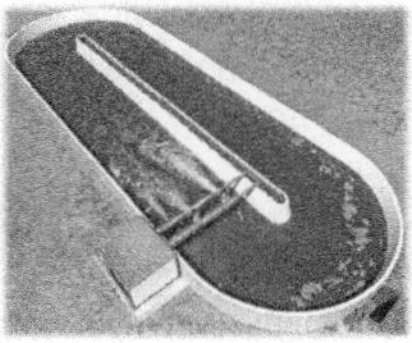	• Low energy utilization • Easy upscale operation • Low operational and principal cost • Reasonable mixing and mass transfer • Non-laborious maintenance and cleaning • Low hydrodynamic cell damage • Simple mixing via paddlewheels • Less prone to photo inactivity • Spacious area to volume ratio for illumination • Small biofilm growth	• Prone to contamination • Species specific; robust microalgae • Low biomass and metabolites yield • Space intensive • Poor light penetration for deep ponds

2.3 Harvesting

In microalgae industry, harvesting is basically a consecutive process upon water removal from culture medium-microalgae mixture. This step is administered by introducing a few downstream processing to get the biomass sludge. A relevant harvesting method is chosen after careful consideration in terms of capital cost and energy usage that mainly relies on microalgal cell density and size [53]. Additionally, harvesting expenditure is said to be one of the primary obstacles for microalgal bioeconomy owing to its extensive energy consumptions [54].

According to Augustine et al. [55] and Muylaert et al. [56], there are challenges during upscaled harvesting because of tiny microalgal cells that usually range from 3 to 30 μm, small concentration of cells which is less than 500 g/m^3, nearly indifferent buoyancy and thin culture. A quintessential biomass recovery technique ought to be discovered that could be utilized for most of the existing microalgal strains in order to recover ultimate amount of biomass. The recovery method must also work optimally with considerable production of energy and cost altogether with low disturbance towards environment.

Jerney and Spilling [41] reported that microalgal harvest processing has become the major bottleneck towards sustainable economy growth of industrial scale microalgae cultivation. The weight of water from even a dense microalgae culture is occupied by water by more than 90%. Nevertheless, there are various methods of harvesting that have been experimented and applied in microalgal biotechnology. In general, methods like sedimentation, filtration, flocculation, centrifugation, flotation, electrical harvesting, electrophoresis, nanomaterial binding, and magnetic nanoparticles are frequently adapted in the processing of microalgae slurry [57, 58]. Figure 1 illustrates a synopsis on several examples of conventional and contemporary methods on microalgae cell harvesting methods.

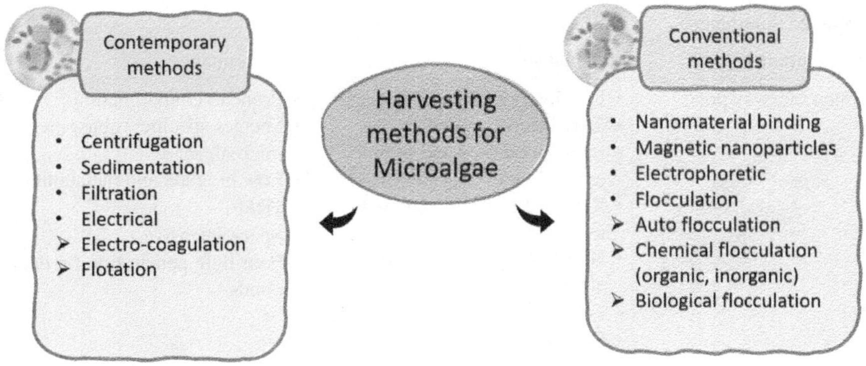

Fig. 1 Microalgal harvesting methods

Singh and Patidar [59] explained that the current trend in microalgal harvest technology revolves around biological, mechanical, electrical, and chemical groundwork. It is essential to take note that in harvesting microalgae, there is no universal method. This downstream step may involve more than one approach to adequately dewater microalgal culture, together with attaining efficient yield of biomass. Enhancement on the productivity of harvesting could be observed with flocculation and coagulation as the primary methods. Through these two procedures, maintenance and manufacturing cost could be scaled down too. The method of harvesting is decided based on several factors, viz. microalgal species, microalgal cell size and density, operational cost and energy, characteristics of end-products, as well as mass scale expediency [59]. Apart from that, microalgal biomass harvesting step may be less strenuous with proper selection of biomass application and culturing system (Jerney and Spilling 2018). For instance, bio-film derived systems like Algal Turf Scrubbers are quite popular for biomass accumulation and downstream processes. This biofilm is convenient since it can be scraped off with ease using mechanical forces as well as via vacuuming. This approach has also been applied in coupled research like wastewater remediation and biofuel production [60]. Table 3 represents an overview on numerous methods of microalgal harvesting.

3 Bioplastic Production from Microalgae

Worldwide demand of plastic is proportional to their increasing use of plastic-based materials and which is increasing and creating more stress to the current waste management organizations [12]. Therefore, the main interest to reduce the dependency on petroleum based plastic or products as they are not sustainable to the environmental concern because they creates a huge amount of environmental pollution. Each year, around eight million metric tons of petroleum-based plastic crap are generated and sent into oceans and other waterbodies, which may be minimized

Table 3 Overview on various harvesting methods of microalgae

Method	Mechanism	Advantages	Disadvantages	References
Sedimentation	Solid and liquid phase separation process via gravitational force	• Cost-effective • Simple operation • No contamination risks • Nutrient medium is recyclable • Suitable for large microalgae like *Spirulina*	• Not reliable due to equivocal density of algae • Time consuming • Low recovery of algae slurry • Possible expensive flocculants • Not suitable for small microalgae like *Dunaliella salina*	• [41] • [61] • [62] • [63]
Filtration	Nutrient medium passes through membrane filter under various forces like gravitational, vacuum, and pressure, to separate it from microalgal culture to form microalgal slurry	• High recovery yield • Low cell damage (shear stress) • Low energy consumption • Chemicals are only necessary for membrane washing (when needed) • Cost-effective • Water is recyclable	• Expensive membrane fabrication cost • Large energy consumption • Limited due to filter size • Replacement of spare parts for membrane (membrane clogging and fouling) and vacuum pump • Time-consuming • Not fitted for small microalgae	• [41] • [63] • [59] • [64] • [65]
Centrifugation	Used for separating samples with different densities	• Time-effective and efficient • Simple operation • High recovery yield • Non species specific • Suitable for laboratory and small-scale cultures	• Expensive operational and maintenance cost • Special equipment is necessary • Cellular biomass is prone to damage • Large energy consumption • Suitable for value-added products recovery • Time-consuming • Expensive for upscale cultures	• [41] • [63] • [59]

(continued)

Table 3 (continued)

Method	Mechanism	Advantages	Disadvantages	References
Flotation	Release of high pressurised air into water column that forms little bubbles that will attach to suspended matter. The suspended matter then will float to surface	• Suitable for upscale cultures • Flexible • Time-effective and efficient • Can be operated within small area • Cost-effective	• Probable issues in further downstream process with supplementary flocculants • Surfactants are necessary • Flocs are prone to breaks due to oversized air bubbles • Large energy consumption • Expensive ozoflotation cost	• [41] • [63] • [59]
Electrophoresis	Hydrolysis releases fine hydrogen bubbles at cathode that detach microalgal biomass from nutrient medium. Hydrogen bubbles adhere to microalgal flocs and float to surface	• Chemicals are unnecessary • Non species specific • Easy operation control • Low energy consumption	• Large energy consumption • Expensive replacement and maintenance cost of electrodes • Increased temperature in microalgal slurry • Changes in pH • Leftover metals in microalgal slurry • Special equipment needed	• [41] • [59]
Auto-flocculation	Natural flocculation without additional chemicals or coagulants responding towards environmental stress like nitrogen change, dissolved O_2 and pH change	• Non-toxic • Cost-effective • Simple operation • Low cell damage (shear stress)	• Time-consuming • Unreliable • Uneconomical commercialisation upon environmental changes	• [41] • [59]
Bio-flocculation	Based on exopolysaccharide (EPS) secretion by microorganisms like microalgae, bacteria, and fungi	• Chemicals are unnecessary • Non cultivation method specific • Suitable for laboratory and small-scale cultures	• Unsuitable for biofuel production due to low lipid yield • Time-consuming • Varying results with varying strains	• [63] • [59] • [66] • [67]

(continued)

Table 3 (continued)

Method	Mechanism	Advantages	Disadvantages	References
Magnetic nanoparticles	In the presence of magnetic field, magnetite (Fe_3O_4) nanoparticles (NPs) directly adhere to microalgal cells and initiates flocculation	• Simple operation • Naturally stable • Recyclable nanoparticles • Uncontaminated algal slurry • Large surface area	• Expensive nanoparticles cost • Special equipment is necessary for nanoparticles recovery	• [63] • [53]

by novel packaging material redesigns [68]. Generally, plastics may be classified according to their two basic properties: petroleum- or bio-based composition, and biodegradability or non-biodegradability. Three types of bio-based plastics have been identified: modified natural polymers, bio-based polymers synthesised from their monomers, and bioplastics derived from waste products [69]. Currently, the global annual production of bioplastics is around 1% only which is too low and need more attention on the production. The major areas in which bioplastics are employed are the packaging industry after the construction, textile industry and the automotive industry [70].

In generic terms, the classification of plastics can be divided into two categories, namely, petroleum-derived, or bio-derived, biodegradable, or non-biodegradable. According to constituents is a big family of materials that comes with different features and applications. Bio-derived plastics are manufactured either partially or fully from renewable sources or biomass and they serve similar purpose as fossil-based plastics [6]. They are further divided into three groups, viz. bio-derived but non-degradable, bio-derived, and degradable, and lastly is petroleum-derived plastics that are non-degradable [71]. Thus far, there are numerous types of bioplastics being manufactured worldwide. Several examples of non-degradable but are bio-derived plastics are bio-derived polyethylene (PE), bio-derived polyethylene terephthalate (PET), bio-derived polytrimethylene terephthalate and bio-derived polyamide (PA) or known as nylon plastic. Next, the second group which is the bio-based and biodegradable plastics include polylactic acid (PLA), polyhydroxyalkanoates (PHA), polybutylene succinate (PBS) and starch blends. Polybutylene adipate terephthalate (PBAT), polycaprolactone (PCL) and PBS are some of the bioplastics that are synthesised from fossil fuel but are biodegradable.

According to Chia et al. (2020), bioplastic manufacturing is made possible with various sources that are mainly from agriculture industry which involves terrestrial crops like soy proteins, corn, and wheat. Natural polymers such as protein and carbohydrates can also be processed for bioplastic making, as well as small molecules like fatty acid, disaccharides, and sugar. The latest innovative way in bioplastic manufacturing is by exploiting microorganisms like microalgae and bacteria [6]. Figure 2 illustrates types of bioplastics that are produced so far, that is adapted from European Bioplastics [72].

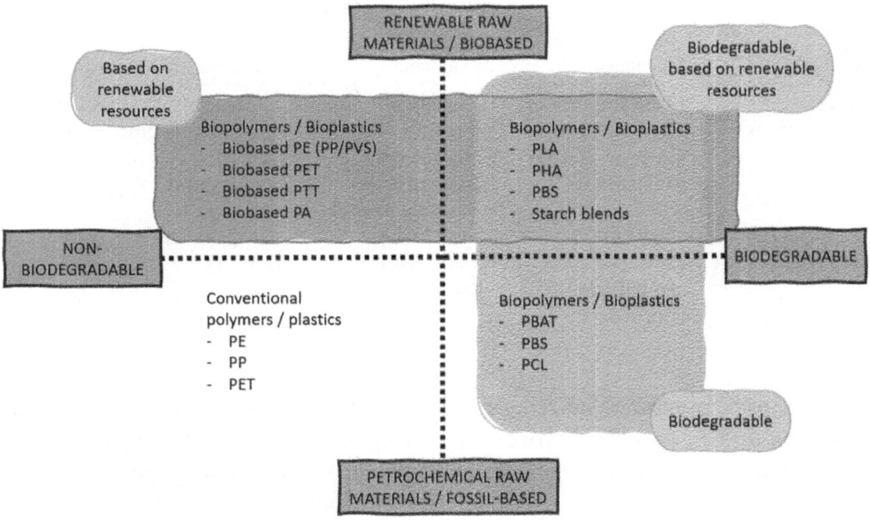

Fig. 2 Different types of bioplastics

Rai et al. [73] stated that the technology of bioplastics are gaining wide attention to replace conventional plastics as they are less likely to have negative impact towards environment. This alternative is also non energy intensive, resource-sufficient, and proficient in cutting down carbon dioxide emission as well as preventing food waste [6]. As defined by European Bioplastics, bioplastics are either produced biologically or they are bio-compostable, or portray both characteristics [71].

Although bioplastics are synthesised from natural materials, not all of them are bio-compostable. Narancic et al. [74] reported that pathway and degree of degradation determine their biodegradability. Similar to traditional petroleum-based plastics, bioplastics can be recycled or reduced to ashes through incineration. However, they are also degradable biologically via microorganism compost at small or large scale, thus creating an archetype of waste-free circular economy. Carbon footprint mitigation could be facilitated through utilisation of microorganisms like microalgae and pivoting agricultural waste towards bioplastic fabrication [73]. Figure 3 exhibits bio-compostable polymer classification according to their origins.

Kartik et al. [75] described that microalgal biopolymers can be processed via three pathways. First route involves microalgal biomass to be fermented by microorganisms and producing synthetic biopolymers. Next, natural biopolymer synthesis can be induced by photosynthetic accumulation by microalgal cells into its biomass (second route). The third pathway is by obtaining bio-composite polymer through microalgal biomass blends with additives. Rahman and Miller [12] also mentioned that to create ideal polymer-synthesising microalgae strains, genetic modification and process of intermediate biorefinery within microalgal cells could be applied. Figure 4 is the summary of routes for biopolymer production through microalgal cells.

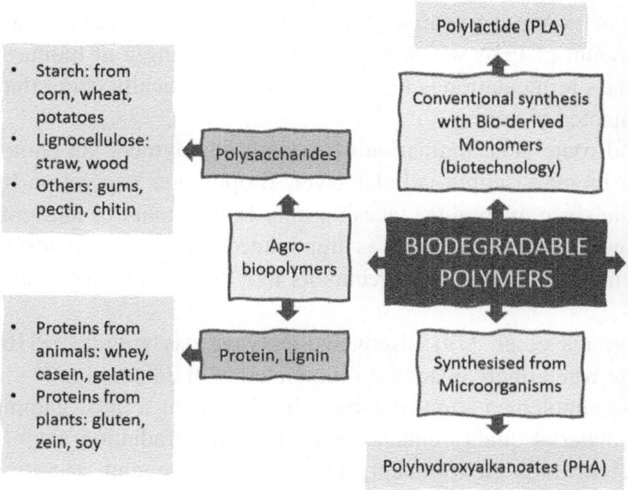

Fig. 3 Biodegradable polymer classification and their origins

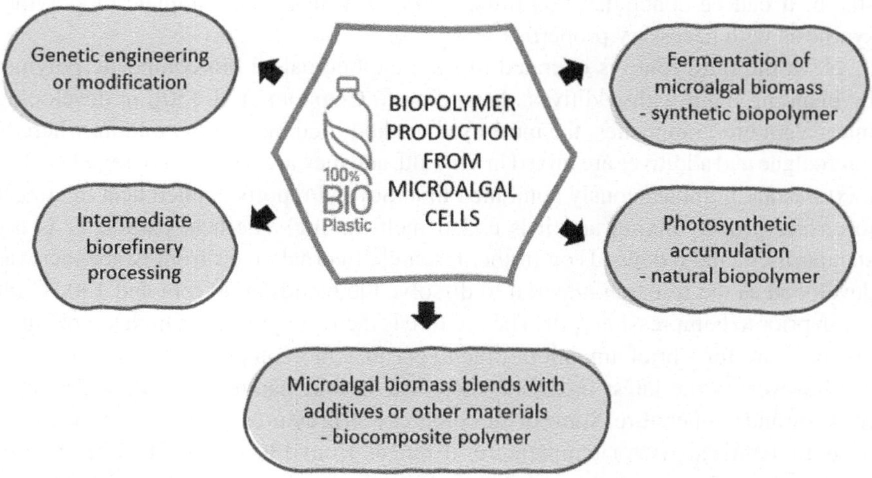

Fig. 4 Routes for biopolymer production through microalgal cells

Khan et al. [76] explained that the first route of microalgal biopolymer production is aided by enzyme-producing algae in fermentation process to transform microalgal biomass into bio-products that contain biopolymers. There has been recent discussion on prior biomass and fundamental macromolecules like lipids, carbohydrates, and proteins defragmentation before proceeding to fermentation. Research by Steinbruch et al. [77] disclosed a novel method to defragment microalgal biomass, which is via subcritical hydrothermal process. The acquired fragments were later introduced to fermentation to attain PHA. The obtained PHA from cellulose and starch fragments were 3.1 and 5.1% mg/g of biomass respectively. In contradiction, untreated biomass created much higher amount of PHA which makes up to 77.8% mg/g of biomass. Henceforth, direct biomass fermentation is regarded to be economically better than fermentation of defragmented biomass.

After that, the second route is accumulation of natural biopolymers via photosynthesis in microalgal biomass within cellular level. Biopolymer generation by microalgae is appropriate since microalgae takes up only little amount of nutrients [78]. Adjustments in growth environments such as illuminance intensity and exposure time can promote increment in biopolymer precursors like lipids, polysaccharides and lignin [79].

The study by Cassuriaga et al. [79] discovered polyhydroxybutyrate (PHB) production at 17.4% rate when microalgae was illuminated with 28 μmol m^{-2} s^{-1} alongside 6 h of xylose supplementation. Additionally, in recent trends, gamma illumination facilitates material quality enhancement. Gamma irradiation also do not have adverse effect towards environment, and it is a convenient operation for biopolymer synthesis. Other than that, biopolymer can possibly be generated from starch via ultraviolet (UV) illuminance, as what was reported by Shahabi-Ghahfarrokhi et al. in 2019 that crosslink chains were created when UV reacts with starch. It can be concluded that utilisation of UV light can stimulate biopolymer synthesis with necessary properties.

Next, the third route is operated to generate microalgal bio-composite polymer by blending them with additives. According to Ciapponi et al. [30], in developing microalgal bio-composites, the most used method is compression molding whereby microalgae and additives are mixed in a mould, and they are compressed together. The mixtures are homogeneously combined first. Several reports applied heat treatment to ensure proper mixing and it is called melt mixing. The heat parameter is not standardised, and it depends on further research. Internal mixer from Brabender was developed as the research material to dissolve the blends at 60 rpm and 130 °C for 4 min prior to compression [80]. The mixture is then compressed at high temperature and pressure for a brief amount of time to produce bio-composites.

However, in the latest literature, there are various parameter settings for time, pressure and temperature. Some of the compression pressures were recorded at 20 kPa up until 10 MPa. As for temperature, it ranges from 130 to 160 °C with time of molding from 3 min to as long as 20 min [20]. Some researchers also experimented with other variations of compression moulding which is without the pressure step like Dianursanti et al. (2018) to generate their bio-composite prototype. Bio-composite prototypes produced from molding method are of various sizes and shapes that relies on the proportion of the mould used and the desired outcome of a study. For instance, Fabra et al. [80] produced bio-composite film for the development of bio-compostable packaging and Ciapponi et al. [30] produced slab form bio-composite of microalgae bio-filler and gluten plasticiser in order to reduce wheat by-products.

Apart from that, solvent casting is typically used too, through microalgae-additives diffusion in a solvent and then left to dry on surfaces to compose films. The framework of this method also differs from one experiment to another. Sabathini et al. [23] had successfully formed bio-composite film upon dissolving microalgae and polyvinyl alcohol (PVA) mixture in water, and then left to air dry for 24 h on a glass plate. Zhang et al. [81] employed an extra step of homogenization unto biomass slurry ahead of polymer addition so that biomass mixture is properly allotted.

On top of that, genetic modification of microalgal cells is also a feasible way of bioplastics manufacture like PHB, which is a biodegradable polymer that can be accumulated in bacteria and microalgae. Rasul et al. [82] described that in the process of producing PHB via genetic engineering, bacterial PHB needs to be inserted into microalgae or macroalgae cells. Consequently, this technique could decrease the synthesising cost of fermentation from bacteria. As an example, microalgae *Chlamydomonas reinhardtii* was genetically tempered to obtain PHB. Two expression vectors that possess genes from bacteria *Ralstonia eutropha* which are phbB and phbC are inserted into the said microalgae [83]. The experiment was successful as PHB granules were observed in transgenic cytoplasm of *C. reinhardtii*. The result was favourable to the target of the research which is to promote high amount of PHB in algae cells and PHB concentration in the chloroplast. In another study by Franziska et al. [84], diatom *Phaeodactylum tricornutum* was injected with bacterial PHB into its cytosolic component for biosynthetic pathway. The obtained PHB was found in the cell's cytosol in granule form in the amount of as high as 10.6% of dry algal biomass.

3.1 Production of Microalgae-Polymer Blends

Utilisation of microalgae as the feedstock for bioplastic manufacture is greatly favourable since microalgae do not impose on human food consumption, able to mitigate carbon emission by using it as source of nutrient, are robust towards extreme growth conditions, and can aid in treating wastewater [85]. Bioplastic making through microalgal biomass is probable because their macromolecular constituents like carbohydrates and protein are able to generate biopolymers. So far, compounds found within microalgal biomass that are feasible for biodegradable polymer productions are cellulose, starch, PLA, PVC, PE, PHB, PHA, and protein-derived polymers [7]. Li et al. [86] stated that several examples of bio-compostable polymers that are broadly applied in PHA blends are PLA, polycaprolactone, amylopectin, amylose, lignin, and cellulose.

Amongst the aforementioned polymers, PHA is widely suggested for bioplastic synthesis due to its ability of enzymatic biodegradation. Properties of PHA are also enhanced upon blending with these materials. Another substantial way to boost bio-derived plastics performance is to blend PHA with plasticisers that ideally needs to be non-toxic, durable and biodegradable [85]. Oxypropylated glycerin or laprol, glycerol, glycerol triacetate, PEG, 4-nonylphenol, acetyl tributyl citrate, dibutyl phthalate, dioctyl phthalate, salicylic ester and acetylsalicylic acid are typical examples of plasticisers [87]. Several characteristics that include thermal and physical qualities, money-saving and crystallinity are paramount for the selection of great plasticisers. PHA that is mixed with 2 wt% of hexagon-shaped boron nitride nanoarchitecture (nanotechnology device) that resembles a ribbon, which act as nanofillers has been proven to have better feature thermally and mechanically. This mixed composite of PHA-boron nitride nanoarchitecture also exhibited boosted pliancy, tensile and yield threshold by 52.3%, 6.01% and 49.4%, respectively [88].

Over the years, several researchers have already succeeded in manufacturing bioplastics. As a proof, duo Dutch designer; Eric Klarenbeek and Maartje Dros [6] produced bioplastic from microalgae that could totally supersede conventional petroleum-based plastics. AlgaeLab was inaugurated by both the designers to culture microalgae to synthesis starch as the raw feedstock of bioplastics. Besides that, biodegradable components that could replace fossil-based plastics originated from microalgae was accomplished by Austeja Platukyte which are composed of agar and calcium carbonate as outer layer. The manufactured bioplastics are water-impermeable, sturdy and stable, yet they are featherweight. Those bioplastics can also be used as organic fertilisers to retain soil moisture. Equivalently, a bottle substitute was created as replacement of conventional plastic bottle by Ari Jónsson [6]. He combined red algae powder with water and this bottle will maintain its bottle form when the bottle is fully filled with water. But once emptied, it will start to decompose and disintegrate. The liquid that is put in the bottle is non-hazardous and safe for drinking. Consumers can even snack on the algae bottle itself.

According to Rahman and Miller [12], in bioplastics industry, the reasons why microalgal biomass are mixed with other components are so that the performance of the bioplastic can be improved. In addition, properties of bioplastics could be boosted, and the lifespan of bioplastics could also be extended. The materials of the blends are varied in terms of the origins like natural products, viz. starch or cellulose, petroleum-based plastics or polymers. There are also blending materials that could be acquired from microalgal biomass such as protein, cellulose, starch, PHA and PLA. To sum up, the production of bioplastic blends that are originated either from microalgae or other higher plants is a sustainable way of reinforcing the tensile strength and mechanical features of the end-products. It is also a feasible solution towards traditional plastic issues that pose adverse impact towards environment. Nevertheless, it is vital to mention that blending materials chosen for bioplastic synthesis must have the ability to be decomposed biologically as to prevent threat unto environment that consequently add up to more pollution and waste management expenses as stated by Rahman and Miller [12].

3.1.1 Proteins from Microalgae

Basically, polypeptide chains are formed through linkages of amide that are connected to 20 varying amino acids and those chains make protein up. The sequences of amino acids would give the ultimate characteristics to a protein. In relevance to Bayón et al. [89], proteins from soy and peanut, wheat gluten and corn zein are no strangers in bioplastic industry as they are frequently found feedstocks.

In spite of that, proteins from plants are in appropriate and unsustainable raw materials because food conflict might rise as proteins are one of human beings' staples. Wang et al. [90] stated that protein content in microalgae is generically high in which it can be exploited for bioplastic and thermoplastic makings and blends.

According to Zhang et al. [85], with polysaccharide as adjuvant, water permeability and mechanical characteristic of bioplastic from protein-loaded microalgae (*Chlorella sorokiniana*) could be increased. In addition, protein modification like chemical or thermal depre-treatments are required for the improvement of bioplastics from proteins regarding its functionality [89]. In a study by Wang et al. [90], they established thermoplastic blends and its bioplastic from *Nannochloropsis* (protein-loaded microalgae) and catfish algae via protein alteration. Nonetheless, the resulted bioplastic turned out different from one another. Catfish algae bioplastic is hard and inflexible, while *Nannochloropsis*-derived bioplastic is pliable and flexible. As for thermoplastic blend, polypropylene (PP) or polyethylene (PE) was mixed with microalgae and this blend could present reasonable mechanical features of for various applications. Apart from that, microalgal bioplastics are also observed as comparable to duckweed, soy protein isolate and feather meal.

Wang [91] had introduced an innovative system of bioplastic blend from microalgae *Spirulina sp.* as well as the strain's protein. The blending system was created to enhance *Spirulina*-derived bioplastic's performance. The whole system is comprised of biomass of *Spirulina* itself, ethylene glycol (EG) as the plasticiser, and polyethylene-graft-maleic anhydride (PE-g-MA) that act as the compatibilizer. These components are mixed with ultrahigh-molecular-weight polyethylene (UHMW-PE). As for estimated molecular weight and protein content determination, sodium dodecyl sulphate–polyacrylamide gel electrophoresis (SDS-PAGE) and Bicinchoninic acid (BCA) assay can be applied, respectively.

3.1.2 Starch from Microalgae

Starch is a typical polysaccharide that is constituted of D-glucose monomers linked by bonds of glycoside. Amylopectin and amylose make up to different percentages of proportions in starch, which are 80–90% and 10–20% respectively. Attributes of starch mostly are varied, according to the different proportions. Starch that contains higher amylopectin possess increased crystallinity property, while amylose-rich starch is stronger in tensile threshold [85].

There are three strategies in processing starch for polymeric components, viz. indirect monomer transformation from starch through dehydration of ethanol like polyethylene from ethylene, and synthesis of polymers like polylactic acid (PLA) from lactic acid. Secondly, starch can be used as the raw resource to generate hydroxylated compounds that have low molecular weight. In polyurethane development, glycolised products and dextrin are examples of polymers. Finally, starch can also act as thermoplastic starch or as a filler for other plastics. The third strategy is the most economic and simple approach to manufacture biopolymers. Macromolecular constitutions are also retained through this approach.

For instance, Mathiot et al. [92] investigated on bioplastics that are derived from starch attained from microalgae *Chlamydomonas reinhadrtii* strain 11e32A. This microalga was selected after being screened amongst nine other microalgae strains, which later was further processed via glycerol plasticisation directly. The resulting bioplastic with starch origins was found out to be a successful plasticisation. In another experiment, starch from marine microalgae *Klebsormidium flaccidum* was examined for bioplastic production [93]. Starch content and quality are studied as to verify the propriety of starch produced from *K. flaccidum*. The qualities that were experimented include: starch granules size, amylopectin-amylose composition, dissolvability, turbidity, and swelling capacity. Corn starch was used as control for the observations because corn starch is extensively employed in bioplastic synthesis. Ramli et al. [93] discovered that starch from *K. flaccidum* is notably tenable in generating bioplastics.

Starch possesses exceptional qualities as polymeric blend and due to this reason that it offers, it has been counted as the primary material of the many polymeric blends available. In spite of being regarded as having low barrier and mechanical features, starch and other kinds of polymer amalgams proffer advantageous traits like biodegradability, viability, and safety [94]. As an instance, polylactic acid (PLA) has great properties mechanically, yet it is not conveniently available and money intensive. Regardless, starch-PLA blends provide better mechanical characteristics and is not costly [95]. This is proven in studies by that involves industry of food packaging production using starch-PLA blends. It is noteworthy to mention that granular starch can be employed as an affordable filler when it is combined with liquified thermoplastic [96].

Referring to Cazón et al. [97], thermoplastic starch is identified as plasticised starch and it is also utilised in polymeric blends owing to its great qualities such as being highly water vapor-permeable and have low strength of mechanics. Water, glycerol, formamide, sorbitol, urea, citric acid, ethylene bis formamide, amino acids, and N-(2-hydroxyethyl) formamide are multiple examples of plasticisers that have been blended with thermoplastics. Furthermore, plasticiser amount gives impact on mechanical features of starch-polycaprolactone thermoplastic blend in which lower plasticiser amount will produce bioplastic that is has lesser elasticity coefficient but enhanced tensile strength [96].

3.2 Isolation of Microalgae Biopolymers

Upon finishing cultivation, microalgae samples are harvested and freeze-dried so that they are ready for isolation, in which it will be subjected to extraction process. Ultrasound extraction, solvent extraction, subcritical water extraction and microwave mediated extraction are several extraction methods that are widely applied. The aforementioned methods are explained in detail as below.

3.2.1 Ultrasound Extraction

Ultrasound extraction is a process on the basis of cavitation activity that is formed via waves from ultrasound. Turbulence is generated from the cavitation, commanding minute particles in microalgal biomass to collide and agitate each other. Vibration energy is transformed from the energy of ultrasound, causing cell walls to be lysed and releasing cellular content. Ultrasound activities promotes better rate of transfer which eventually alleviates biopolymer withdrawal from the cells [98, 99]. Ultrasound extraction is favoured due to its convenience, eco-friendly, and timesaving. Besides that, ultrasound extraction does not necessitate membrane separation step and can be operated at normal room temperature without influencing the resulting yield. Material waste could also be reduced via ultrasound extraction [100].

In reference to Kartik et al. [75], extraction yield from ultrasound was discovered to be higher by 33% more efficient as compared to traditional method. This statement is proven in study by Flórez-Fernández et al. [101] whereby biopolymer alginate was extracted from brown seaweed *Sargassum muticum* via ultrasound process and the experiment had cut down the extraction time by four-fold. The isolation process is highly impacted by several factors such as ultrasound wave frequency, temperature, and duration of sonication. Biomass yield will be enhanced with higher settings of frequency and operational time. Flórez-Fernández et al. [101] identified that yield of biomass is amplified from 5.7% to 15% upon longer sonication time which is 30 min.

3.2.2 Solvent Extraction

Apart from ultrasound assisted extraction, extraction via solvents is another method that can be employed for biopolymer production from microalgal biomass. Although this technique necessitates usage of chemicals, this method is simpler and does not involve many biorefinery processes in contrast to fermentation. As mentioned by Roja et al. [102], microalgal biomass can be combined with chemical agents in producing polymer precipitates. Physical and chemical factors that are optimised could boost higher synthesis of biopolymer. Faidi et al. [103] explained that extraction process is essentially influenced by mechanical processing like centrifugation, sifting and filtration. In their research whereby brown macroalgae *Padina pavonica* was utilised in extracting alginate biopolymer, the sample was pre-treated with formaldehyde for depigmentation. Then, the sample was further processed for precipitation by using mineral acids (pH as 1.5) to excerpt alginate biopolymer.

Downstream operations are laborious and unquestionably would incur extra expenses on production cost. Therefore, the strenuousness of unit operations could possibly be overcome by screening and selecting the most suitable microalgae amongst several candidates according to their efficiencies. Eventually, the best suited microalgae could be exploited for biopolymer manufacture and discarding unnecessary tests and experiments, altogether will save time and money. Morales-Jiménez et al. [104] conducted a screening experiment that involved six different microalgae strains for the best biopolymer-producing microalgae. Among the six strains, three microalgae strains were selected as the probable candidates, namely, *Synechocystis sp.* (Sy), *Porphyridium purpureum* (Pp), and *Nostoc sp.* (No). They discovered that yield of biopolymer accumulated in Sy, Pp and No is 204, 83, and 323 mg/L respectively.

3.2.3 Subcritical Water Extraction

Subcritical water extraction is an up-and-coming extraction technique that withdraws and isolates bioactive compounds from biomass. In this procedure, the water is subjected to pressure of not more than 22.12 MPa, which means less than critical pressure, and subjected to extreme heat of up to 373.99 °C (beyond boiling point) [105]. Subcritical water extraction offers several benefits over traditional methods. In this approach, water acts as solvent that dismisses the usage of toxic chemicals because water acts as solvent. In comparison to conventional extraction courses, subcritical water extraction requires short operational time, hence uses less energy and power. Apart from that, the yield collected through this extraction process is of high quality and quantity. These advantages provide subcritical water extraction an upper hand on other traditional methods [106].

There are several experiments that represents feasibility and practicality of subcritical water extraction. For example, Saravana et al. [107] exhibited a novel strategy of extracting biopolymer (fucoidan) from brown seaweed *Saccharina japonica* via subcritical water extraction in which accumulated fucoidan was in the amount of 4.85%, whereas the yield from traditional method was merely 2.47% of fucoidan. Correspondingly, Alboofetileh et al. [106] highlighted the significance of subcritical water extraction whereby they successfully removed 25.98% of fucoidan from another brown seaweed, *Nizamuddinia zanardinii*. The optimum settings studied were at temperature of 150 °C and 29 min of retention time. The result is highly contradicting to conventional method in which only 5.2% of fucoidan was extracted via conventional approach.

Apart from that, there are also some studies that developed combined extraction methods. Subcritical water extraction was coupled with ionic liquid catalyst to obtain biopolymer κ-carrageenan from red seaweed *Kappaphycus alvarezii*. The added ionic liquid catalyst highly influences biopolymer dissolution, making it easier to be isolated from the cells. Ionic liquid catalyst is chosen over organic solvents because ionic liquids have better thermal and chemical stability, high dissolution power, and only slight vapour pressure [105]. Besides that, there is another group of researchers that combined subcritical water extraction with deep eutectic solvents in isolating two types of biopolymers from *Saccharina japonica* which are fucoidan and alginate [108]. This study provided successful evidence of efficiently using this approach, as they obtained 14.93% and 28.1% of fucoidan and alginate respectively.

3.2.4 Microwave Mediated Extraction

Next approach is microwave mediated extraction that uses electromagnetic radiations on both ions and dipoles. This method is an innovative and green strategy to remove value-added compounds from biomass. The importance of using microwave assisted extraction include its conciseness, swift and constant procedure, time-saving steps, zero energy needed, and usage of solvents is little [109]. Improved yield of extraction was proven in an investigation by Ponthier et al. [110], whereby hybrid carrageenan was isolated from red seaweed *Mastocarpus stellatus*. The boosted biopolymer yield was attained via optimal conditions at 6 min retention time and at temperature of 150 °C. On top of that, it was reported that increment in temperature would induce better tensile strength in the extracted biopolymers.

Moreover, this approach suggests bioplastic extraction at short time duration which is advantageous in upscale operations. Additionally, industrial mass extraction via microwave mediation is highly advantageous due to dearth of biopolymeric gel syneresis. Syneresis is the process of gels contracting and loosing overall liquid inside them after being left for a long time, which acts like spontaneous ageing of the biopolymeric gels. This hinders progressive solvent expulsion and usability of the polymeric gels. Therefore, biopolymer making via microwave mediated extraction could be well underway for novel approach in achieving elevated yield that is also economical, at the same time [75]. Table 4 demonstrates isolation strategies and parameters of biopolymer synthesis in assorted microorganisms.

Figure 5 illustrates an integrated summary of algal biopolymers and their applications that are currently being studied and developed. There is a myriad of types of biopolymers synthesised from algal cells and the applications are certainly various as well. Brief explanations on biopolymer types and the applications are as follows.

4 Types of Biopolymers Produced from Algal Biomass and Their Applications

4.1 Polyhydroxyalkanoates (PHAs)

PHAs are the most sought-after bioplastic owing to its great properties and credibility that is equivalent to petroleum-based plastics but environmentally friendly [79]. PHAs do not pose negative impact towards nature because they are biodegradable and biocompatible [113]. Originally, PHAs are obtained from plant-based raw materials like fruit wastewater, wheat bran, glycerol waste, waste cooking oil and cheese whey but due to possible issues of food scarcity and land requirement, researchers investigated on microorganisms like microalgae, cyanobacteria and bacteria for PHA production [114, 115].

To date, the feedstocks of PHAs are numerous, that include microalgae *Spirulina sp.*, as well as cyanobacteria *Calothrix scytonemicola* and *Synechococcus subsalsus*.

Table 4 Isolation strategies and parameters of bioplastic synthesis in assorted microorganisms

Microorganism	Method and parameters	Bioplastic, % of recovery	References
Cyanobacteria	Solvent extraction • Chloroform and benzoic acid • Methanol and sulfuric acid • 100 °C, 300 min	PHB, 4.5%	[111]
Microalgae • *Nostoc sp.* • *Synechocystis sp.* • *Porphyridium purpuruem*	Solvent extraction • Sodium chloride and glycerol as solvents • 80 °C, 10 min	PHB, 47.5%	[104]
Microalgae • *Chlorella sp.* • *Scenedesmus sp.*	Solvent extraction • Sodium hypochlorite and chloroform as solvents • 32 °C, 180 min	PHB, 63%	[112]
Green seaweed *Ulva sp.*	Solvent extraction • Dimethyl sulphoxide as solvent • 180 °C, 40 min	PHA, 77.80%	[77]
Red seaweed *Mastocarpus stellatus*	Microwave assisted extraction • 150 °C, 3 min	Carrageenan, 40%	[110]
Red seaweed *Kappaphycus alvarezii*	Subcritical water extraction with ionic liquid (IL) • 1-butyl-3-methylimidazolium acetate (IL) as solvent • 150 °C, 30–40 min	κ-carrageenan, 78.75%	[105]
Brown seaweed *Saccharica japonica*	Subcritical water extraction • Water as solvent • 127 °C, 11.98 min	Fucoidan, 13.56%	[108]
Brown seaweed *Saccharica japonica*	Subcritical water and deep eutectic solvent extraction • Deep eutectic solution mixed with water as solvents • 150 °C, 25 min	Fucoidan, 14.93% Alginate, 28.1%	[107]
Brown seaweed *Nizamuddinia zanardinii*	Subcritical water extraction • Water as solvent • 150 °C, 29 min	Fucoidan, 25.98%	[106]
Brown seaweed *Sargassum muticum*	Ultrasound mediated extraction • Water as solvent • 25 °C, 30 min	Alginate, 15%	[98]

Several bacterial species like *Streptomyces*, *Pseudomonas* and *Bacillus* are amongst the microorganisms that can synthesis PHAs [75, 116]. It can be manufactured in research laboratories by manipulating the growth conditions. According to Magagula et al. [115], *Escherichia coli* can be genetically manipulated for PHA accumulations. Basically, microalgae and bacteria are induced for PHA production through stressed

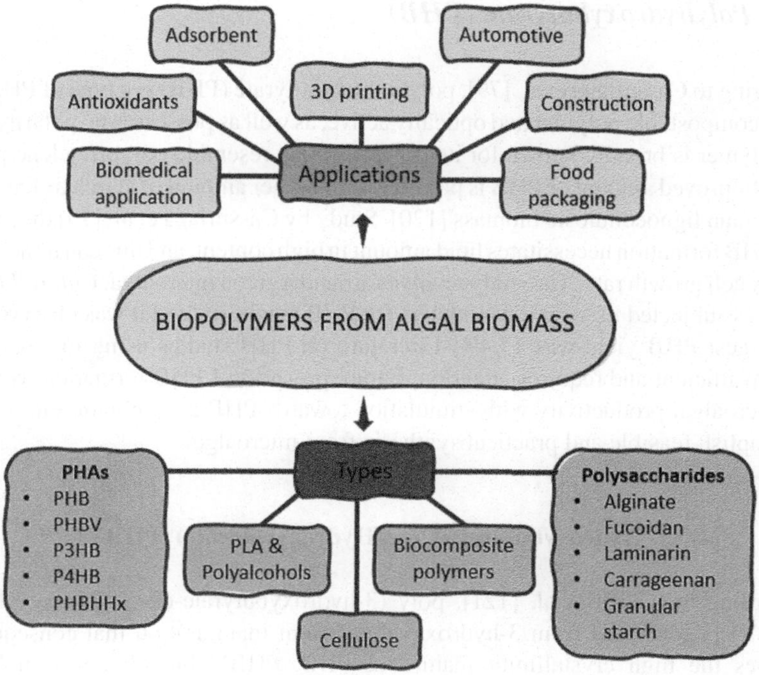

Fig. 5 Integrated summary of biopolymers produced from algal biomass and their applications

environmental conditions such as nitrogen, magnesium as well phosphorus depletion with excessive carbon. The cultivation medium is assembled in such a way that C/N ratio is kept consistent., and PHA synthesis occurs during stationary phase of cultivation [117, 118]. According to García et al. [119], PHAs production in microalgae can also be accumulated by subjecting thermal and mechanical polymerisation process towards the protein accumulated in microalgal biomass.

PHAs come in various types whereby its monomers compositions are generated based on strains of microorganisms [78]. Several types of PHAs include polyhydroxybutyrate (PHB), poly(3-hydroxybutyrate-co-3-hydroxyvalerate) (PHBV) and poly(3-hydroxybutyrate) (P3HB). Poly-4-hydroxybutyrate (P4HB) and poly(3-hydroxybutyrate-co-3-hydroxyhexanoate) (PHBHHx) are another two examples of PHAs [116]. Despite intensive experiments on PHAs at laboratory scale, industrial scale PHAs manufacturing is still in the pipeline, and it requires big attention for in-depth studies.

4.2 *Polyhydroxybutyrate (PHB)*

Referring to Cassuriaga et al. [79], polyhydroxybutyrate (PHB) is a type of PHA that is bio compostable polymer and optically active, as well as possesses no polarity. This biopolymer is broadly known for its properties that resemble polypropylene plastic (PP). Improved amount of PHB is possible with higher amount of starch as feedstock rather than lignocellulosic biomass [120]. Study by Cassuriaga et al. [79] discovered that PHB formation necessitates lipid amount in high content, and this can be achieved by low cell growth rate. The study revolves around a green microalga, *Chlorella fusca* that was subjected to several parameters for PHB synthesis, and it was observed that the largest PHB yield was 17.4%. Literature on PHB studies using microalgae is still insufficient and requires attention. Future prospect of PHB formation would be on microalgal productivity with stimulation towards PHB accumulation in order to accomplish feasible and practical synthesis from microalgae.

4.2.1 Poly (3- Hydroxybutyrate-Co-3-Hydroxyvalerate) (PHBV)

According to Ghosh et al. [121], poly (3-hydroxybutyrate-co-3-hydroxyvalerate) (PHBV) is generated from 3-hydroxyvalerate unit inauguration that consequently jumbles the high crystallinity feature of PHB. PHBV biopolymer is a highly functional means in drug delivery processes owing to its slow rate of degradation and outstanding physiochemical characteristics. Tebaldi et al. [122] executed in-depth research on PHBV formation and its biomedical applications in which PHBV nanoparticles as the tool for tumour-targeting is particularly progressive and intriguing. Up-to-date techniques in securing nanoparticle-based PHBVs were also explained.

Despite PHBV biopolymer being a promising future of bioplastic, literature of microalgae exploitation of PHBV is still lacking. Akdoğan and Çelik [123] administered research on a recombinant bacterial strain which is *Bacillus megaterium* in improving PHBV production properties from glucose. The study was conducted without precursor addition and involved two types of bioreactors; batch and fed-batch. The result indicated that fed-batch cultivation is a better strategy because nearly 80% PHBV biopolymer was generated as compared to only 46% from batch cultivation. Besides that, Talan et al. [124] suggested that cultivation of pure microbes in food wastes promotes higher microbial cell growth and leads to higher PHA accumulation. It was also proposed that utilisation of halophilic microorganism like *Haloferax mediterranei* (a salt-loving bacteria) might stimulate biosynthesis of PHBV and PHB polymers. It is also highly convenient because no sterilisation is required. With the aforementioned approaches, one can opt for algae as precursor and exploit it in similar approach for up-scale production of PHBV in future.

4.2.2 Polylactic Acid (PLA) and Polyalcohols

Polylactic acid (PLA) is a biopolymer that has vast probable applications in biomedical scope like braces, sutures, bone screws, sewing needles and bandages. As of now, algae-PLA composite is being broadly studied [125]. Wound dressing, tissue regeneration and augmentation are some of the areas that can make use of biomaterials generated from algae-based PLA.

Other than that, polyalcohols is another biopolymer that has gained much attention to produce bio-composite for numerous applications. Examples of advantageous qualities of polyalcohols are its biodegradability, water solubility, and great tensile strength. Polyvinyl alcohol (PVA), a type of polyalcohols is often implied as protective films, emulsifiers and sizing agents. Exploration in PVA bio-composites could develop novel material for bio-composite. According to Tran et al. [126], microalgal lipid from *Nannochloropsis salina* was extracted to be used as a filler for PVA bio-composite. The report described that microalgal filler strengthen the thermal stability of PVA-microalgal filler but elongation at break and tensile strength was decreased. This issue was solved with addition of polydiallyldimethylammonium chloride (polyDADMAC) as the plasticiser.

4.3 Polysaccharides

Kumar et al. [127] reported that long chain polysaccharides can be studied to produce biodegradable polymers that are compatible and works well with human systems. As such, algae can be manipulated to accumulate wide examples of polysaccharides including carrageenan, laminarin, fucoidan, and alginate. These polysaccharides vary in terms of their structure and morphology. Algae-derived polysaccharides can be applied in industries like cosmetics surgery, tissue engineering and cosmetics production.

Carrageenan is typically isolated from red macroalgae, and it is broadly employed as drug delivery tool and prohibits clogging in membranes [128, 129]. Extraction yield of carrageenan is highly dependent of the extraction methods. As an instance, carrageenan that is isolated from *Kappaphycus alvarezii* via ultrasound process for 15 min at temperature of 90 °C exhibited 56% extraction yield.

According to Bouanati et al. [130], the extracted carrageenan would have slight dissemination and decreased molecular weight. A type of carrageenan, which is κ-carrageenan or also known as Kappa-carrageenan is a water-soluble ionic polysaccharide from red seaweed [131]. κ-carrageenan displays great features of high viscosity and strength of gel. The quality of viscosity relies on extraction temperature, whereas gel strength is dependent on critical gel temperature, purity as well as contents of monosaccharide [132].

Another type of long chain polysaccharide is laminarin which is a non-hydrocolloid polysaccharide with low polydispersity [133]. Laminarins can be identified from seaweed extracts that they have low molecular weight. They are productively used in drug delivery system, carbon cycles of marine life, altogether with direct ethanol production [134]. Zargazadeh et al. (2020) mentioned that processes of oxidation and reduction intensify anticancer, anti-inflammatory and antioxidant activities of laminarin. Rajauria et al. [133] examined a study and disclosed that laminarin yield with high anisometric stretching intensity and high antioxidant properties can be improved via purification.

Furthermore, fucoidan is another beneficial polysaccharide that can be withdrawn from brown algae and have diverse structure. Alongside cancer treatments, fucoidan is utilised widespread in pharmaceutical industry and for different health products. Etman et al. [135] stated fucoidans can excellently be used as coating material and ligand targeting nanocarrier. Fucoidan nanoparticles are also useful because they could control drug deliver minus the toxicity. The diverse structure of fucoidans can be substantially determined through the feedstock and means of extraction. Extracted fucoidans that possess high contents of sulphur with low molecular weight portrays great antitumor property (Etman et al. 2020). On top of that, brown seaweed *Fucus evanescens* was examined for fucoidan characterisation by Hmelkov et al. [100]. The extracted fucoidan has low molecular weight which is 188 kDa with up to 96.1% of monosaccharide composition. It also had high anticomplement property and high sulfation degree which is at 0.5. This research concluded that the isolated fucoidan exhibited high inhibition towards neutrophil emigration at 93% and high antitumor properties as compared to conventional method.

Alginate is another biopolymer that can be generated from brown algae and marine seaweed. This biopolymer is safe for environment as well as biocompatible [136]. Specifically in brown algae *Undaria pinnatifida*, it is freely available in the amount of about 50%. Alginate is found beneficial in a myriad of applications such as in drug delivery system, tissue engineering, biomedicine and food industry [8]. Hydrogels with qualities like rigid, flexible yet elastic can be formed with a combination of alginates and cations. Alginate-cation hydrogels are also found to have great magnitude of water retention and adsorption criteria [85].

Alginate manufacturing can also benefit industries that produce biomedical devices and hydrogel beads [137, 138]. According to Martău et al. [139], alginates are also widely used as gelling, thickening and stabilising agents. It is worth mentioning that types of alginates extracted highly relies on pH during isolation process. Insoluble alginic acids will be generated from alginate salts through extraction process that has acidic pH whereas extraction process that is resulted from a hindered extraction process. Meanwhile, basic extraction process will obtain alginates with high extraction capacity [140].

4.4 Applications

Microalgal biopolymers are composed of units that are repeated which is advantageous for molecular level interactions with great criteria. Biopolymers from algae find useful applications in a myriad of areas like 3D printing, biomedical engineering, nutraceuticals (antioxidants) and environmental bioremediation (adsorbent).

4.4.1 3D Printing

3D printing is one of the budding applications for microalgal biopolymers since complex configurations can be produced within short amount of time with great properties. Microalgae *Nannochloropsis salina* was exploited in study by Ponthier et al. [110] and the produced PVA bio-composite was used as a filler for 3D printing. In addition, biopolymers are highly versatile that they can be used and adjusted for multidisciplinary applications only by changing the properties of materials. Biopolymers that are printed through 3D technology can be employed in environmental bioremediation and biomedicine. For instance, Sangiorgi et al. [141] utilised polylactic acid that was tempered with titanium dioxide (TiO_2) to synthesise filler for methyl orange disintegration. Whereas, with TiO_2 at 30 wt% in the composite allows for successful methyl orange full degradation within 24 h. Moreover, nano-sized ceramics that are infused with 3D printed scaffolds were examined for stem cell viability and behaviour [142].

4.4.2 Biomedical Engineering

Biopolymers also find its efficacious application in biomedical engineering notably in tissue engineering, drug delivery system and regenerative drugs. Biopolymers promote functional properties like non-toxicity, biodegradability, water holding capacity and high tensile strength. As an example, Sathiyavimal et al. [143] revealed a novel study on hydroxyapatite (HAp) and chitosan bio-composite manufacture due to significance of HAp being the essential mineral for human bones. Additionally, soybean oil epoxidized acetate (SOEA) was utilised alongside HAp nanoparticles to determine investigate orthotropic activities for bone-inspired composites [144].

Several marine algae were studied for collagen membrane, and they were reinforced with polylactic acid (PLA) composites. The alga strains the were used were *Stypopodium schimperi*, *Cystoseria compressa*, *Corallina elongata*, *Sargassum vulgare* and *Galaxaura oblongata*. The purpose of this study is to improve its characteristics for skin grafting and amongst the five strains, *S. vulgare* was observed to possess the best properties of cell proliferation and viability. Considering the result, this approach could aid in clinical practices for guided wound and tissue regeneration, as well as connective tissue extension [125].

4.4.3 Nutraceuticals (Antioxidants)

Next, biopolymers that act as antioxidants are widely known in nutraceutical and food industry and they are superior compared to other antioxidants as they are non-toxic and safe. Sivakanthan et al. [145] stated that biopolymers do not pose negative effects as compared to synthetic antioxidants. Unsaturated lipids in food tend to be exposed to oxidation and biopolymers can be employed to inhibit oxidation for food preservation [146]. Therewith, a comparative experiment was conducted by Córdoba and Sobral [147] that involved pure gelatine (G), gelatine-chitosan composite (G-Ch) and gelatine-sodium caseinate composite (G-C). These bio-composites were reinforced with active compounds via water nano-emulsification. As a result, G-C composite that is impregnated with active compounds demonstrated high antioxidant ability.

Furthermore, bovine and goat milk were applied for bioactive peptides generation via protease enzymes from fungi *Aspergillus flavipes* and *Aspergillus oryzae*. Antioxidant potency was executed through 2,2-diphenyl-1-picrylhydrazyl (DPPH) scavenging activity. The DPPH test depicted that antioxidant activities were present of up to 92.5%, resulting it to be crucial in food and nutraceutical industries [148]. Besides that, Gopu and Selvam [149] explored the potential of red algae *Amphira rigida* for antioxidants. A novel polysaccharide was successfully extracted via ultrasonic mediated extraction to be used as antioxidants. The extracted polysaccharide was subjected to DPPH and (2,2'-azino-bis(3-ethylbenzothiazoline-6-sulfonic acid)) (ABTS) tests and the result portrayed high antioxidant activities. It was also observed that the polysaccharide has great anticancer and bacteriostatic properties, as well as apoptotic and cytotoxicity potential.

4.4.4 Environmental Remediation (Adsorbent)

In environmental bioremediation, biological approach for adsorption plays an important role of being effective and efficient and do not pose any consequent pollution. It is also recyclable and easy to be retrieved [150]. Heavy metal remediation and dye removal involves biopolymer-clay mixture for adsorption purposes. Biopolymers are combined with clays because natural clays are inadequate in eliminating hydrophobic pollutants. Polypeptides and polysaccharide biopolymers are selected for this combination, and it has been proven to be successful in a scientific report by del Mar Orta et al. [151]. Biopolymer-clay bio-composite exhibits enhanced characteristics if compared to pure biofilm usage. These enhanced qualities include better wettability, resistant towards changes in pH, and higher specificity. Research by Xia et al. [152] described about effective mercury removal by two bacterial strains, i.e., *Bacillus sp.* (EPS-B) and *Klebsiella sp.* (EPS-K) that were isolated from activated sludge in a wastewater treatment plant.

On top of that, there are plenty of adsorption studies that utilised tempered cellulose owing to its abundance, benefits and its convenience in modification due to hydroxyl group (OH) interaction. Silva et al. [153] conducted experiment on organic pollutants (Amitriptyline and Rhodamine B) removal by biopolymer that is incorporated with phosphate group (PCel). The said biopolymer is generated with the usage of sodium tripolyphosphate and phosphoric acid in order to temper the cellulose surface. It was discovered that the highest adsorption level was accomplished at acidic to neutral pH value for both organic pollutants. Adsorption capacity was achieved at 45.52 mg/g and 47.58 mg/g for Amitriptyline and Rhodamine B, respectively.

Besides that, biopolymer chitosan is the next extensively used chitosan. Chitosan is often chosen because its constituents are made up of amenable functional groups like primary and secondary hydroxyl groups (OH) and primary amine (NH). These groups that make chitosan up leaves chitosan to be easily amended without jeopardising polymerisation degree for better efficiency and improved targeted characteristics [154]. Shariful et al. [155] unearthed a novel study on divalent heavy metal ion via chitosan–polyethylene oxide bio composite in the form of nanofibers. This biocomposite is well known in remediation of heavy metals, namely, copper(II) oxide (Cu(II)), lead(II) nitrate (Pb(II)), zinc (Zn(II)) and hexavalent chromium or chromium 6 (Cr(VI)) [156].

As a conclusion, literature on microalgal polysaccharides as modified biopolymer for applications like bioremediation and nutraceuticals are still insufficient and in need of further studies. One can imply above-mentioned studies by using algae as feedstock because adsorption through tempered biopolymers. This approach is promising owing to its advantages like cost-effective feedstock, effective pollutants elimination, high recyclability, and green strategy.

5 Conclusion

In regard of global circular bioeconomy, microalgae are regarded as one of the best suited candidates for biomass feedstock for the mass generation of biopolymers. The chapter provides an insight to the cultivation of microalgae using closed and open systems with the common harvesting techniques adopted. The chapter also incorporated the production of microalgae polymer blends, isolation of microalgae biopolymers, different types of biopolymers produced and their potential applications. The chapter gives a comprehensive account to obtain biopolymers from microalgal biomass.

References

1. Sharuddin, S. D. A., Abnisa, F., Daud, W. M. A. W., & Aroua, M. K. (2016). A review on pyrolysis of plastic wastes. *Energy Conversion and Management, 115*, 308–326.
2. Horton, A. A. (2022). Plastic pollution: When do we know enough? *Journal of Hazardous Materials, 422*, 126885.
3. Muneer, F. (2015). Biocomposites from natural polymers and fibers.

4. Bai, B., Jin, H., Fan, C., Cao, C., Wei, W., & Cao, W. (2019). Experimental investigation on liquefaction of plastic waste to oil in supercritical water. *Waste Management, 89*, 247–253.
5. Geyer, R., Jambeck, J. R., & Law, K. L. (2017). Production, use, and fate of all plastics ever made. *Science Advances, 3*(7), e1700782.
6. Chia, W. Y., Tang, D. Y. Y., Khoo, K. S., Lup, A. N. K., & Chew, K. W. (2020). Nature's fight against plastic pollution: Algae for plastic biodegradation and bioplastics production. *Environmental Science and Ecotechnology, 4*, 100065.
7. Karan, H., Funk, C., Grabert, M., Oey, M., & Hankamer, B. (2019). Green bioplastics as part of a circular bioeconomy. *Trends in Plant Science, 24*(3), 237–249.
8. Devadas, V. V., Khoo, K. S., Chia, W. Y., Chew, K. W., Munawaroh, H. S. H., Lam, M.-K., Lim, J.-W., Ho, Y.-C., Lee, K. T., & Show, P. L. (2021). Algae biopolymer towards sustainable circular economy. *Bioresource Technology, 325*, 124702.
9. Das, S. K., Sathish, A., & Stanley, J. (2018). Production of biofuel and bioplastic from Chlorella pyrenoidosa. *Materials Today: Proceedings, 5*(8), 16774–16781.
10. Garbowski, T., Richter, D., & Pietryka, M. (2019). Analysis of changes of particle size distribution and biological composition of flocs in wastewater during the growth of algae. *Water, Air, & Soil Pollution, 230*(6), 1–13.
11. Hamid, S. S., Wakayama, M., Ichihara, K., Sakurai, K., Ashino, Y., Kadowaki, R., Soga, T., & Tomita, M. (2019). Metabolome profiling of various seaweed species discriminates between brown, red, and green algae. *Planta, 249*(6), 1921–1947.
12. Rahman, A., & Miller, C. (2017). Microalgae as a source of bioplastics. Algal Green Chemistry. *Recent Progress in Biotechnology* 121–138.
13. Tang, D. Y. Y., Khoo, K. S., Chew, K. W., Tao, Y., Ho, S.-H., & Show, P. L. (2020). Potential utilization of bioproducts from microalgae for the quality enhancement of natural products. *Bioresource Technology, 304*, 122997.
14. Koyande, A. K., Chew, K. W., Rambabu, K., Tao, Y., Chu, D.-T., & Show, P.-L. (2019). Microalgae: A potential alternative to health supplementation for humans. *Food Science and Human Wellness, 8*(1), 16–24.
15. Mehta, P., Singh, D., Saxena, R., Rani, R., Gupta, R. P., Puri, S. K., & Mathur, A. S. (2018). High-value coproducts from algae—An innovational way to deal with advance algal industry. In *Waste to wealth* (pp. 343–363). Springer.
16. Elrayies, G. M. (2018). Microalgae: Prospects for greener future buildings. *Renewable and Sustainable Energy Reviews, 81*, 1175–1191.
17. Amini, E., Babaei, A., Mehrnia, M. R., Shayegan, J., & Safdari, M.-S. (2020). Municipal wastewater treatment by semi-continuous and membrane algal-bacterial photo-bioreactors. *Journal of Water Process Engineering, 36*, 101274.
18. Madadi, R., Tabatabaei, M., Aghbashlo, M., Zahed, M. A., & Pourbabaee, A. A. (2018). Biodiesel from microalgae. In *Waste to wealth* (pp. 277–318). Springer.
19. Beckstrom, B. D., Wilson, M. H., Crocker, M., & Quinn, J. C. (2020). Bioplastic feedstock production from microalgae with fuel co-products: A techno-economic and life cycle impact assessment. *Algal Research, 46*, 101769.
20. Onen Cinar, S., Chong, Z. K., Kucuker, M. A., Wieczorek, N., Cengiz, U., & Kuchta, K. (2020). Bioplastic production from microalgae: A review. *International Journal of Environmental Research and Public Health, 17*(11), 3842.
21. Vander Wiel, J. B., Mikulicz, J. D., Boysen, M. R., Hashemi, N., Kalgren, P., Nauman, L. M., Baetzold, S. J., Powell, G. G., He, Q., & Hashemi, N. N. (2017). Characterization of Chlorella vulgaris and Chlorella protothecoides using multi-pixel photon counters in a 3D focusing optofluidic system. *RSC Advances*, *7*(8), 4402–4408.
22. Sakarika, M., & Kornaros, M. (2019). Chlorella vulgaris as a green biofuel factory: Comparison between biodiesel, biogas and combustible biomass production. *Bioresource Technology, 273*, 237–243.
23. Sabathini, H., Windiani, L., & Gozan, M. (2018). Mechanical Physicial properties of chlorella-PVA based bioplastic with ultrasonic homogenizer. In *E3S Web of Conferences*. EDP Sciences.

24. Rout, N. P., Khandual, S., Gutierrez-Mora, A., Ibarra-Montoya, J. L., & Vega-Valero, G. (2015). Divergence in three newly identified Arthrospira species from Mexico. *World Journal of Microbiology and Biotechnology, 31*(7), 1157–1165.
25. Alagawany, M., Taha, A. E., Noreldin, A., El-Tarabily, K. A., & Abd El-Hack, M. E. (2021). Nutritional applications of species of Spirulina and Chlorella in farmed fish: A review. *Aquaculture , 542*, 736841.
26. Dianursanti, C. N., Windiani, L., & Gozan, M. (2019). Effect of compatibilizer addition in Spirulina platensis based bioplastic production. In *AIP Conference Proceedings*. AIP Publishing LLC.
27. Feng, H., Zhang, B., He, Z., Wang, S., Salih, O., & Wang, Q. (2018). Study on co-liquefaction of Spirulina and Spartina alterniflora in ethanol-water co-solvent for bio-oil. *Energy, 155*, 1093–1101.
28. Khalis, S. (2018). The effect of compatibilizer addition on Chlorella vulgaris microalgae utilization as a mixture for bioplastic. In *E3S Web of Conferences*. EDP Sciences.
29. Gifuni, I., Olivieri, G., Krauss, I. R., D'Errico, G., Pollio, A., & Marzocchella, A. (2017). Microalgae as new sources of starch: Isolation and characterization of microalgal starch granules. *Chemical Engineering Transactions, 57*, 1423–1428.
30. Ciapponi, R., Turri, S., & Levi, M. (2019). Mechanical reinforcement by microalgal biofiller in novel thermoplastic biocompounds from plasticized gluten. *Materials, 12*(9), 1476.
31. Zhu, N., Ye, M., Shi, D., & Chen, M. (2017). Reactive compatibilization of biodegradable poly (butylene succinate)/Spirulina microalgae composites. *Macromolecular Research, 25*(2), 165–171.
32. Simonič, M., & Zemljič, F. L. (2021). Production of bioplastic material from algal biomass. *Chemical Industry and Chemical Engineering Quarterly, 27*(1), 79–84.
33. Rocha, C. J. L., Álvarez-Castillo, E., Yáñez, M. R. E., Bengoechea, C., Guerrero, A., & Ledesma, M. T. O. (2020). Development of bioplastics from a microalgae consortium from wastewater. *Journal of Environmental Management, 263*, 110353.
34. Monshupanee, T., Nimdach, P., & Incharoensakdi, A. (2016). Two-stage (photoautotrophy and heterotrophy) cultivation enables efficient production of bioplastic poly-3-hydroxybutyrate in auto-sedimenting cyanobacterium. *Scientific Reports, 6*(1), 1–9.
35. Johnsson, N., & Steuer, F. (2018). Bioplastic material from microalgae: Extraction of starch and PHA from microalgae to create a bioplastic material.
36. Chakraborty, S., Karmaker, D., Rahman, M. A., Bali, S. C., Das, S. K., & Hossen, R. (2021). Impacts of pH and salinity on community composition, growth and cell morphology of three freshwater phytoplankton. *Plant Science Today, 8*(3), 655–661–655–661.
37. Ahmad, I., Abdullah, N., Koji, I., Yuzir, A., & Muhammad, S. E. (2021). Evolution of photobioreactors: A review based on microalgal perspective. In *IOP Conference Series: Materials Science and Engineering*. IOP Publishing.
38. Rasel, I., & Karim, S. (2021). A mini review on microalgae biomass production: Recent progress in cultivation systems.
39. Apel, A., Pfaffinger, C., Basedahl, N., Mittwollen, N., Göbel, J., Sauter, J., Brück, T., & Weuster-Botz, D. (2017). Open thin-layer cascade reactors for saline microalgae production evaluated in a physically simulated Mediterranean summer climate. *Algal Research, 25*, 381–390.
40. Min, M., Hu, B., Mohr, M. J., Shi, A., Ding, J., Sun, Y., Jiang, Y., Fu, Z., Griffith, R., & Hussain, F. (2014). Swine manure-based pilot-scale algal biomass production system for fuel production and wastewater treatment—A case study. *Applied Biochemistry and Biotechnology, 172*(3), 1390–1406.
41. Jerney, J., & Spilling, K. (2018). Large scale cultivation of microalgae: Open and closed systems.: In: Biofuels from algae (pp. 1–8). Springer.
42. Lutzu, G. A., Ciurli, A., Chiellini, C., Di Caprio, F., Concas, A., & Dunford, N. T. (2021). Latest developments in wastewater treatment and biopolymer production by microalgae. *Journal of Environmental Chemical Engineering, 9*(1), 104926.

43. Lim, Y. A., Chong, M. N., Foo, S. C., & Ilankoon, I. (2021). Analysis of direct and indirect quantification methods of CO_2 fixation via microalgae cultivation in photobioreactors: A critical review. *Renewable and Sustainable Energy Reviews, 137*, 110579.
44. Pham, H.-M., Kwak, H. S., Hong, M.-E., Lee, J., Chang, W. S., & Sim, S. J. (2017). Development of an X-Shape airlift photobioreactor for increasing algal biomass and biodiesel production. *Bioresource Technology, 239*, 211–218.
45. Wen, X., Du, K., Wang, Z., Peng, X., Luo, L., Tao, H., Xu, Y., Zhang, D., Geng, Y., & Li, Y. (2016). Effective cultivation of microalgae for biofuel production: A pilot-scale evaluation of a novel oleaginous microalga Graesiella sp. WBG-1. *Biotechnology for Biofuels, 9*(1), 1–12.
46. Cheng, Y. W., Lim, J. S. M., Chong, C. C., Lam, M. K., Lim, J. W., Tan, I. S., Foo, H. C. Y., Show, P. L., & Lim, S. (2021). Unravelling CO_2 capture performance of microalgae cultivation and other technologies via comparative carbon balance analysis. *Journal of Environmental Chemical Engineering, 9*(6), 106519.
47. Qin, C., Lei, Y., & Wu, J. (2018). Light/dark cycle enhancement and energy consumption of tubular microalgal photobioreactors with discrete double inclined ribs. *Bioresources and Bioprocessing, 5*(1), 1–10.
48. Kothari, R., Pandey, A., Ahmad, S., Kumar, A., Pathak, V. V., & Tyagi, V. (2017). Microalgal cultivation for value-added products: A critical enviro-economical assessment *3 Biotech, 7*(4), 1–15.
49. Sheng, A., Bilad, M., Osman, N., & Arahman, N. (2017). Sequencing batch membrane photobioreactor for real secondary effluent polishing using native microalgae: Process performance and full-scale projection. *Journal of Cleaner Production, 168*, 708–715.
50. Arcigni, F., Friso, R., Collu, M., & Venturini, M. (2019). Harmonized and systematic assessment of microalgae energy potential for biodiesel production. *Renewable and Sustainable Energy Reviews, 101*, 614–624.
51. Vo, H. N. P., Ngo, H. H., Guo, W., Nguyen, T. M. H., Liu, Y., Liu, Y., Nguyen, D. D., & Chang, S. W. (2019). A critical review on designs and applications of microalgae-based photobioreactors for pollutants treatment. *Science of the Total Environment, 651*, 1549–1568.
52. Jiménez-González, A., Adam-Medina, M., Franco-Nava, M., & Guerrero-Ramírez, G. (2017). Grey-box model identification of temperature dynamics in a photobioreactor. *Chemical Engineering Research and Design, 121*, 125–133.
53. Vasistha, S., Khanra, A., Clifford, M., & Rai, M. (2021). Current advances in microalgae harvesting and lipid extraction processes for improved biodiesel production: A review. *Renewable and Sustainable Energy Reviews, 137*, 110498.
54. Sutherland, D. L., McCauley, J., Labeeuw, L., Ray, P., Kuzhiumparambil, U., Hall, C., Doblin, M., Nguyen, L. N., & Ralph, P. J. (2021). How microalgal biotechnology can assist with the UN sustainable development goals for natural resource management. *Current Research in Environmental Sustainability, 3*, 100050.
55. Augustine, A., Tanwar, A., Tremblay, R., & Kumar, S. (2019). Flocculation processes optimization for reuse of culture medium without pH neutralization. *Algal Research, 39*, 101437.
56. Muylaert, K., Bastiaens, L., Vandamme, D., & Gouveia, L. (2017). Harvesting of microalgae: Overview of process options and their strengths and drawbacks. *Microalgae-Based Biofuels and Bioproducts*, 113–132.
57. Zou, X., Li, Y., Xu, K., Wen, H., Shen, Z., & Ren, X. (2018). Microalgae harvesting by buoy-bead flotation process using Bioflocculant as alternative to chemical Flocculant. *Algal Research, 32*, 233–240.
58. Xu, K., Zou, X., Wen, H., Xue, Y., Zhao, S., & Li, Y. (2018). Buoy-bead flotation harvesting of the microalgae Chlorella vulgaris using surface-layered polymeric microspheres: A novel approach. *Bioresource Technology, 267*, 341–346.
59. Singh, G., & Patidar, S. (2018). Microalgae harvesting techniques: A review. *Journal of Environmental Management, 217*, 499–508.
60. DeRose, K., DeMill, C., Davis, R. W., & Quinn, J. C. (2019). Integrated techno economic and life cycle assessment of the conversion of high productivity, low lipid algae to renewable fuels. *Algal Research, 38*, 101412.

61. Watsuji, T.-O., Naka, A., Morita, Y., & Kurahashi, M. (2021). Effect of temperature and dissolved oxygen on gravity sedimentation of the unicellular alga Dunaliella salina. *Annals of Microbiology, 71*(1), 1–7.
62. Ortiz, A., García-Galán, M. J., Garcia, J., & Diez-Montero, R. (2021). Optimization and operation of a demonstrative full scale microalgae harvesting unit based on coagulation, flocculation and sedimentation. *Separation and Purification Technology, 259*, 118171.
63. Li, S., Hu, T., Xu, Y., Wang, J., Chu, R., Yin, Z., Mo, F., & Zhu, L. (2020). A review on flocculation as an efficient method to harvest energy microalgae: Mechanisms, performances, influencing factors and perspectives. *Renewable and Sustainable Energy Reviews, 131*, 110005.
64. Razak, N. N. A. N., & Bilad, M. R. (2021). Progress in development of Membrane Fouling Control for Microalgae Filtration: A Review. *Jurnal Penelitian dan Pengkajian Ilmu Pendidikan: e-Saintika, 5*(1), 67–91.
65. Eliseus, A., Bilad, M., Nordin, N., Putra, Z. A., & Wirzal, M. (2017). Tilted membrane panel: A new module concept to maximize the impact of air bubbles for membrane fouling control in microalgae harvesting. *Bioresource Technology, 241*, 661–668.
66. Salim, S., Bosma, R., Vermuë, M. H., & Wijffels, R. H. (2011). Harvesting of microalgae by bio-flocculation. *Journal of Applied Phycology, 23*(5), 849–855.
67. Mathimani, T., & Mallick, N. (2018). A comprehensive review on harvesting of microalgae for biodiesel-key challenges and future directions. *Renewable and Sustainable Energy Reviews, 91*, 1103–1120.
68. MacArthur, E. (2017). *Beyond plastic waste* (pp. 843–843). American Association for the Advancement of Science.
69. Thielen, M. (2014). Bioplastics MAGAZINE In: Bioplastics–Plants and crops, raw materials. Published in Berlin: Fachagentur Nachwachsende Rohstoffe eV (FNR). *Order*, 237.
70. Alaerts, L., Augustinus, M., & Van Acker, K. (2018). Impact of bio-based plastics on current recycling of plastics. *Sustainability, 10*(5), 1487.
71. Bioplastics, E. (2014). What are bioplastics. *Material Types, Terminology and Labels, an Introduction. European Bioplastic Fact Sheet*.
72. Sheet, M. R. F. (2015). European bioplastics. *Life Cycle Assessment of Biopolymers–Position Paper-2008*.
73. Rai, P., Mehrotra, S., Priya, S., Gnansounou, E., & Sharma, S. K. (2021). Recent advances in the sustainable design and applications of biodegradable polymers. *Bioresource Technology, 325*, 124739.
74. Narancic, T., Cerrone, F., Beagan, N., & O'Connor, K. E. (2020). Recent advances in bioplastics: Application and biodegradation. *Polymers, 12*(4), 920.
75. Kartik, A., Akhil, D., Lakshmi, D., Gopinath, K. P., Arun, J., Sivaramakrishnan, R., & Pugazhendhi, A. (2021). A critical review on production of biopolymers from algae biomass and their applications. *Bioresource Technology, 329*, 124868.
76. Khan, M. I., Shin, J. H., & Kim, J. D. (2018). The promising future of microalgae: Current status, challenges, and optimization of a sustainable and renewable industry for biofuels, feed, and other products. *Microbial Cell Factories, 17*(1), 1–21.
77. Steinbruch, E., Drabik, D., Epstein, M., Ghosh, S., Prabhu, M. S., Gozin, M., Kribus, A., & Golberg, A. (2020). Hydrothermal processing of a green seaweed Ulva sp. for the production of monosaccharides, polyhydroxyalkanoates, and hydrochar. *Bioresource Technology*, 318, 124263.
78. Costa, S. S., Miranda, A. L., Andrade, B. B., de Jesus Assis, D., Souza, C. O., de Morais, M. G., Costa, J. A. V., & Druzian, J. I. (2018). Influence of nitrogen on growth, biomass composition, production, and properties of polyhydroxyalkanoates (PHAs) by microalgae. *International Journal of Biological Macromolecules, 116*, 552–562.
79. Cassuriaga, A., Freitas, B., Morais, M., & Costa, J. (2018). Innovative polyhydroxybutyrate production by Chlorella fusca grown with pentoses. *Bioresource Technology, 265*, 456–463.
80. Fabra, M. J., Martínez-Sanz, M., Gómez-Mascaraque, L. G., Gavara, R., & López-Rubio, A. (2018). Structural and physicochemical characterization of thermoplastic corn starch films containing microalgae. *Carbohydrate Polymers, 186*, 184–191.

81. Zhang, C., Wang, C., Cao, G., Wang, D., & Ho, S.-H. (2020). A sustainable solution to plastics pollution: an eco-friendly bioplastic film production from high-salt contained Spirulina sp. residues. *Journal of Hazardous Materials, 388*, 121773.
82. Rasul, I., Azeem, F., Siddique, M. H., Muzammil, S., Rasul, A., Munawar, A., Afzal, M., Ali, M. A., & Nadeem, H. (2017). Algae biotechnology: A green light for engineered algae. *Algae Based Polymers, Blends, and Composites* (pp. 301–334). Elsevier.
83. Chaogang, W., Zhangli, H., Anping, L., & Baohui, J. (2010). Biosynthesis of poly-3-hydroxybutyrate (phb) in the transgenic green alga Chlamydomonas reinhardtii 1. *Journal of Phycology, 46*(2), 396–402.
84. Franziska, H., Andrew, B., Nicole, L., Andreas, K., & Stefan, Z. (2011). Microalgae as bioreactors for bioplastic production.
85. Zhang, C., Show, P.-L., & Ho, S.-H. (2019). Progress and perspective on algal plastics—A critical review. *Bioresource Technology, 289*, 121700.
86. Li, Z., Yang, J., & Loh, X. J. (2016). Polyhydroxyalkanoates: Opening doors for a sustainable future. *NPG Asia Materials, 8*(4), e265–e265.
87. Mukheem, A., Hossain, M. M., Shahabuddin, S., Muthoosamy, K., Manickam, S., Sudesh, K., Saidur, R., & Sridewi, N. (2018). Bioplastic Polyhydroxyalkanoate (PHA): Recent advances in modification and medical applications.
88. Zhao, Y., Liu, Z., Cao, C., Wang, C., Fang, Y., Huang, Y., Yu, C., Zhang, J., Li, L., & Hu, L. (2017). Self-sacrificed template synthesis of ribbon-like hexagonal boron nitride nano-architectures and their improvement on mechanical and thermal properties of PHA polymer. *Scientific Reports, 7*(1), 1–6.
89. Bayón, B., Berti, I. R., Gagneten, A. M., & Castro, G. R. (2018). Biopolymers from wastes to high-value products in biomedicine. In *Waste to wealth* (pp. 1–44). Springer.
90. Wang, K., Mandal, A., Ayton, E., Hunt, R., Zeller, M., & Sharma, S. (2016). Modification of protein rich algal-biomass to form bioplastics and odor removal. In *Protein byproducts* (pp. 107–117). Elsevier.
91. Wang, K. (2014). *Bio-plastic potential of spirulina microalgae*. University of Georgia.
92. Mathiot, C., Ponge, P., Gallard, B., Sassi, J.-F., Delrue, F., & Le Moigne, N. (2019). Microalgae starch-based bioplastics: Screening of ten strains and plasticization of unfractionated microalgae by extrusion. *Carbohydrate Polymers, 208*, 142–151.
93. Ramli, R., Lee, C., & Kassim, M. (2020). Extraction and characterization of starch from microalgae and comparison with commercial corn starch. In *IOP Conference Series: Materials Science and Engineering*. IOP Publishing.
94. Kamaruddin, Z. H., Jumaidin, R., Ilyas, R. A., Selamat, M. Z., Alamjuri, R. H., & Yusof, F. A. M. (2022). Biocomposite of cassava starch-cymbopogan citratus fibre: Mechanical, thermal and biodegradation properties. *Polymers, 14*(3), 514.
95. Ilyas, R., Zuhri, M., Aisyah, H., Asyraf, M., Hassan, S., Zainudin, E., Sapuan, S., Sharma, S., Bangar, S., & Jumaidin, R. (2022). Natural fibre-reinforced polylactic acid, polylactic acid blends and their composites for advanced applications. *Polymers, 14*(1), 202.
96. Madadi, R., Maljaee, H., Serafim, L. S., & Ventura, S. P. (2021). Microalgae as contributors to produce biopolymers. *Marine Drugs, 19*(8), 466.
97. Cazón, P., Velazquez, G., Ramírez, J. A., & Vázquez, M. (2017). Polysaccharide-based films and coatings for food packaging: A review. *Food Hydrocolloids, 68*, 136–148.
98. Flórez-Fernández, N., López-García, M., González-Muñoz, M. J., Vilariño, J. M. L., & Domínguez, H. (2017). Ultrasound-assisted extraction of fucoidan from Sargassum muticum. *Journal of Applied Phycology, 29*(3), 1553–1561.
99. Jain, A., Sarsaiya, S., Awasthi, M. K., Singh, R., Rajput, R., Mishra, U. C., Chen, J., & Shi, J. (2022). Bioenergy and bio-products from bio-waste and its associated modern circular economy: Current research trends, challenges, and future outlooks. *Fuel, 307*, 121859.
100. Hmelkov, A. B., Zvyagintseva, T. N., Shevchenko, N. M., Rasin, A. B., & Ermakova, S. P. (2018). Ultrasound-assisted extraction of polysaccharides from brown alga Fucus evanescens. Structure and biological activity of the new fucoidan fractions. *Journal of Applied Phycology, 30*(3), 2039–2046.

101. Flórez-Fernández, N., Domínguez, H., & Torres, M. (2019). A green approach for alginate extraction from Sargassum muticum brown seaweed using ultrasound-assisted technique. *International Journal of Biological Macromolecules, 124*, 451–459.
102. Roja, K., Sudhakar, D. R., Anto, S., & Mathimani, T. (2019). Extraction and characterization of polyhydroxyalkanoates from marine green alga and cyanobacteria. *Biocatalysis and Agricultural Biotechnology, 22*, 101358.
103. Faidi, A., Lassoued, M. A., Becheikh, M. E. H., Touati, M., Stumbé, J.-F., & Farhat, F. (2019). Application of sodium alginate extracted from a Tunisian brown algae Padina pavonica for essential oil encapsulation: Microspheres preparation, characterization and in vitro release study. *International Journal of Biological Macromolecules, 136*, 386–394.
104. Morales-Jiménez, M., Gouveia, L., Yáñez-Fernández, J., Castro-Muñoz, R., & Barragán-Huerta, B. E. (2020). Production, preparation and characterization of microalgae-based biopolymer as a potential bioactive film. *Coatings, 10*(2), 120.
105. Gereniu, C. R. N., Saravana, P. S., & Chun, B.-S. (2018). Recovery of carrageenan from Solomon Islands red seaweed using ionic liquid-assisted subcritical water extraction. *Separation and Purification Technology, 196*, 309–317.
106. Alboofetileh, M., Rezaei, M., Tabarsa, M., You, S., Mariatti, F., & Cravotto, G. (2019). Subcritical water extraction as an efficient technique to isolate biologically-active fucoidans from Nizamuddinia zanardinii. *International Journal of Biological Macromolecules, 128*, 244–253.
107. Saravana, P. S., Tilahun, A., Gerenew, C., Tri, V. D., Kim, N. H., Kim, G.-D., Woo, H.-C., & Chun, B.-S. (2018). Subcritical water extraction of fucoidan from Saccharina japonica: Optimization, characterization and biological studies. *Journal of Applied Phycology, 30*(1), 579–590.
108. Saravana, P. S., Cho, Y.-N., Woo, H.-C., & Chun, B.-S. (2018). Green and efficient extraction of polysaccharides from brown seaweed by adding deep eutectic solvent in subcritical water hydrolysis. *Journal of Cleaner Production, 198*, 1474–1484.
109. Mirzadeh, M., Arianejad, M. R., & Khedmat, L. (2020). Antioxidant, antiradical, and antimicrobial activities of polysaccharides obtained by microwave-assisted extraction method: A review. *Carbohydrate Polymers, 229*, 115421.
110. Ponthier, E., Domínguez, H., & Torres, M. (2020). The microwave assisted extraction sway on the features of antioxidant compounds and gelling biopolymers from Mastocarpus stellatus. *Algal Research, 51*, 102081.
111. Rueda, E., García-Galán, M. J., Ortiz, A., Uggetti, E., Carretero, J., García, J., & Díez-Montero, R. (2020). Bioremediation of agricultural runoff and biopolymers production from cyanobacteria cultured in demonstrative full-scale photobioreactors. *Process Safety and Environmental Protection, 139*, 241–250.
112. Kumar, A. N., Chatterjee, S., Hemalatha, M., Althuri, A., Min, B., Kim, S.-H., & Mohan, S. V. (2020). Deoiled algal biomass derived renewable sugars for bioethanol and biopolymer production in biorefinery framework. *Bioresource Technology, 296*, 122315.
113. Novelli, L. D. D., Sayavedra, S. M., & Rene, E. R. (2021). Polyhydroxyalkanoate (PHA) production via resource recovery from industrial waste streams: A review of techniques and perspectives. *Bioresource Technology, 331*, 124985.
114. Kalia, V. C., Patel, S. K. S., Shanmugam, R., & Lee, J.-K. (2021). Polyhydroxyalkanoates: Trends and advances toward biotechnological applications. *Bioresource Technology, 326*, 124737.
115. Magagula, S., Mohapi, M., Sefadi, J., & Mochane, M. (2021). The production and applications of microbial-derived polyhydroxybutyrates. In *Microbial polymers* (pp. 3–43). Springer.
116. Awasthi, M. K., Kumar, V., Yadav, V., Sarsaiya, S., Awasthi, S. K., Sindhu, R., Binod, P., Kumar, V., Pandey, A., & Zhang, Z. (2021). Current state of the art biotechnological strategies for conversion of watermelon wastes residues to biopolymers production: A review. *Chemosphere* , 133310.
117. Müller-Santos, M., Koskimäki, J. J, Alves, L. P. S., de Souza, E. M., Jendrossek, D., & Pirttilä, A. M. (2021). The protective role of PHB and its degradation products against stress situations in bacteria. *FEMS Microbiology Reviews* , *45*(3), fuaa058.

118. Sarsaiya, S., Awasthi, S. K., Awasthi, M. K., Awasthi, A. K., Mishra, S., & Chen, J. (2018). The dynamic of cellulase activity of fungi inhabiting organic municipal solid waste. *Bioresource Technology, 251*, 411–415.
119. García, G., Sosa-Hernández, J. E., Rodas-Zuluaga, L. I, Castillo-Zacarías, C., Iqbal, H., & Parra-Saldívar, R. (2021). Accumulation of PHA in the microalgae Scenedesmus sp. under nutrient-deficient conditions. *Polymers, 13*(1), 131.
120. Banu, J. R., Kavitha, S., Kannah, R. Y., Devi, T. P., Gunasekaran, M., Kim, S.-H., & Kumar, G. (2019). A review on biopolymer production via lignin valorization. *Bioresource Technology, 290*, 121790.
121. Ghosh, S., Gnaim, R., Greiserman, S., Fadeev, L., Gozin, M., & Golberg, A. (2019). Macroalgal biomass subcritical hydrolysates for the production of polyhydroxyalkanoate (PHA) by Haloferax mediterranei. *Bioresource Technology, 271*, 166–173.
122. Tebaldi, M. L., Maia, A. L. C., Poletto, F., de Andrade, F. V., & Soares, D. C. F. (2019). Poly (-3-hydroxybutyrate-co-3-hydroxyvalerate)(PHBV): Current advances in synthesis methodologies, antitumor applications and biocompatibility. *Journal of Drug Delivery Science and Technology, 51*, 115–126.
123. Akdoğan, M., & Çelik, E. (2021). Enhanced production of poly (3-hydroxybutyrate-co-3-hydroxyvalerate) biopolymer by recombinant Bacillus megaterium in fed-batch bioreactors. *Bioprocess and Biosystems Engineering, 44*(2), 403–416.
124. Talan, A., Pokhrel, S., Tyagi, R., & Drogui, P. (2022). Biorefinery strategies for microbial bioplastics production: Sustainable pathway towards Circular Bioeconomy. *Bioresource Technology Reports, 17*, 100875.
125. Sayin, S., Kohlhaas, T., Veziroglu, S., Okudan, E. Ş, Naz, M., Schröder, S., Saygili, E., Açil, Y., Faupel, F., & Wiltfang, J. (2020). Marine Algae-PLA composites as de novo alternative to porcine derived collagen membranes. *Materials Today Chemistry, 17*, 100276.
126. Tran, D.-T., Lee, H. R., Jung, S., Park, M. S., & Yang, J.-W. (2018). Lipid-extracted algal biomass based biocomposites fabrication with poly (vinyl alcohol). *Algal Research, 31*, 525–533.
127. Kumar, L., Brice, J., Toberer, L., Klein-Seetharaman, J., Knauss, D., & Sarkar, S. K. (2019). Antimicrobial biopolymer formation from sodium alginate and algae extract using aminoglycosides. *PLoS ONE, 14*(3), e0214411.
128. Qureshi, D., Nayak, S. K., Maji, S., Kim, D., Banerjee, I., & Pal, K. (2019). Carrageenan: A wonder polymer from marine algae for potential drug delivery applications. *Current Pharmaceutical Design, 25*(11), 1172–1186.
129. Kim, D., & Kang, S. M. (2020). Red algae-derived carrageenan coatings for marine antifouling applications. *Biomacromolecules, 21*(12), 5086–5092.
130. Bouanati, T., Colson, E., Moins, S., Cabrera, J.-C., Eeckhaut, I., Raquez, J.-M., & Gerbaux, P. (2020). Microwave-assisted depolymerization of carrageenans from Kappaphycus alvarezii and Eucheuma spinosum: Controlled and green production of oligosaccharides from the algae biomass. *Algal Research, 51*, 102054.
131. Krempel, M., Griffin, K., & Khouryieh, H. (2019). Hydrocolloids as emulsifiers and stabilizers in beverage preservation. In *Preservatives and preservation approaches in beverages* (pp. 427–465). Elsevier.
132. Bui, V. T., Nguyen, B. T., Renou, F., & Nicolai, T. (2019). Structure and rheological properties of carrageenans extracted from different red algae species cultivated in Cam Ranh Bay, Vietnam. . *Journal of Applied Phycology, 31*(3), 1947–1953.
133. Rajauria, G., Ravindran, R., Garcia-Vaquero, M., Rai, D. K., Sweeney, T., & O'Doherty, J. (2021). Molecular characteristics and antioxidant activity of laminarin extracted from the seaweed species Laminaria hyperborea, using hydrothermal-assisted extraction and a multi-step purification procedure. *Food Hydrocolloids, 112*, 106332.
134. Becker, S., Scheffel, A., Polz, M. F., & Hehemann, J.-H. (2017). Accurate quantification of laminarin in marine organic matter with enzymes from marine microbes. *Applied and Environmental Microbiology, 83*(9), e03389-e3416.

135. Etman, S. M., Elnaggar, Y. S., & Abdallah, O. Y. (2020). Fucoidan, a natural biopolymer in cancer combating: From edible algae to nanocarrier tailoring. *International Journal of Biological Macromolecules, 147*, 799–808.
136. Wang, B., Wan, Y., Zheng, Y., Lee, X., Liu, T., Yu, Z., Huang, J., Ok, Y. S., Chen, J., & Gao, B. (2019). Alginate-based composites for environmental applications: A critical review. *Critical Reviews in Environmental Science and Technology, 49*(4), 318–356.
137. Murujew, O., Whitton, R., Kube, M., Fan, L., Roddick, F., Jefferson, B., & Pidou, M. (2021). Recovery and reuse of alginate in an immobilized algae reactor. *Environmental Technology, 42*(10), 1521–1530.
138. Ahmad, A., Mubarak, N., Jannat, F. T., Ashfaq, T., Santulli, C., Rizwan, M., Najda, A., Bin-Jumah, M., Abdel-Daim, M. M., & Hussain, S. (2021). A critical review on the synthesis of natural sodium alginate based composite materials: An innovative biological polymer for biomedical delivery applications. *Processes, 9*(1), 137.
139. Martău, G. A., Mihai, M., & Vodnar, D. C. (2019). The use of chitosan, alginate, and pectin in the biomedical and food sector—Biocompatibility, bioadhesiveness, and biodegradability. *Polymers, 11*(11), 1837.
140. Charoensiddhi, S., Lorbeer, A. J., Lahnstein, J., Bulone, V., Franco, C. M., & Zhang, W. (2016). Enzyme-assisted extraction of carbohydrates from the brown alga Ecklonia radiata: Effect of enzyme type, pH and buffer on sugar yield and molecular weight profiles. *Process Biochemistry, 51*(10), 1503–1510.
141. Sangiorgi, A., Gonzalez, Z., Ferrandez-Montero, A., Yus, J., Sanchez-Herencia, A. J., Galassi, C., Sanson, A., & Ferrari, B. (2019). 3D printing of photocatalytic filters using a biopolymer to immobilize TiO_2 nanoparticles. *Journal of the Electrochemical Society, 166*(5), H3239.
142. Shishkovsky, I., Sherbakov, V., Ibatullin, I., Volchkov, V., & Volova, L. (2018). Nano-size ceramic reinforced 3D biopolymer scaffolds: Tribomechanical testing and stem cell activity. *Composite Structures, 202*, 651–659.
143. Sathiyavimal, S., Vasantharaj, S., LewisOscar, F., Selvaraj, R., Brindhadevi, K., & Pugazhendhi, A. (2020). Natural organic and inorganic–hydroxyapatite biopolymer composite for biomedical applications. *Progress in Organic Coatings, 147*, 105858.
144. Bahmani, A., Comeau, P. A., Montesano, J., & Willett, T. L. (2019). Extrudable hydroxyapatite/plant oil-based biopolymer nanocomposites for biomedical applications: Mechanical testing and modeling. *Materials & Design, 174*, 107790.
145. Sivakanthan, S., Rajendran, S., Gamage, A., Madhujith, T., & Mani, S. (2020). Antioxidant and antimicrobial applications of biopolymers: A review. *Food Research International, 136*, 109327.
146. Gu, L., Peng, N., Chang, C., McClements, D. J., Su, Y., & Yang, Y. (2017). Fabrication of surface-active antioxidant food biopolymers: Conjugation of catechin polymers to egg white proteins. *Food Biophysics, 12*(2), 198–210.
147. Córdoba, L. J. P., & Sobral, P. J. (2017). Physical and antioxidant properties of films based on gelatin, gelatin-chitosan or gelatin-sodium caseinate blends loaded with nanoemulsified active compounds. *Journal of Food Engineering, 213*, 47–53.
148. Zanutto-Elgui, M. R., Vieira, J. C. S., do Prado, D. Z., Buzalaf, M. A. R., de Magalhães Padilha, P., de Oliveira, D. E., & Fleuri, L. F. (2019). Production of milk peptides with antimicrobial and antioxidant properties through fungal proteases. *Food Chemistry, 278*, 823–831.
149. Gopu, M., & Selvam, K. (2020). Polysaccharides from marine red algae Amphiroa rigida and their biomedical potential: An in-vitro study. *Biocatalysis and Agricultural Biotechnology, 29*, 101769.
150. Upadhyay, U., Sreedhar, I., Singh, S. A., Patel, C. M., & Anitha, K. (2021). Recent advances in heavy metal removal by chitosan based adsorbents. *Carbohydrate Polymers, 251*, 117000.
151. del Mar Orta, M., Martín, J., Santos, J. L., Aparicio, I., Medina-Carrasco, S., & Alonso, E. (2020). Biopolymer-clay nanocomposites as novel and ecofriendly adsorbents for environmental remediation. *Applied Clay Science , 198*, 105838.
152. Xia, L., Tan, J., Wu, P., He, Q., Song, S., & Li, Y. (2020). Biopolymers extracted from Klebsiella sp. and Bacillus sp. in wastewater sludge as superb adsorbents for aqueous Hg (II) removal from water. *Chemical Physics Letters, 754*, 137689.

153. Silva, M. S., Silva, L. S., Ferreira, F. J., Bezerra, R. D., Marques, T. M., Meneguin, A. B., Barud, H. S., Osajima, J. A, & Silva Filho, E. C. (2020). Study of interactions between organic contaminants and a new phosphated biopolymer derived from cellulose. *International Journal of Biological Macromolecules, 146*, 668–677.
154. Negm, N. A., Hefni, H. H., Abd-Elaal, A. A., Badr, E. A., & Abou Kana, M. T. (2020). Advancement on modification of chitosan biopolymer and its potential applications. *International Journal of Biological Macromolecules , 152*, 681–702.
155. Shariful, M. I., Sharif, S. B., Lee, J. J. L., Habiba, U., Ang, B. C., & Amalina, M. A. (2017). Adsorption of divalent heavy metal ion by mesoporous-high surface area chitosan/poly (ethylene oxide) nanofibrous membrane. *Carbohydrate Polymers, 157*, 57–64.
156. Zhang, C., Liu, S., Li, S., Tao, Y., Wang, P., Ma, X., & Chen, L. (2019). Enahanced biosorption of Cu (II) by magnetic chitosan microspheres immobilized Aspergillus sydowii (MCMAs) from aqueous solution. *Colloids and Surfaces A: Physicochemical and Engineering Aspects, 581*, 123813.

Starch/Carrageenan Blend-Based Biocomposites as Packaging Materials

Heru Suryanto, Uun Yanuhar, Aminnudin, Yanuar Rohmat Aji Pradana, and Redyarsa Dharma Bintara

1 Introduction

Nowadays, millions of tons of plastic are produced and applied in various industries that generate excessive plastic wastes causing massive global problems. Such large amounts of synthetic plastic have caused severe ecological problems because of the non-renewable and non-degradable nature of the material [1]. Even so, plastic is still used as a packaging material because of several advantages: cheap, light, inert, good heat insulator, easy to print, and flexible to fabricate various shapes [2]. Fortunately, new bio-based materials have been developed to replace non-degradable materials for food packaging to avoid pollution problems in the last decades [3].

The primary function of packaging in the food industries is to maintain the safety and quality of food products such as chemical contaminants, spoilage microorganisms, oxygen, moisture, light, high mechanical strength, heat resistance, chemical

H. Suryanto (✉)
Center of Excellence for Cellulose Composite (CECCom), Department of Mechanical Engineering, Universitas Negeri Malang, Jl. Semarang 5, Malang 65145, Indonesia
e-mail: heru.suryanto.ft@um.ac.id

H. Suryanto · Aminnudin · Y. R. A. Pradana · R. D. Bintara
Deparment of Mechanical Engineering, Faculty of Engineering, Universitas Negeri Malang, Jl. Semarang 5, Malang 65145, East Java, Indonesia
e-mail: aminnudin.ft@um.ac.id

Y. R. A. Pradana
e-mail: yanuar.rohmat.ft@um.ac.id

R. D. Bintara
e-mail: redyarsa.dharma.ft@um.ac.id

U. Yanuhar
Department of Aquatic Resources Management, Faculty of Fisheries and Marine Sciences, Universitas Brawijaya, Jl. Veteran, Malang, East Java, Indonesia
e-mail: doktoruun@ub.ac.id

Sapuan S. M. and I. Ahmad (eds.), *Composites from the Aquatic Environment*, Composites Science and Technology, https://doi.org/10.1007/978-981-19-5327-9_6

and renewable [4]. Microbial contamination can occur in many traditional paper-based food packaging products [5]. In this case, bioplastic as active packaging can prevent bacterial contamination by inhibiting the growth of food pathogens, either by producing an active atmosphere or by direct contact aging [6]. Efforts to reduce microbial contamination have been carried out by using antibacterial materials containing various natural and synthetic antibacterial ingredients. Antibacterial agents commonly use heavy metals such as mercury, cadmium, lead, chromium [7], copper [8], ZnO nanoparticles [1], silver nanoparticles [9], which are, of course, dangerous and also expensive. Other types of antibacterial are benzoic acid, sodium propionate, potassium sorbate [10], essential and volatile oils [11], proteins from gluten [12], and peptides [13] which have limited resources. Some bioplastic materials such as egg and soy protein [14] and chitosan from marine animal shells [15] have antibacterial ability. In this case, natural ingredients from seaweed through carrageenan compounds also can be antibacterial [16, 17] and can be used as bioplastics [17].

Natural starch is hardly suitable for packaging because the granules are easily degraded. Efforts to overcome these weaknesses are carried out by thickening starch products using various methods such as thermo-mechanical treatment. Its chemical modification by cross-linking is conducted to maintain the granular structure of starch. Wax starch can prevent the release of starch molecular components during the pasting process. High viscosity and relatively low flow properties possessed by starch-based materials raise complications during injection molding; a typical method for making synthetic thermoplastic polymer products [18]. Many approaches have been conducted to increase the starch film properties by making starch-based biocomposites using reinforcement of natural fibers [19], Zinc nanoparticles [20], nanoclay [21], and chitosan [22]. The wide starch-based bioplastics application is limited by their brittleness, moisture sensitivity, low tensile strength [23] and modest processability in the processing equipment [19]. The starch blended with polycaprolactone, gelatin, and polylactic acid has been reported [20–22] Polymer blend gives better influence on their physicochemistry characteristics. This chapter provides a general overview of the biocomposite based cassava starch-carrageenan blend, including its properties and prospective application.

2 Starch

Starch is the main stock of polysaccharides in food plants. In 2019, cassava was produced about 303.57 million tons, where Africa is the largest producer region with 45.74% of the total production [24]. Nigeria takes the first position in the country with the largest cassava production, followed by the Republic of Congo, Thailand, Ghana, Brazil, and Indonesia producing 19.5, 13.19, 10.24, 7.39, 5.76, and 4.81% of worldwide production, respectively [24]. Although cassava production is growing and starch from cassava is competitive, corn is still the primary starch source globally (>70%). Starch extracted from both tubers and roots has specific physical and rheological properties (lower retrogradation, high viscosity, and clear gel). These properties are required to formulate certain products [25].

As one of the natural polymer sources, starch can be developed extensively. Starch is a raw material that has the potential as a biocomposite material. They have advantages such as being cheap, flexible, wide availability, tasteless, odorless, resistant to O_2, semipermeable to CO_2, and able to be degraded without forming toxic residues [10, 26].

2.1 *Structure of Starch*

Starch is a class of organic compounds containing carbohydrates, a glucose polymer consisting of amylose and amylopectin, and a semi-crystalline phase structure [27]. The chemical structure of starch is composed of carbon, hydrogen, and oxygen with a proportion of 6:10:5 [$C_6H_{10}O_5$], as shown in Fig. 1. Amylose has a linear molecule chain of –(1–4) linked D-glucose units with a degree of polymerization (DP) at a range of 500–6000 glucose units. Amylopectin is a branched-chain molecule with a DP ranging from 3.0×10^5 to 3.0×10^6 glucose units. Also, it is composed of linear –(1–4) glucan chains connected at –(1–6) branching points. One molecule contains 4–5% of all bonds in amylopectin. Other ingredients that have minor value in starch are lipids (up to 1%), protein, phosphorus, and other minerals [28]. Lipids, especially phospholipids, have a great tendency to form helical complexes with starch (especially with amylose) to form very strongly bound amylose-phospholipid complexes. The amylopectin content in starch ranges from 50 to 58%, while the amylose content in tapioca flour is 20–27%. Amylose and amylopectin have different physical characteristics. High amylose content can improve mechanical properties because the bioplastic development process occurs limitedly. In contrast, amylopectin content provides optimal stickiness [29]. Amylose tends to form long and flexible helical chains and always moves in a circle to be more easily dissolved in solution.

2.2 *Plasticization Mechanisms*

Plasticizers are non-volatile compounds with low molecular weight largely used as additives in polymer industries. The main role of plasticizers is to improve the processability and flexibility of polymers. The plasticizers are grouped into two types external and internal plasticizers. The external plasticizers are low volatility molecules added for interacting with the polymer, resulting in polymer swelling without chemical reaction. Intermolecular forces among plasticiser molecules, between polymer and plasticizer, such as hydrogen bonds, dipole–dipole interactions, dispersion forces, and induction forces, have a significant role in external plasticization. Besides, an internal plasticizer is a polymer molecule that is reacted or grafted with the original polymer into the polymer structure. This reaction makes

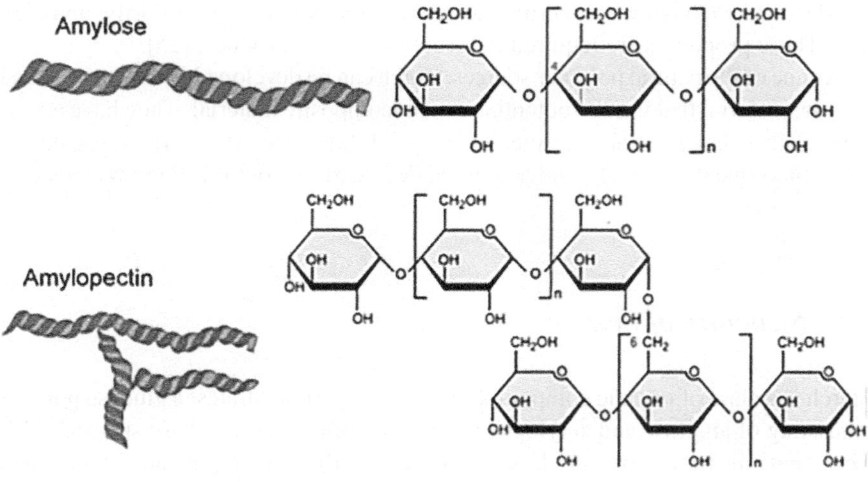

Fig. 1 Molecular structure of starch [30]

the polymer chains tightly compact and more difficult to fit. They also reduce the elastic modulus and lower the glass transition temperature (T_g) [31].

The other categories of plasticizers are namely primary and secondary plasticizers. The group is called a primary plasticizer if the polymer dissolves in the plasticizer at high polymer concentrations. These plasticizers have a single element of plasticizer as the main component of the plasticizer. They make rapidly gel the polymer at the normal processing temperature. On the contrary, the secondary plasticizers are less compatible with the original polymer and have lower gelation capacity. They are mostly mixed with main plasticizers to reduce cost and improve product properties [32].

Theories about the action mechanism of plasticizers within polymer networks that have been accepted include lubrication theory, gel theory, and free volume theory. Lubrication theory is identical to the lubrication mechanism of metal parts using a liquid lubricant. The plasticizers act as a lubricant for reducing friction and facilitating the mobility of the polymer chains across each other to reduce deformation. On the other hand, gel theory expands the lubrication theory and shows that plasticizers interfere with and displace interactions of polymer–polymer (ionic forces, van der Waals or hydrogen bonds). Lastly, free volume theory is the internal space available in the polymer for chain movement. Flexible resin has relatively large free volumes, while rigid resins have a limited free volume. The plasticizer increases the resin-free volume, maintains the free volume after post-processing the polymer-plasticizer mixture, and reduces the glass transition temperature [33].

Processing of thermoplastic starch is usually carried out between temperatures of 70 and 90 °C using a plasticizer. Granular starch has a semi-crystalline structure in its natural form. If dried starch is heated, thermal degradation occurs under the melting point of the granule crystals. It makes starch not able to be processed in its original form. Hydrogen bonds holding the starch molecules reduce the melting process of the original starch. These bonds can be reduced from the starch using a solvent such as water. The phase transition state changes from regular to disordered in an aqueous medium when starch is heated to a critical temperature. The solvent interacts with the hydroxyl groups in the starch, called the gelatinization process [34]. This process removes polymer crystallinity because of the loss of double helix and lamellar crystal structures. Gelatinization temperature can be determined using Differential Scanning Calorimetry (DSC) with the temperature range of 52–75 °C, depending on the starch type [35].

Thermoplastic Starch (TPS) forms the plasticised moldable thermoplastic material by structuring native starch at high temperatures under high shear conditions in little water. The properties of starch in water are alkaline, where starch can undergo a melting process. Several substances used as plasticizers for processing the TPS are polyol compounds such as propylene glycol butyladiol, ethylene glycol, maltitol, xylitol, sorbitol, glycol, glycerol: fatty acids (such as myristic or palmitic), mannose, fructose, and sucrose. It is known that water is a softening agent of starch; however, alone, water action is not preferred because it can produce a brittle film. Single starch chains move relatively freely with respect to other chains, thereby allowing the starch to be melt-processed. These changes, in turn, facilitate their molecular mobility in the amorphous region and allow swelling of the granular. The plasticizer penetrates the starch grains, breaks the starch's internal hydrogen bonds, and eliminates the starch–starch interaction replaced by the starch-plasticizer interaction. Starch granular transformation is affected by process conditions like temperature and plasticizer contents. TPS is suitable for blow molding, extrusion or injection molding, similar to the process of synthetic thermoplastic polymer. Dissolved amorphous starches tend to return to states as insoluble, aggregated, or crystalline when stored at temperatures above their glass transition temperatures. This phenomenon is called retrogradation. Plasticizers have a function as lubricants that facilitate the mobility of polymer chains and reduce the retrogradation of TPS products [31, 34, 36].

The starch plastic properties can be changed by adjusting the processing temperature, type and amounts of plasticizers, and moisture content. Starch sources are also important to produce the desired TPS properties. The content of starch is depended on the plant producer type. So that, starch can have different content such as granular size, crystallinity, molecular weight dan its distributions, and also amylose/amylopectin ratios. These contents affect gelatinization and glass transition temperature that is directly correlated with TPS thermoplasticity. The brittle starch film can be converted to an elastic film with modulation of plasticizer or its plasticizer blends. However, plasticized starch has some drawbacks, such as lower mechanical properties than synthetic polymers, high affinity to water, and unstable properties over time [37].

TPS can be improved their properties using reinforcement with nanomaterial such as nanoclay, nanocellulose, nanochitosan, silver nanoparticles with the characteristic of composite listed in Table 1. Properties of TPS composite depend on the source of starch, type and content of plasticizer, processing route, and type of reinforcement.

Table 1 Characteristics of starch-based biocomposites

Starch biocomposite	Plasticizer	Main thermal degradation (°C)	Tensile Strength (MPa)	Elongation at break (%)	References
Cassava film	Glycerol	337	21.7 ± 6.08	5.2 ± 1.89	[38, 39]
Cassava starch/nanoclay	Glycerol	362	10.8–28	2.1–3.5	[39–41]
Corn starch/lignocellulose	Glycerol	300	5.26	110	[42]
Wheat starch/kaolin	Glycerol	400	1.2–3.48	123.28	[43]
Sago starch/nano chitosan	Glycerol	385	1.2–2.8	20.0–53.8	[44]
Corn starch/sisal fibre	Glycerol	334	14.15	31.8	[45]
Potato starch/nanoclay	Gliserol and sorbitol	511	2.28–10.78	26.1–43.3	[46, 47]
Yam bean/nanocellulose WHF	Glycerol	341–371	1.8–5.8	27.4–69.1	[48]
Sugar palm/AgNPs	Glycerol and Sorbitol	320	16.5–40.8	38.84–197	[49]

3 Carrageenan

Carrageenan is a generic name for the linear polysaccharides family consisting of repeating disaccharide units. It is a group of linear galactan with an ester sulphate. They are obtained from red seaweed as the major source of carrageenan, including *Eucheuma, Furcellaria, Iridaea, Hypnea, and Chondrus* [50]. This material is composed of D-galactose and 3,6-anhydrous-D-galactose units linked by α-1,3 and β-1,4-glycosidic linkages [51]. Showing a molecular weight ranging from 100,000 to 1,000,000 Dalton, carrageenan forms a gel in the presence of calcium or potassium ions. There are three common carrageenan types, such as lambda (λ), kappa (κ), and iota (ι), with chemical structure, as shown in Fig. 2. These carrageenan types differ in the chemical composition and sulfation degree at certain locations in the polymer and do not indicate definite chemical structures. This carrageenan name is based on sulphate groups ranging from 22 to 35% and its solubility in KCl solution. The position and number of ester sulphate groups in carrageenan structure make its variety and properties [52].

The properties of carrageenan types are very different, for instance, kappa-carrageenan can form a gel film in the presence of potassium ions. On the other hand, Iota-carrageenan can react with calcium ions resulting in an elastic gel, while lambda-carrageenan forms a gel at high salt concentrations. The characteristics of carrageenan are shown in Table 2. The properties of the chemical content of carrageenan are determined by its solubility, viscosity, gel strength, pH and stability. The decrease

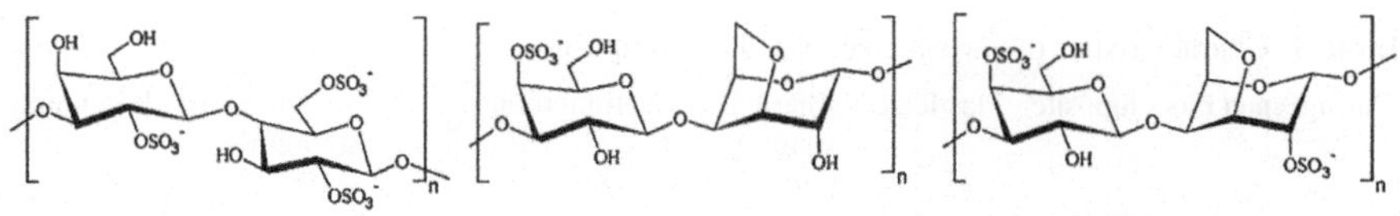

Fig. 2 Chemical structures of carrageenans

in pH causes viscosity loss and the potential to form gels because of the hydrolysis of the glycosidic bonds. Moreover, the process is accelerated by the presence of heat. Carrageenan dissolves easily in hot water, even at ambient temperature [50]. Carrageenan can form gel and thickening properties. It can form thermo-reversible hydrogels, widely applied as gelling agents in the pharmaceutical and food industries. Oxidation processes and acidic conditions easily damage carrageenan. Likewise, an increase in temperature and time can break the glycosidic bond.

The main property for packaging products is barrier property. High film barrier property prevents the migration of low molecular weight chemical compounds such as vapours and gases. Like almost bio-based films, carrageenan film has low water vapour permeability due to a large number of sulphate groups and hydroxyl in their structures. Consequently, the inherent hydrophilic nature of these films gives poor water resistance and moisture barrier properties [53]. On the contrary, carrageenan-based films show several desirable properties for developing packaging films because of their ability to form high-strength films with better homogeneity, excellent oxygen barrier properties, and good transparency properties. Several researchers

Table 2 The carrageenan properties (adopted from [52])

Parameter	Description
Composition	Structured of α-d-1,3 and β-d-1,4 galactose that is sulphated at up to 40% of the total weight
Solubility	Lambda is dissolved in hot or cold water; Kappa and Iota is dissolved in hot water (80 °C)
Gel formation	KCl promote gel formation of Kappa; iota form right-handed helices and calcium ion form a gel, and Lambda does not form gels
Source	Red algae; gigartina, chondrus, and various Eucheuma species
Molecular weight	Native carrageenan: 1.5×10^6 to 2.0×10^7 Dalton; food-grade carrageenan: 1.0×10^5 to 8.0×10^5 or 2.0×10^5 to 4.0×10^5 Dalton, degraded carrageenan: 2.0×10^4 to 3.0×10^4 Dalton
Viscosity	5 to 800 cps for 1.5% solution at 75 °C
Properties	Serve as a stabilizer, emulsifier, thickening agent
Texture	Kappa is brittle, Iota is elastic, and lambda is not forming a gel
Major application	Infant formula, skin preparations, cosmetics, processed meats, dietetic formulations, milk products, toothpaste, pesticides, laxative

Table 3 Characteristics of carrageenan based biocomposites

Carrageenan biocomposite	Plasticizer	Thermal stability (°C)	Antimicrobial	Tensile strength (MPa)	References
Carrageenan film	Glycerol	290	–	48.95	[54]
Carrageenan/nanosilver	Glycerol	~290	*L. mono-cytogenes* and *E. coli*	64.6	[55–57]
Carrageenan/zinc oxide	Glycerol	~230	*L. mono-cytogenes* and *E. coli*	12.3	[57]
Carrageenan/nanoparticles (ZnO, CuO)	Glycerol	222.5–228	*L. mono-cytogenes* and *E. coli*	30.4–55.2	[58]
κ-carrageenan/nanosilica	Glycerol	272–304	–	32–38	[59]
κ-carrageenan/nanoclay	Glycerol	262.6	*L. mono-cytogenes*, *E. coli* and *S. enterica*	26.9–33.8	[60–62]
κ-carrageenan/copper sulphide nanoparticles	Glycerol	240	*L. mono-cytogenes* and *E. coli*	54–66.5	[61]
κ-carrageenan/cellulose nanowhisker	Glycerol	201–222.5	–	38.33 ± 3.79	[63–65]
κ-carrageenan/melanin nanoparticle	Glycerol	228	*L. mono-cytogenes* and *E. coli*	46.2–62.9	[64]

have strengthened its characteristics by involving reinforcement into composites, as listed in Table 3.

4 Starch/Carrageenan Blend Biocomposite

Biopolymers have distinct structures showing scattered molecules and long worm-like chains or twisted ropes when the materials are captured by microscopic observation. The poor performance of biopolymers can be overcome by adapting composite/blend technology. Biopolymers can be mixed with other biopolymers or polymers to form blends. They can be combined with various compatible reinforcing materials called fillers to make biopolymer composites. The properties of the composite/mixture are strongly influenced by the dimensions of the constituent phases and the content ratio of the two phases. The characteristics of the composite/mixture depend on the properties of the constituents used and the synthesis procedure.

Blending polymers with other polymeric materials is a more simple and effective method for obtaining desirable film properties. The performance of the polymer is based on the characteristics of each polymer blend, its structure, and its morphology. Biopolymers can also be mixed with other bio/synthetic polymers to produce biopolymer blends. It will improve the properties compared to the parent polymer. Solubility, a function of molecular weight, copolymer composition, temperature, and mixture composition, is the main factor determining the mixture's formation. Two polymers can mix if they have a negative mixed free energy. Solubility is polymer blends main requirement; however, interfacial adhesion between polymer components is highly desirable to enhance the specific properties of the mixture. Suppose two immiscible polymers are to be mixed. In this case, adding a substance located at the polymer–polymer interface can increase the degree of mixture, ultimately resulting in superior properties.

The various approaches followed for the synthesis of polymer blends are [65]:

1. Melt mixing: each constituent for the blend are mixed in the molten state in batch mixers or extruders.
2. Solution blending: each constituent is dissolved in the solvent, stirred and then precipitated or evaporated.
3. Graft co-polymerization: monomers are used as a solvent for another component, followed by polymerization.
4. Interpenetrating polymer networks: At least one of the blend components is synthesized and/or cross-linked in the immediate presence of the other(s).

The composition of carrageenan and starch in the blending process results in different film properties. Their interaction was also affected by the ratio of the components. Carrageenan inhibits the molecular structure rearrangement of starch by limiting the water molecules. Higher levels of carrageenan increase the gelatinization onset temperature of starch. Also, carrageenan is able to restrain the starch granules swelling from forming a film with a homogenous network structure [66]. Starch can increase gel strength and reduce kappa-carrageenan dispersion during storage. The stiffness of the gelatinized starch granules plays an important role in the gel properties of the carrageenan starch mixture. Starch with high granular stiffness is suitable for combining kappa-carrageenan to increase gel strength and reduce dispersion syneresis [67].

The main concerns in developing starch/carrageenan blends are starch/carrageenan mixtures and gelling properties. The flexibility of the film related to the interaction of carrageenan chains with starch granules is influenced by the bond chain length of the constituents and the small molecular weight of the chain, which can be the main supporting factor. The presence of carrageenan in the continuous phase of the starch gel causes the development of the surface size of the starch granules. As the swelling increases, the starch granules can be softer, resulting in decreased resistance to mechanical treatment, leading to higher amounts of amylopectin in the system. Carrageenan is adsorbed on the starch when interacting with starch after or before swelling. The degree of adsorption depends on the molecular weight and the charge of the carrageenan. The lower molecular weight and the higher charge lead to a higher adsorption rate [68].

Additional carrageenan into starch forming bioplastic blend results in various physical effects. Blend of starch and kappa-carrageenan forms an edible film with good mechanical properties and structure that varies according to the carrageenan content [69]. Interactions between starch and carrageenan determine the phase behavior of the constituent polymers, either phase separation or miscibility. Starch and carrageenan have good miscibility because the carrageenan has a double-helical structure that traps starch in the coiled structure and acts as a protector for the starch molecules. Strong interactions with starch result in a structure with high crystalline, which strengthens the polymeric chain integrity of bioplastic [70].

5 Characteristic of Starch/Carrageenan Blend Biocomposite

Mechanical Properties

Blending carrageenan with cassava starch through the casting process can increase in strength of up to 200% at the addition of 20% carrageenan. The synthesis and optimization of bionanocomposite based on the nanoclay-reinforced starch increase tensile strength by up to 344% after optimizing the sonication process for up to 1 h [40]. The addition of nanoclay to starch-based bioplastics increases the optimum tensile strength by adding 5% nanoclay. During biocomposite synthesis, amyloses form an amylose-glycerol complex. The blend polymer penetrates nanoclay galleries and separates their layer disperse uniformly in the polymer blend. Synergetic interaction and compatibility of starch/carrageenan blend with nanoclay on the surface layer are important to form intercalated or exfoliated starch/carrageenan blend biocomposites. The mechanical properties of the biocomposite were also explained by its fracture surface. At the higher content of clay addition, the agglomeration of nanoclay characterized the rougher fracture surface, indicating lower compatibility between clay and its matrix. As a critical point, the 5% nanoclay addition provided the restriction means of bioplastic chain mobility at the interface region, enabling effective interaction between polymer chain and nanoclay surface. As the excessive nanoclay was added, the restriction ability of nanoclay was subsequently decreased, followed by the drop of tensile strength [71].

A similar circumstance on carrageenan involvement was also found on bioplastic fabricated by hot extrusion. By adding the different amounts of carrageenan content into cassava starch, the tensile strength of bioplastic showed a significant uptrend to 1.43 MPa in 10% (w/w) addition. It is almost seven times that without carrageenan content. Similarly, the elastic modulus significantly increased from 0.77 to 9.65 MPa after a similar carrageenan addition. Consequently, the elongation was dramatically reduced into half of that on the pure extruded bioplastic counterpart with 14.78% on 10% carrageenan addition [72]. Also, the role of carrageenan is obvious in terms of the mechanical properties of edible films. Starch/carrageenan blend film with different ratios had considerable tensile strength and elongation when the carrageenan content was increased from 50 to 75% for all starch contents. These findings can be explained due to the higher hygroscopic properties of carrageenan than starch. Therefore, the higher amount of water absorption in higher carrageenan content induced the plasticization behavior during tensile loading and led to the larger range of elongation until the breaking phase [69].

Physical Properties

Initially, a clear-transparent appearance was shown by starch-made bioplastic. Meanwhile, in terms of extruded-bioplastic resulted by a subsequent process, voids were visible, indicating gas entrapped during extrusion. Therefore, a vacuum environment was needed during fabrication. In the case of the carrageenan addition effect, the pure bioplastic exhibited an obviously rough surface, while the surface quality rose gradually by adding a higher amount of carrageenan, indicating the better affinity with the starch [72].

The XRD analysis reveals that the peak of bioplastics with no carrageenan content was identified at diffraction angles of 17.5°, 19.4°, and 22.2°. Although the peak located at 19.4° shifted to different positions of 19.9° and 20.8° after 5% carrageenan addition, more than 2.5% peaks at the range of 17.5° and 22.2° were reduced and disappeared. However, by adding higher carrageenan content until 10%, the peak at 19.9° was gradually disappeared, leading to a stronger peak of 20.8°. In parallel, the peak of amylose remained stagnant, while new peaks at 20.8° raised using 5% carrageenan. It indicates that adding a carrageenan of 5% successfully made a considerable starch/carrageenan blend. Still, with more concentration of carrageenan, the structural change of bioplastic was not significantly indicated by the relatively similar shape of XRD curves [73].

On the other hand, nanoclay addition performs the different states in their structure (Fig. 3). Crystal planes of (103), (004), and (220) are indicated by the XRD peaks at the 2θ of 16.9°, 19.6° and 21.7° on pure biocomposite, respectively. These peaks indicate the presence of cassava B-crystal type mainly contains amylose characterized at 17°, 22°, and 24°. The effect of nanoclay in the starch/carrageenan blend composite is indicated through the decrease in peaks intensity of 16.9° and 21.7°. These peaks disappear after adding 10% nanoclay, indicating a fully or partially exfoliated biocomposite blend structure caused by the movement restriction of bioplastic molecules by nanoclay. Henceforth, the optimum interactions and compatibility between starch/carrageenan blend matrix and nanoclay are important in forming its composites with intercalated or exfoliated nanoclay-layered [71].

Chemical Properties

The functional groups of nanoclay-reinforced starch-carrageenan blend composite were analysed using FTIR (Fig. 4). The OH stretching in the starch was indicated at the broad wavelength ranging from 3000 to 3670 cm^{-1}. The OH stretching observed in this blend biocomposite is caused by the presence of non-bound or bound water molecules due to the humidity. The C–H stretching of alkanes group was observed at wave number ranging from 2800 to 3000 cm^{-1} with intensity decreased at increasing nanoclay content, did not appear by 5% of nanoclay. Nanomaterial such as nanoclay can break the blend matrix, then bond with nanoclay through the mechanism of exfoliating. The peaks at the wavenumber of 1405–1435 cm^{-1} indicate the CH_2–bending vibrations and C–H deformation. The low transmittance at the wavenumber of 1360–1390 cm^{-1} indicated CH_3 symmetric bending with CH_3 vibration, suggesting acetates' presence. Nanoclay reinforcement in composite absorbs the acetate into the

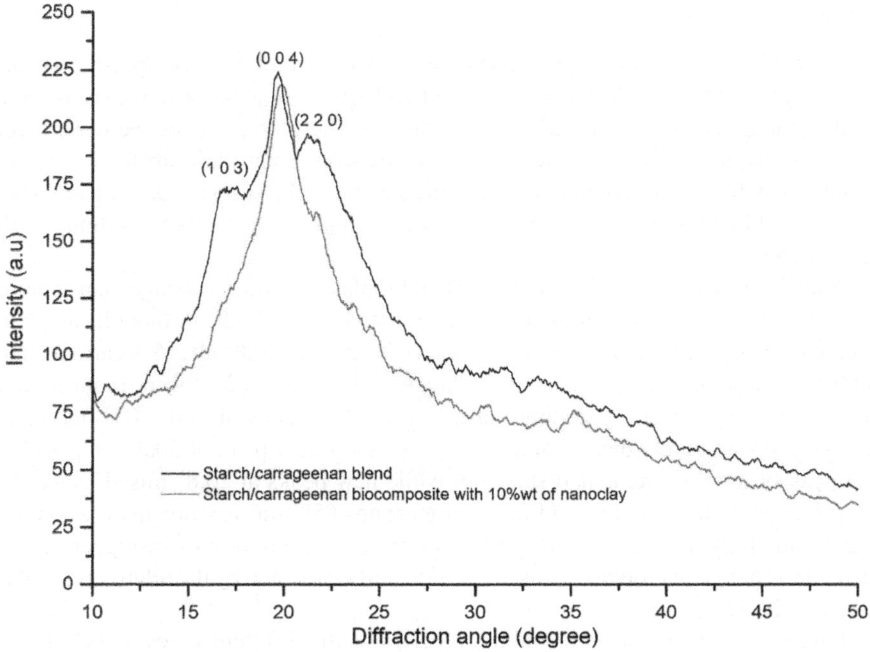

Fig. 3 Diffractogram of starch/carrageenan blend biocomposite reinforced by 10 wt% of nanoclay

interlamellar of the clay, causing disappearing the peak of 1361. Wavenumber in the range from 800 to 1200 cm^{-1} indicates that the main polysaccharide is present in the polysaccharides blend. Wavenumber in the range from 915 to 955 cm^{-1} attributed to C–O–C out of plane stretching. These functional groups indicate a carboxylic acid group and pyranose type sugars. The amylose was identified at 400 and 700 cm^{-1} [71].

In the case of bacterial inhibition performance of starch-carrageenan bioplastic in packaging application, the durian starch-carrageenan/carvacrol film showed considerable anti-microbial properties when applied on food samples made from durian. The durian starch of 11.42% was successfully extracted from durian seeds using 0.5% $NaHSO_3$ aqueous treatment and resulted in starch with a purity of 42%. By using 8% carvacrol content, the inhibition region increased from 15.89 to 22.45 mm after 24 h of storage. Similarly, durian starch-carrageenan blend film exhibited an effective inhibition performance against *S. aureus* count in spike-inoculated food sample with the reduction of 83.6% under 4 °C storage temperature for 24 h, while total inhibition was performed using carvacrol concentration of 10% under a similar storage period. Even though showing a relatively similar level of bacterial inhibitability with durian starch-carrageenan, the addition of carvacrol successfully restricted the bacteria counts with the higher amount at an early stage of inhibition (8 h) [74].

Similarly, for oxygen (O_2) scavenging performance investigation, a 1-mm-thick extruded thermoplastic starch oxygen scavenging film was fabricated with iron

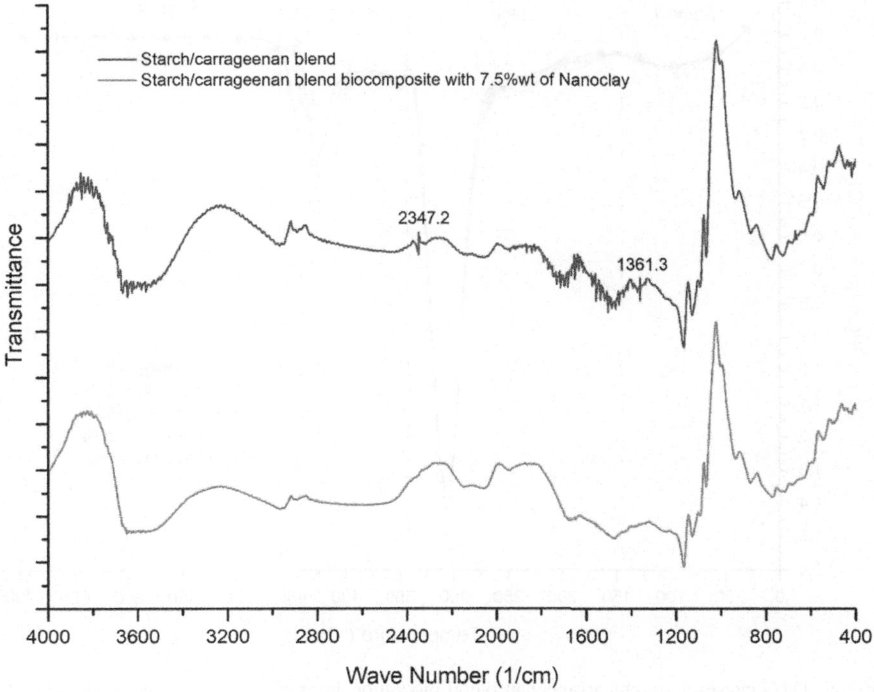

Fig. 4 Functional group analysis of starch/carrageenan blend biocomposite reinforced by 7.5 wt% of nanoclay

powder and ascorbic acid. The film showed exceptional performance in terms of oxygen scavenging when increasing the water content (humidity) could improve the scavenging ability, with the oxygen reduction from 20.9 to 1% proportional with the absorbing capacity of 13.5 mL of oxygen for each gram of dry film after 15 days under the 80% relative humidity environment. Moreover, the mechanical properties improved when the oxygen scavenging film made from thermoplastic starch was blended with 20% polycaprolactone (PCL), but only at humidities below 72% [75].

Thermal Properties

Blend biocomposite characteristics in the presence of elevated temperature can be analyzed using Thermal Gravimetric Analysis (TGA) equipment. Derivative TGA results were analysed to show the decomposition process at elevated temperatures (Fig. 5). After being reinforced by nanoclay, starch/carrageenan blend biocomposite exhibited a multi-stage thermal decomposition process. It is distributed into different four stages temperature observed at stage 1 (below 145 °C), stage 2 (146–340 °C), stage 3 (341–475 °C), stage 4 (476–1000 °C). Firstly, stage 1 represented the evaporation process of water and plasticizer (glycerol) devolatilization. Stage 2 then illustrated the matrix depolymerization with a relatively high mass loss of 61–67.5%. Following this, stage 3 was defined as the transition process becoming char with

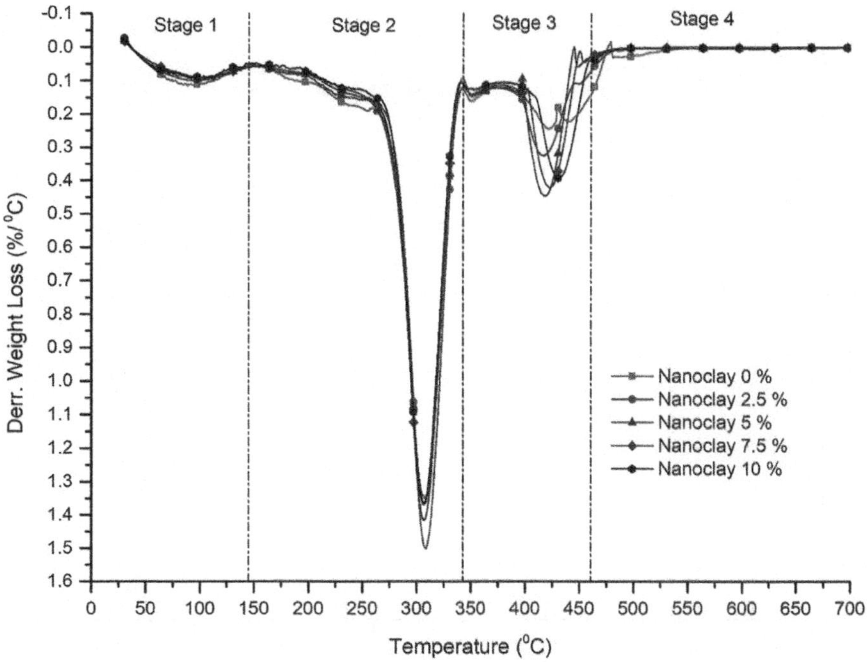

Fig. 5 DTG curve of starch/carrageenan blend biocomposite with *Nanoclay* reinforcement of 0, 2.5, 5, 7.5 and 10 wt%

approximately mass loss of 21%. Final stage, all samples decomposed to the ash with the content of 2.2, 2.2, 4.1, 5.8, and 9.9% for pure starch/carrageenan blend and its composite with nanoclay content of 2.5, 5.0, 7.5, and 10%, respectively [71].

Several researchers have synthesised starch/carrageenan biocomposite with properties shown in Table 4.

6 Prospective Application in the Packaging Industry

Biopolymers are now widely used in food industries and packaging. Several studies have developed starch-based bioplastic because starch is the most abundant biopolymer in nature. It has several advantages: being widely available, low-cost material, edible, biodegradable, colorless, tasteless, and easy to use [86]. The cassava starch-based plastic has good properties in forming film to obtain a flexible material of homogeneous and smooth surfaces [87]. It can be an alternative in coating or packaging material in the food industry [88]. Figure 6 shows the process of making starch/carrageenan-based nanocomposite film for food packaging application. Starch/carrageenan blend bioplastic is already produced as a film with highly

Table 4 Characteristics of starch/carrageenan blend-based biocomposites

Biocomposite	Thermal degradation	XRD peaks at 2θ	Tensile strength (MPa)	Elongation at break (%)	Water solubility (%)	Water vapour permeability (10^{-10} g/m Pa s)	Refs.
Starch/carrageenan blends	–	13.5°, 17.5°, 18.1°, 22° and 22.4°	1.43	20.22	71.3 ± 2.2	4.6	[69, 72, 76]
Starch/carrageenan blends with nanoclay	Start at 146 °C	16.9°, 19.6° and 21.7°	2.48	17.91	–	–	[77–79]
Corn starch/carrageenan with chitosan	Start at 146 °C	–	44	13	–	1.25	[80]
Starch/carrageenan with SiO_2–ZnO nanoparticle	Start at 150 °C	–	7.323 ± 0.595	27.962 ± 5.768	81.381 ± 4.340	3.847 ± 0.552	[81]
Starch/LDPE/carrageenan with cotton fiber	–	21.3°, 23.6° and 36.0°	9–10	–	–	7	[18]
Starch/carrageenan with nanocellulose	–	19.8° and 22.6°	7.10	–	98	2 ± 0.05	[82, 83]
Starch/carrageenan with carboxymethyl cellulose	120 °C	–	15.7	40	–	–	[84]
Starch/carrageenan with ZnO nanoparticle	–	–	10.78–35.71	41.42–68.79	74.69	1.031 ± 0.047	[76, 85]
Starch/carrageenan with SiO_2 nanoparticle	–	–	19.37 ± 0.52	10.92 ± 1.91	85.88	1.035 ± 0.074	[85]

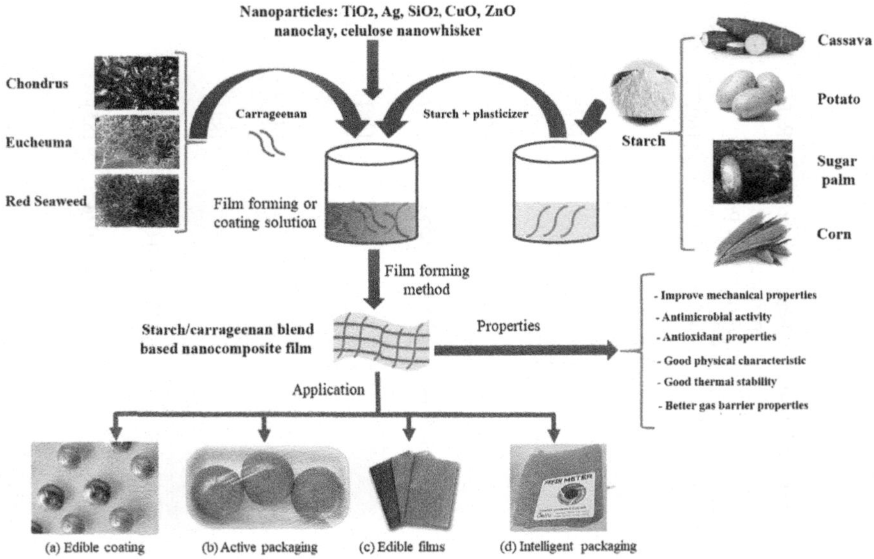

Fig. 6 Prospective applications of starch/blend based biocomposite

thermal stability in packaging application. κ-carrageenan can make a stiff film modulated by adjusting the weight ratio of κ-carrageenan and starch in the polymer blend [51].

Nanotechnology offers food safety in packaging for avoiding spoilage of food quality. Nanoparticles of titanium, zinc oxide, dioxide, copper oxide, and copper had been added in polymer composite due to their antimicrobial properties [89]. Moreover, in self-cleaning surfaces application, SiO_2- and TiO_2-based nanofillers can be used [89]. The antibacterial activity of metal nanostructures is mostly dependent on various factors such as size, chemical functionalization, particle internalization, large surface area, and particle shape. The nanostructures can also penetrate inner and outer bacterial membranes [90]. Starch/carrageenan blend based biocomposite properties can increase by additional content of nanoclay. The nanoclay content affects their tensile strength, and a concentration of 5% showed the optimum tensile strength. In addition, the structure of biocomposite was thermally stable after being reinforced by nanoclay. These results show that the starch/carrageenan blend-based biocomposite, which is improved using nanoclay reinforcement, have the potential to enhance the properties of packaging materials [71]. The starch/carrageenan-based biocomposite also can be modified as an edible coating. The biopolymer coatings can combine food additives, such as antioxidants, antimicrobials, flavors, and it still allows for expansion of its applications [91]. It would be applied as a coating for vegetables, fruits, and others, effectively delaying the dehydration process that causes weight loss [92].

Active packaging has a good role in food preservation and provides an inert barrier to external conditions. Active packaging is mainly developing a packaging system that can respond to environmental changes. It acts by releasing active molecules that act as enzyme immobilization, oxygen scavengers, antimicrobials, and antioxidants to improve food stability. Nanocomposites in active packaging can also be used as delivery systems, thereby aiding the migration of functional additives, such as probiotics and vitamins, into food. Silver nanoparticles can be implemented as packaging films to preserve food for longer periods which is used for killing microorganisms in 6 min [93].

7 Conclusions and Future Direction

Starch has received considerable attention for biodegradable film formulation because of its biodegradability, edible, low cost, easy to use, and thermo-processable nature. Carrageenan is a biopolymer with some desirable properties for packaging films development because of its ability to increase film's tensile strength, high thermal stability, antifungal activity, excellent oxygen barrier properties, and good transparency properties. Starch/carrageenan blend-based films have been considered an alternative for conventional packaging in improving food safety and quality. Additionally, starch/carrageenan-based blends biocomposite are applied as carriers of functional ingredients to prepare active, antioxidant, and intelligent packaging by incorporating nutraceutical, antibrowning, and antimicrobial agents for improving shelf-life and quality.

However, improving the delivery properties of packaging needs an innovative technology and still requires future research. Mostly, the applied packaging materials have been produced on a laboratory scale and need to be developed for commercially packaged products. Further studies should optimize the blend film formulation and processing conditions for improving film properties and subsequently applied into specific applications.

References

1. Diez-Pascual, A. M., & Diez-Vicente, A. L. (2014). Poly(3-hydroxybutyrate)/ZnO bionanocomposites with improved mechanical, barrier and antibacterial properties. *International Journal of Molecular Sciences, 15*, 10950–10973. https://doi.org/10.3390/ijms150610950
2. Mose, B. R., & Maranga, S. M. (2011). A review on starch based nanocomposites for bioplastic materials. *Journal of Materials Science and Engineering B, 1*, 239–245.
3. López, O. V., Castillo, L. A., García, M. A., Villar, M. A., & Barbosa, S. E. (2015). Food packaging bags based on thermoplastic corn starch reinforced with talc nanoparticles. *Food Hydrocolloids, 43*, 18–24. https://doi.org/10.1016/j.foodhyd.2014.04.021
4. Rhim, J.-W.W., Park, H.-M.M., & Ha, C.-S.S. (2013). Bio-nanocomposites for food packaging applications. *Progress in Polymer Science, 38*, 1629–1652. https://doi.org/10.1016/j.progpolymsci.2013.05.008

5. Rana, M. S., Mahmud, S., Hossain, M. A., Rana, M., Kabir, E., Das, A. K., & Roy, R. K. (2019). Bacteriological load in traditional food packaging paper. *Journal of Advance in Microbiology, 15*(2), 1–9. https://doi.org/10.9734/jamb/2019/v15i230085
6. Cano, A., Cháfer, M., Chiralt, A., & González-Martínez, C. (2015). Physical and antimicrobial properties of starch-PVA blend films as affected by the incorporation of natural antimicrobial agents. *Foods, 5*, 3. https://doi.org/10.3390/foods5010003
7. Parolo, M., Fernandez, L., Zajonkovsky, I., Sanchez, M., & Baschini, M. (2011). Antibacterial activity of materials synthesized from clay minerals. In *Science against microbial pathogens: Communicating current research and technological advances* (pp. 144–151).
8. Bruna, J. E., González, V., Rodríguez, F., Guarda, A., & Galotto, M. J. (2011). Eco-nanocomposites films containing copper as potential antimicrobial active packaging. In *11th Congresso Brasileiro de Polimeros* (pp. 3770–3775).
9. Abreu, A. S., Oliveira, M., de Sá, A., Rodrigues, R. M., Cerqueira, M. A., Vicente, A. A., & Machado, A. V. (2015). Antimicrobial nanostructured starch based films for packaging. *Carbohydrate Polymers, 129*, 127–134. https://doi.org/10.1016/j.carbpol.2015.04.021
10. Chowdhury, T., & Das, M. (2013). Effect of antimicrobials on mechanical, barrier and optical properties of corn starch based self-supporting edible film. *International Journal of Food Studies, 2*(2), 212–223.
11. Nairetti, D., Mironescu, M., & Tita, O. (2014). Antimicrobial activity of active biodegradable starch films on pathogenic microorganisms. *Annals of the Romanian Society for Cell Biology, XIX*(1), 73–78.
12. Gómez-Heincke, D., Martínez, I., Partal, P., Guerrero, A., & Gallegos, C. (2015). Development of antimicrobial active packaging materials based on gluten proteins. *Journal of the Science of Food and Agriculture*. https://doi.org/10.1002/jsfa.7525
13. de Azeredo, H. M. C. (2009). Nanocomposites for food packaging applications. *Food Research International, 42*(9), 1240–1253.
14. Jones, A., Mandal, A., & Sharma, S. (2015). Protein-based bioplastics and their antibacterial potential. *Journal of Applied Polymer Science, 132*, 1–11. https://doi.org/10.1002/app.41931
15. Nadrajah, K. (2005). *Development and characterization of nanocomposite materials*. Louisiana State University.
16. Natrah, F. M. I., Harah, Z. M., Sidik, B. J., Izzatul, N. M. S., & Syahidah, A. (2015). Antibacterial activities of selected seaweed and seagrass from port dickson coastal water against different aquaculture pathogens.
17. Gade, R., Siva Tulasi, M., & Aruna Bhai, V. (2013). Seaweeds: A novel biomaterial. *International Journal of Pharmacy and Pharmaceutical Sciences, 5*, 40–44.
18. Prachayawarakorn, J., & Pomdage, W. (2014). Effect of carrageenan on properties of biodegradable thermoplastic cassava starch/low-density polyethylene composites reinforced by cotton fibers. *Materials and Design, 61*, 264–269. https://doi.org/10.1016/j.matdes.2014.04.051
19. Lopez-Gil, A., Rodriguez-Perez, M. A., De Saja, J. A., Bellucci, F. S., & Ardanuy, M. (2014). Strategies to improve the mechanical properties of starch-based materials: Plasticization and natural fibers reinforcement. *Polimeros, 24*, 36–42. https://doi.org/10.4322/polimeros.2014.054
20. Fadeyibi, A., Osunde, Z. D., Egwim, E. C., & Idah, P. A. (2017). Performance evaluation of cassava starch-zinc nanocomposite film for tomatoes packaging. *Journal of Agricultural Engineering, 48*, 137–146. https://doi.org/10.4081/jae.2017.565
21. Suryanto, H., Hutomo, P. T., Wanjaya, R., & Puspitasari, P. (2016). The structure of bioplastic from cassava starch with nanoclay reinforcement. In *AIP Int. Proceeding Int. Mech. Eng. Eng. Educ.* (Vol. 030027, pp. 1–5). https://doi.org/10.1063/1.4965761
22. Akter, N., Khan, R. A., Tuhin, M. O., Haque, M. E., Nurnabi, M., Parvin, F., & Islam, R. (2014). Thermomechanical, barrier, and morphological properties of chitosan-reinforced starch-based biodegradable composite films. *Journal of Thermoplastic Composite Materials, 27*(7), 933–948. https://doi.org/10.1177/0892705712461512
23. Jagadeesh, D., Prashantha, K., Mithil Kumar Nayunigari, N., & Maity, A. (2016). Effect of gelatin content on potato starch green composite films. *Indian Journals of Advances in Chemical*

Science, 4(4), 355–361. Retrieved June 11, 2018, from http://www.ijacskros.com/. 4 Volume 4 Issue/IJACS-M207.pdf

24. Tridge. (2022). Cassava global production and top producing countries. Retrieved Jan 26, 2022, from https://www.tridge.com/intelligences/mandioca/production
25. Bertolini, A. (2010). *Starches characterization, properties, and applications*. CRC Press.
26. Sulaiman, S., Manut, A., & Nur Firdaus, A. R. (2009). Starch plastic packaging and agriculture applications. In *ICIMT'09. International Conference, IEEE* (pp. 513–516). http://digitalcommons.unl.edu/usdaarsfacpub/1459
27. Bertoft, E. (2017). Understanding starch structure: Recent progress. *Agronomy, 7*(3), 1–29. https://doi.org/10.3390/agronomy7030056
28. Jacobs, H., & Delcour, J. A. (1998). Hydrothermal modifications of granular starch, with retention of the granular structure: A review. *Journal of Agricultural and Food Chemistry, 46*(8), 2895–2905. American Chemical Society. https://doi.org/10.1021/jf980169k
29. Hardwianti, R., Primaniyarta, M., & Palupi, N. S. (2014). Konsistensi Mutu Pilus Tepung Tapioka: Identifikasi Parameter Utama Penentu Kerenyahan. *J. Mutu Pangan, 1*(2), 91–99. Retrieved Aug 05, 2018, http://jurnalmutupangan.com/files/JMP-00000001/JMP-02-15-002/naskah-JMP-02-15-002.pdf
30. Madhumitha, G., Fowsiya, J., Mohana Roopan, S., & Thakur, V. K. (2018). Recent advances in starch–clay nanocomposites. *International Journal of Polymer Analysis and Characterization, 23*(4), 331–345. https://doi.org/10.1080/1023666X.2018.1447260
31. Mekonnen, T., Mussone, P., Khalil, H., & Bressler, D. (2013). Progress in bio-based plastics and plasticizing modifications. *Journal of Materials Chemistry A, 1*(43), 13379. https://doi.org/10.1039/c3ta12555f
32. Vieira, M. G. A., Da Silva, M. A., Dos Santos, L. O., & Beppu, M. M. (2011). Natural-based plasticizers and biopolymer films: A review. *European Polymer Journal, 47*(3), 254–263. Pergamon. https://doi.org/10.1016/j.eurpolymj.2010.12.011
33. G. Wypych, *Handbook of plasticizer* (3rd ed.) (Vol. 1). ChemTec Publishing.
34. Nafchi, A. M., Moradpour, M., Saeidi, M., & Alias, A. K. (2013). Thermoplastic starches: Properties, challenges, and prospects. *Starch/Staerke, 65*(1–2), 61–72. https://doi.org/10.1002/star.201200201
35. Coral, D., Pineda-Gomez, P., & Rosales-Rivera, A. (2009). Determination of the gelatinization temperature of starch presented in maize flours. *Journal of Physics: Conference Series*, 012057. https://doi.org/10.1088/1742-6596/167/1/012057
36. Jane, J. (1995). Starch properties, modifications, and applications. *Journal of Macromolecular Science, Part A, 32*(4), 751–757. https://doi.org/10.1080/10601329508010286
37. Mikus, P. Y., Alix, S., Soulestin, J., Lacrampe, M. F., Krawczak, P., Coqueret, X., & Dole, P. (2014). Deformation mechanisms of plasticized starch materials. *Carbohydrate Polymers, 114*, 450–457. https://doi.org/10.1016/j.carbpol.2014.06.087
38. Alves, V. D., Mali, S., Beléia, A., & Grossmann, M. V. E. (2007). Effect of glycerol and amylose enrichment on cassava starch film properties. *Journal of Food Engineering, 78*(3), 941–946. https://doi.org/10.1016/j.jfoodeng.2005.12.007
39. Ren, J., Dang, K., Pollet, E., & Avérous, L. (2018). Preparation and characterization of thermoplastic potato starch/halloysite nano-biocomposites: Effect of plasticizer nature and nanoclay content. *Polymers (Basel), 10*(8), 808. https://doi.org/10.3390/polym10080808
40. Suryanto, H., Jenuardy, S., Kustono, D., Puspitasari, P., & Lubis, D. Z. (2018). Sonication assisted synthesis of biocomposite from starch/nanoclay and its properties. *International Review of Mechanical Engineering, 12*(3), 272–278. https://doi.org/10.15866/ireme.v12i3.13641
41. Monteiro, M. K. S., de Oliveira, V. R. L., dos Santos, F. K. G., de Leite, R. H., Aroucha, E. M. M., da Silva, R. R., & de Silva, K. N. (2017). Analysis of water barrier, mechanical and thermal properties of nanocoposites based on cassava starch and natural clay or modified by anionic exchange. *Materials Research, 20*(suppl2), 69–76. https://doi.org/10.1590/1980-5373-mr-2016-1087

42. Yin, P., Zhou, W., Zhang, X., Guo, B., & Li, P. (2020). Bio-based thermoplastic starch composites reinforced by dialdehyde lignocellulose. *Molecules, 25*(14). https://doi.org/10.3390/molecules25143236
43. Tuntachon, S., Sukolrat, A., Numnuam, A., & Kaewtatip, K. (2019). Effect of kaolin content and sonication on the properties of wheat gluten composites. *Powder Technology, 351*, 66–70. https://doi.org/10.1016/j.powtec.2019.04.007
44. Azadi, A., Supriyadi, S., & Herawati, H. (2020). Starch based biocomposite from sago (Metroxylon sagu) with nano-chitosan reinforcement: Mechanical and thermal characteristics. *The Journal of Pure and Applied Chemistry Research, 9*(2), 98–107. https://doi.org/10.21776/ub.jpacr.2020.009.02.516
45. de Freitas, R. R. M., do Carmo, K. P., de Souza Rodrigues, J., de Lima, V. H., Osmari da Silva, J., & Botaro, V. R. (2021). Influence of alkaline treatment on sisal fibre applied as reinforcement agent in composites of corn starch and cellulose acetate matrices. *Plastics, Rubber and Composites, 50*(1), 9–17. https://doi.org/10.1080/14658011.2020.1816119
46. Ren, J., Dang, K. M., Pollet, E., & Avérous, L. (2018). Preparation and characterization of thermoplastic potato starch/halloysite nano-biocomposites: Effect of plasticizer nature and nanoclay content. *Polymers (Basel), 10*(8). https://doi.org/10.3390/polym10080808
47. Suryanto, H., Rahmawan, A., Solichin, S., Sahana, R., & Yanuhar, U. (2018). Characteristics of nanoclay reinforced starch biocomposites through the extrusion process. In *MATEC web of conferences* (in Press).
48. Asrofi, M., Abral, H., Kasim, A., Pratoto, A., Mahardika, M., & Hafizulhaq, F. (2018). Characterization of the sonicated yam bean starch bionanocomposites reinforced by nanocellulose water hyacinth fiber (WHF): The effect of various fiber loading. *Journal of Engineering Science and Technology, 13*(9), 2700–2715.
49. Rozilah, A., Aiza Jaafar, C. N., Sapuan, S. M., Zainol, I., & Ilyas, R. A. (2020). The effects of silver nanoparticles compositions on the mechanical, physiochemical, antibacterial, and morphology properties of sugar palm starch biocomposites for antibacterial coating. *Polymers (Basel), 12*(11), 1–21. https://doi.org/10.3390/polym12112605
50. Sudhakar, Y. N., Selvakumar, M., & Bhat, D. K. (2018). Biopolymer electrolytes for solar cells and electrochemical cells. In *Biopolymer electrolytes* (pp. 117–149). Elsevier.
51. de Lima Barizão, C., Crepaldi, M. I., de O.S. Junior, O., de Oliveira, A. C., Martins, A. F., Garcia, P. S., & Bonafé, E. G. (2020). Biodegradable films based on commercial κ-carrageenan and cassava starch to achieve low production costs. *International Journal of Biological Macromolecules, 165*, 582–590. https://doi.org/10.1016/j.ijbiomac.2020.09.150
52. Necas, J., & Bartosikova, L. (2013). Carrageenan: A review. *Veterinarni medicina (Praha), 58*(4), 187–205. Retrieved Jan 18, 2018, from http://www.agriculturejournals.cz/publicFiles/91236.pdf
53. Sedayu, B. B., Cran, M. J., & Bigger, S. W. (2019). A review of property enhancement techniques for carrageenan-based films and coatings. *Carbohydrate Polymers, 216*, 287–302. Elsevier Ltd. https://doi.org/10.1016/j.carbpol.2019.04.021
54. Kassab, Z., Aziz, F., Hannache, H., Ben Youcef, H., & El Achaby, M. (2019). Improved mechanical properties of k-carrageenan-based nanocomposite films reinforced with cellulose nanocrystals. *International Journal of Biological Macromolecules, 123*, 1248–1256. https://doi.org/10.1016/j.ijbiomac.2018.12.030
55. Rhim, J. W., & Wang, L. F. (2014). Preparation and characterization of carrageenan-based nanocomposite films reinforced with clay mineral and silver nanoparticles. *Applied Clay Science, 97–98*(August), 174–181. https://doi.org/10.1016/j.clay.2014.05.025
56. Fouda, M. M. G., El-Aassar, M. R., El Fawal, G. F., Hafez, E. E., Masry, S. H. D., & Abdel-Megeed, A. (2015). K-Carrageenan/poly vinyl pyrollidone/polyethylene glycol/silver nanoparticles film for biomedical application. *International Journal of Biological Macromolecules, 74*, 179–184. https://doi.org/10.1016/j.ijbiomac.2014.11.040
57. Kanmani, P., & Rhim, J. W. (2014). Properties and characterization of bionanocomposite films prepared with various biopolymers and ZnO nanoparticles. *Carbohydrate Polymers, 106*(1), 190–199. https://doi.org/10.1016/j.carbpol.2014.02.007

58. Oun, A. A., & Rhim, J. W. (2017). Carrageenan-based hydrogels and films: Effect of ZnO and CuO nanoparticles on the physical, mechanical, and antimicrobial properties. *Food Hydrocolloids, 67*, 45–53. https://doi.org/10.1016/j.foodhyd.2016.12.040
59. Rane, L. R., Savadekar, N. R., Kadam, P. G., & Mhaske, S. T. (2014). Preparation and characterization of K-carrageenan/nanosilica biocomposite film. *Journal of Materials, 2014*, 1–8. https://doi.org/10.1155/2014/736271
60. Martins, J. T., Bourbon, A. I., Pinheiro, A. C., Souza, B. W. S., Cerqueira, M. A., & Vicente, A. A. (2013). Biocomposite films based on κ-carrageenan/locust bean gum blends and clays: Physical and antimicrobial properties. *Food and Bioprocess Technology, 6*(8), 2081–2092. https://doi.org/10.1007/s11947-012-0851-4
61. Roy, S., & Rhim, J. W. (2020). Fabrication of copper sulfide nanoparticles and limonene incorporated pullulan/carrageenan-based film with improved mechanical and antibacterial properties. *Polymers (Basel), 12*(11), 1–14. https://doi.org/10.3390/polym12112665
62. Sánchez-García, M. D., Hilliou, L., & Lagarón, J. M. (2010). Morphology and water barrier properties of nanobiocomposites of κ/l-hybrid carrageenan and cellulose nanowhiskers. *Journal of Agriculture and Food Chemistry, 58*(24), 12847–12857. https://doi.org/10.1021/jf102764e
63. Yadav, M., & Chiu, F. C. (2019). Cellulose nanocrystals reinforced κ-carrageenan based UV resistant transparent bionanocomposite films for sustainable packaging applications. *Carbohydrate Polymers, 211*, 181–194. https://doi.org/10.1016/j.carbpol.2019.01.114
64. Roy, S., & Rhim, J. W. (2019). Preparation of carrageenan-based functional nanocomposite films incorporated with melanin nanoparticles. *Colloids Surfaces B Biointerfaces, 176*, 317–324. https://doi.org/10.1016/j.colsurfb.2019.01.023
65. Sreekumar, K., & Bindhu, B. (2020). An overview on biopolymers and biopolymer blends/composites. *Journal of Xidian University, 14*(9), 686–691. https://doi.org/10.37896/jxu14.9/076
66. Liu, B., Zhu, S., Zhong, F., Yokoyama, W., Huang, D., & Li, Y. (2021). Modulating storage stability of binary gel by adjusting the ratios of starch and kappa-carrageenan. *Carbohydrate Polymers, 268*, 118264. https://doi.org/10.1016/j.carbpol.2021.118264
67. Lin, J. H., Liang, C. W., & Chang, Y. H. (2016). Effect of starch source on gel properties of kappa-carrageenan-starch dispersions. *Food Hydrocolloids, 60*, 509–515. https://doi.org/10.1016/j.foodhyd.2016.04.024
68. Zia, K. M., Tabasum, S., Nasif, M., Sultan, N., Aslam, N., Noreen, A., & Zuber, M. (2017). A review on synthesis, properties and applications of natural polymer based carrageenan blends and composites. *International Journal of Biological Macromolecules, 96*, 282–301. Elsevier B.V. https://doi.org/10.1016/j.ijbiomac.2016.11.095
69. Abdou, E. S., & Sorour, M. A. (2014). Preparation and characterization of starch/carrageenan edible films. *International Food Research Journal, 21*(1), 189–193. Retrieved July 02, 2018 from http://www.ifrj.upm.edu.my/21. (01) 2014/27 IFRJ 21 (01) 2014 Manal 348.pdf
70. Thakur, R., Saberi, B., Pristijono, P., Golding, J., Stathopoulos, C., Scarlett, C., Bowyer, M., & Vuong, Q. (2016). Characterization of rice starch-ι-carrageenan biodegradable edible film. Effect of stearic acid on the film properties. *International Journal of Biological Macromolecules, 93*(Pt A), 952–960. https://doi.org/10.1016/j.ijbiomac.2016.09.053
71. Suryanto, H., Mahera, L. P., Kharismawan, F. A., Saragih, S. M., Solichin, S., & Yanuhar, U. (2019). Properties of starch/carrageenan blend biocomposite with nanoclay reinforcement. *Journal of Engineering Science & Technology Review, 12*(5), 122–126. https://doi.org/10.25103/jestr.125.14
72. Suryanto, H., Rahmawan, A. W., Solichin, Sahana, R. T., Muhajir, M., & Yanuhar, U. (2019). Influence of carrageenan on the mechanical strength of starch bioplastic formed by extrusion process. In *IOP Conference Series: Materials Science and Engineering* (Vol. 494, no. 1). https://doi.org/10.1088/1757-899X/494/1/012075
73. Suryanto, H., Sahana, R. T., Aminnudin, Suyetno, A., Widiyanti, & Yanuhar, U. (2019). Effect of carrageenan on the structure of cassava starch bioplastic after extrusion process. In *AIP Conference Proceedings* (Vol. 2120). https://doi.org/10.1063/1.5115761

74. Velasco, E. M. Z., & Fundador, N. G. V. (2020). Development and use of antimicrobial durian starch-carrageenan/carvacrol films. *Mindanao Journal of Science and Technology, 18*(1), 118–128.
75. Mahieu, A., Terrié, C., & Youssef, B. (2015). Thermoplastic starch films and thermoplastic starch/polycaprolactone blends with oxygen-scavenging properties: Influence of water content. *Industrial Crops and Products, 72*, 192–199. https://doi.org/10.1016/j.indcrop.2014.11.037
76. Shahbazi, M., Majzoobi, M., & Farahnaky, A. (2018). Physical modification of starch by high-pressure homogenization for improving functional properties of κ-carrageenan/starch blend film. *Food Hydrocolloids, 85*, 204–214. https://doi.org/10.1016/j.foodhyd.2018.07.017
77. Goonoo, N., Bhaw-Luximon, A., Passanha, P., Esteves, S., Schönherr, H., & Jhurry, D. (2017). Biomineralization potential and cellular response of PHB and PHBV blends with natural anionic polysaccharides. *Materials Science and Engineering C, 76*, 13–24. https://doi.org/10.1016/j.msec.2017.02.156
78. Suryanto, H., Kharismawan, F. A., Solichin, Rahmawan, A. W., Sahana, R. T., Muhajir, M., & Yanuhar, U. (2019). Influence of nanoclay on thermal decomposition of biocomposite matrix starch/carrageenan blend. In *IOP Conference Series: Materials Science and Engineering* (Vol. 494, no. 1). https://doi.org/10.1088/1757-899X/494/1/012077
79. Suryanto, H., Fitrasakti, D. A. D., Ramadhani, A. R., Suyetno, A., & Aminnudin. (2020). The effect of extrusion speed on mechanical properties of starch-based biocomposite. In *AIP Conference Proceedings* (Vol. 2231). https://doi.org/10.1063/5.0002524
80. Panatarani, C., Rochima, E., Ayunani, Yoga, S., & Joni, I. M. (2020). Reinforcement of carrageenan/starch based bio-composite by beads-milled chitosan. In *5th International Conference on Food, Agriculture and Natural Resources (FANRes 2019)* (Vol. 194, no. FANRes 2019, pp. 272–276). https://doi.org/10.2991/aer.k.200325.054.
81. Praseptiangga, D., Widyaastuti, D., Panatarani, C., & Joni, I. M. (2021). Development and characterization of semi-refined iota carrageenan/sio2-zno bionanocomposite film with the addition of cassava starch for application on minced chicken meat packaging. *Foods, 10*(11). https://doi.org/10.3390/foods10112776
82. Lopes, T., Bufalino, L., Júnior, M., Tonoli, G., & Mendes, M. (2018). Eucalyptus wood nanofibrils as reinforcement of carrageenan and starch biopolymers for improvement of physical properties. *Journal of Tropical Forest Science, 30*(3), 292–303. Retrieved Jan 29, 2022 from https://www.jstor.org/stable/26512514?read-now=1&refreqid=excelsior%3Ade72e33b55906386426dd98513abfb55&seq=10#page_scan_tab_contents
83. Chen, Q. J., Zong, Z. Y., Gao, X., Zhao, Y. L., & Wang, J. H. (2021). Preparation and characterization of nanostarch-based green hard capsules reinforced by cellulose nanocrystals. *International Journal of Biological Macromolecules, 167*, 1241–1247. https://doi.org/10.1016/j.ijbiomac.2020.11.078
84. Abdullah, A., Firdiana, B., Nissa, R., Satoto, R., Karina, M., & Fransiska, D. (2021). Effect of K-carrageenan on mechanical, thermal and biodegradable properties of starch—carboxymethyl cellulose (CMC) bioplastic. *Cellulose Chemistry and Technology, 55*(9–10), 1109–1117.
85. Praseptiangga, D., Mufida, N., Panatarani, C., & Joni, I. M. (2021). Enhanced multi functionality of semi-refined iota carrageenan as food packaging material by incorporating SiO2 and ZnO nanoparticles. *Heliyon, 7*(5), e06963. https://doi.org/10.1016/j.heliyon.2021.e06963
86. Basiak, E., Galus, S., & Lenart, A. (2014). Characterisation of composite edible films based on wheat starch and whey-protein isolate. *International Journal of Food Science & Technology*, 1–9. https://doi.org/10.1111/ijfs.12628
87. Orozco-Parra, J., Mejía, C. M., & Villa, C. C. (2020). Development of a bioactive synbiotic edible film based on cassava starch, inulin, and Lactobacillus casei. *Food Hydrocolloids, 104*, 105–754. https://doi.org/10.1016/J.FOODHYD.2020.105754
88. Gutiérrez, T. J., Morales, N. J., Pérez, E., Tapia, M. S., & Famá, L. (2015). Physico-chemical properties of edible films derived from native and phosphated cush-cush yam and cassava starches. *Food Packaging and Shelf Life, 3*, 1–8. https://doi.org/10.1016/j.fpsl.2014.09.002
89. Keun Taik, L. (2010). Quality and safety aspects of meat products as affected by various physical manipulations of packaging materials. *Meat Science, 86*(1), 50–138. https://doi.org/10.1016/j.meatsci.2010.04.035. Epub 2010 Apr 28.

90. Pathakoti, K., Manubolu, M., & Hwang, H. M. (2017). Nanostructures: Current uses and future applications in food science. *Journal of Food and Drug Analysis, 25*(2), 245–253. https://doi.org/10.1016/j.jfda.2017.02.004
91. Jaramillo, C. M., González Seligra, P., Goyanes, S., Bernal, C., & Famá, L. (2015). Biofilms based on cassava starch containing extract of yerba mate as antioxidant and plasticizer. *Starch/Staerke, 67*, 780–789. https://doi.org/10.1002/star.201500033
92. Cortés-Rodríguez, M., Villegas-Yépez, C., Gil González, J. H., Rodríguez, P. E., & Ortega-Toro, R. (2020). Development and evaluation of edible films based on cassava starch, whey protein, and bees wax. *Heliyon, 6*(9), 2–7. https://doi.org/10.1016/j.heliyon.2020.e04884
93. Ahmad, S. S., Yousuf, O., Islam, R. U., & Younis, K. (2021). Silver nanoparticles as an active packaging ingredient and its toxicity. *Packaging Technology and Science*, 1–11. https://doi.org/10.1002/pts.2603

Chitosan Composites for the Removal of Pollutants in Aqueous Environment

A. H. Nordin, N. Ngadi, R. A. Ilyas, and M. L. Nordin

1 Introduction

Aquatic organisms are available in a huge amount in the environment and have been discovered to contain many futuristic potential biomolecules for diverse scientific areas. Biopolymers derived from marine sources have been widely used in commercial applications and product development. Chitin and chitosan are two of the most common biopolymers found as supporting materials in a variety of aquatic organisms, including shrimp, crabs, lobsters, crayfish, and krill. In Southeast Asia, biowaste from the aquatic organisms is used to produce a significant amount of chitin and chitosan in a commercial-scale [1]. This is due to the unique characteristics of chitin and chitosan such as biocompatibility, non-toxicity and biodegradability which are beneficial to be used a lot in environmental applications. Fig. 1 shows the production of chitosan from aquatic crustaceans using sodium hydroxide solution and hydrochloric acid. Generally, to make crustacean shells easier to handle, they are usually crushed into small size. Then, minerals such as $CaCO_3$, $Ca_2(PO_4)$, and protein must be removed from the crustacean skeleton using alkali and acid, respectively. Subsequently, the isolated chitin was dried and stored. Finally, chitosan is made via de-acetylation of chitin with a concentrated NaOH solution at high temperature [2].

A. H. Nordin · N. Ngadi (✉) · R. A. Ilyas
School of Chemical and Energy Engineering, Faculty of Engineering, Universiti Teknologi Malaysia, 81310 Skudai, Johor, Malaysia
e-mail: norzita@utm.my

R. A. Ilyas
Centre for Advanced Composite Materials (CACM), Universiti Teknologi Malaysia, UTM Johor Bahru, 81310 Johor, Malaysia

M. L. Nordin
Department of Clinical Studies, Faculty of Veterinary Medicine, Universiti Malaysia Kelantan, Pengkalan Chepa, 16100 Kota Bharu, Kelantan, Malaysia

Sapuan S. M. and I. Ahmad (eds.), *Composites from the Aquatic Environment*, Composites Science and Technology, https://doi.org/10.1007/978-981-19-5327-9_7

Fig. 1 Production of chitosan from aquatic crustacean shells

The utilization of chitosan has gained much attention because of the properties (the presence of amine and hydroxyl groups as displayed in Fig. 2) which makes it feasible for the physical and chemical modifications. Development in chitosan composites is a new field of material science that has the potential to meet the needs of future generations [3, 4]. Among the applications is in the removal of pollutants from wastewater [5, 6].

Wastewater treatment using chitosan composites is a significant application associated with their ability to treat pollutants from wastewater effectively. Previously, removal of pollutants such as heavy metals, dye, fluoride, herbicides, pesticides and pharmaceuticals from wastewater using chitosan composites have been studied extensively. Various substances, such as carboxylation, amination, inclusion of magnetic particles, hydroxyapatites, multi–walled carbon nanotubes and montmorillonite have been utilized to produce composites with chitosan.

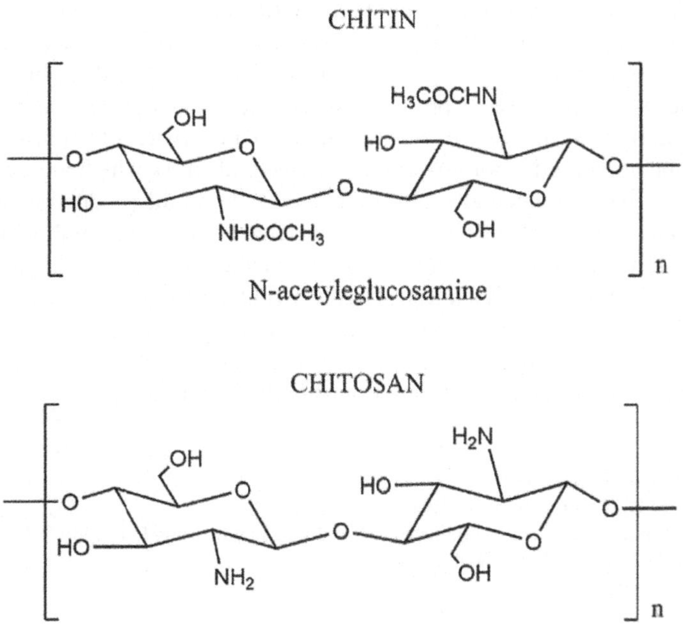

Fig. 2 Molecular structure of chitin and chitosan

2 Chitosan and Its Global Market

The global market for chitin and chitosan has seen phenomenal growth, owing to the expansion of the application sector. The global market for chitin and chitosan derivatives is expected to reach $4.2 billion by 2021 and the market volume is expected to project to be more than 155 thousand metric tons by the year 2022. From 2016 to 2021, the market for chitin and chitosan grew at a compound annual growth rate (CAGR) of 15.4% [7]. Chitosan is the chitin derivative with the highest growth potential among the chitin derivatives. In countries of South East Asia, the demand of biobased products is increased significantly. For example, chitooligosaccharides and glucosamine are considered safe for use in dietary supplements by the Korea Food & Drug Administration (FDA), the Japan Food & Drug Administration (FDA), and the European Food Safety Authority (EFSA). Japan is a major producer and consumer of chitosan, which is used in water treatment, wound dressing, and the production of artificial skin [8]. Meanwhile, in Latin America and the Middle East, the agrochemicals and health-care industries are growing and expanding, resulting in increased chitosan demand. As new value-added products enter the market, the demand for a dependable source of high–quality chitosan is rapidly growing.

3 Different Forms of Chitosan

Chitosan comes in a variety of forms, including powder [9], flakes [10], fibers [11, 12], films/membranes [13], composites [14], depending on the applications. Furthermore, chitosan can be transformed into nanoparticles [15, 16] and nanofibers [17] at nanoscale. Fig. 3 depicts the various types of chitosan and their applications.

3.1 Flakes

Chitosan flakes are more convenient to work with than gelled materials. Iqbal et al. [18] have reported the usage of chitosan flakes for the removal of acid yellow dye. Authors achieved adsorption capacity of 127.0 mg/g dry mass of chitosan [19]. Chitosan in flake for also has been used for the removal of other pollutants such as heavy metals [20] and residual oil from palm oil effluent [21].

3.2 Powder

Many studies have demonstrated the effectiveness of chitosan powder in removing dyes [9], heavy metals [22] and residual oil from oil mill effluent [21]. Chitosan

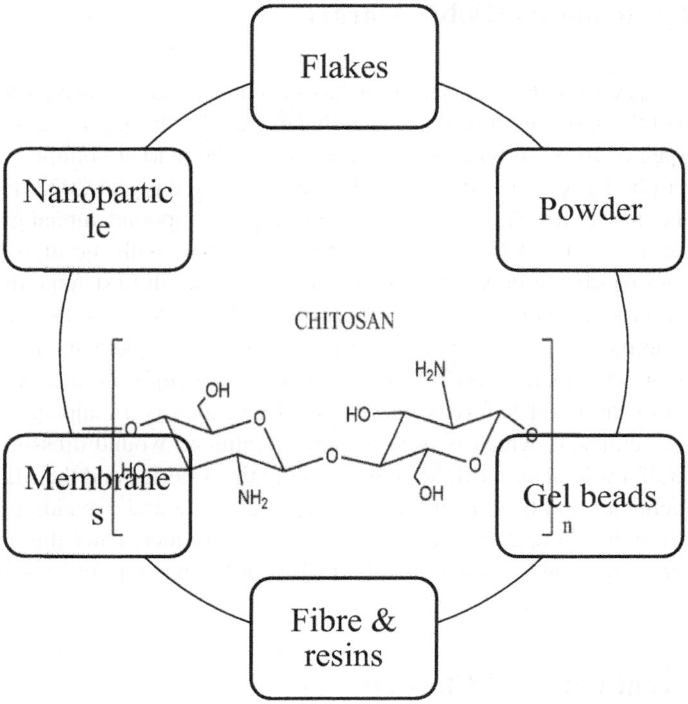

Fig. 3 Different forms of chitosan

powder can also be used as a flocculating and coagulating agent in addition to its adsorptive properties. The use of chitosan is less harmful than other chemicals as it is a green polymer in nature [22]. Jagaba et al. [23] reported that chitosan powder showed better removal of total suspended solids and turbidity from palm oil mill wastewater compared to conventional alum. Moreover, the dosage of chitosan (400 mg/l) used was much lower than conventional alum (4000 mg/l) for 96.4% color removal. Rao [24] investigated the potential and efficiency of chitosan as a coagulant in the treatment of industrial textile wastewater. The work found that chitosan was not only effective in reducing COD by 63–64%, but it also helped to clear the cloudy sample completely.

3.3 *Gel Beads*

Chitosan beads are made in a variety of ways, depending on the structure required, including via emulsion crosslinking, template sacrifice, ion imprinting, extrusion and etc. [25]. Table 1 depicts the various structures of chitosan beads as well as their preparation methods. Ngah and Fatinathan [26] synthesized chitosan beads,

Table 1 Preparation of different form of chitosan beads using different method

Chitosan beads form	Synthesis method	References
Micro-/nano-beads	Emulsion crosslinking, Electrospray, Extrusion with filtration, Ultrasonic-assisted extrusion, Inverse emulsion	[28–31]
Molecularly imprinted beads	immobilization of chitosan beads—template imprinting—cross-linking—removal of the template	[32–34]
Core-shell beads	Complex coacervation, Extrusion crosslinking	[35, 36]
Multi-porous beads	Combination of in situ co-precipitation and sodium citrate cross-linking, Combination of electrospray layer by layer and lyophilization	[37–39]

chitosan crosslink with glutaraldehyde (GTA) and chitosan-alginate beads for the adsorption of Cu(II) in aqueous solution. They compared the adsorption capacity of chitosan beads, chitosan–GTA and chitosan–alginate beads with the maximum capacity obtained was 64.62, 31.20 and 67.66, respectively. In another study done by Muedas-Taipe et al. [27], magnetized chitosan beads were prepared to remove azo dyes from wastewater. It was found that the maximum adsorption capacity for adsorption of were 131.58 and 526.32 mg/g, respectively.

3.4 *Chitosan Fibers and Resins*

Chitosan fibres were first discovered in 1926, but commercial production was prohibitively expensive at the time. Treatment of chitin with alkaline results in highly deacetylated chitin, which is used to make fibres. Fibers can be classified into three types based on their crystalline structure, degree of crystallinity, and average lateral crystallite sizes [6]. Chitosan resins can be utilized for removing pollutants such as metal ions [11] and dyes [12]. Because chitosan can be modified in a variety of ways, cationic groups are added through the quaternization process to create anion exchange resins that can replace current anion exchange resins [11].

3.5 *Membranes*

Chitosan is used as a membrane matrix material for affinity separations because of its high content of amino groups [16, 40, 41]. Electrospinning methods such as multilayering electrospinning, needle-less electrospinning, template-assisted collection, porogen-added electrospinning, and three-fluid electrospinning are the most effective technologies for preparing nanofiber because of its versatility, efficiency, and ease of use [42]. The synthesised chitosan membranes had excellent removal capacities for heavy metals, dyes, and organic pollutants.

3.6 Nanoparticles

A nanoparticle is a particulate dispersion with a size between 1 and 100 nm [43]. Nanoparticles are used in a variety of fields, including medicine, drug delivery, enzyme immobilisation, and so on. [44]. Zhao et al. [43] has described two methods for preparing chitosan nanoparticles: ionic gelation and reverse micellar methods. Ion gelation technique was employed to form nanoparticles through linkages between negatively charged tripolyphosphate and positively charged chitosan. According to de Pinho Neves et al. [45], the best conditions to prepare the particles were at pH 4.4 and the ratio between chitosan and tripolyphosphate was 3:0.8. On the other hand, Orellano et al. [46] synthesized chitosan nanoparticles via reverse micellar approach. The procedure involves the chitosan crosslinking reaction into polar cores of reverse micelles (RMs). In this study, benzyl-n-hexadecyldimethylammonium chloride (BHDC) and sodium 1,4 bis-2-ethylhexylsulfosuccinate (AOT) RMs were used as nanoreactors. The results showed crosslinking reaction takes place more effective in AOT RMs.

4 Surface Area and Particle Size

Porosity, pore volume, and pore size distribution are important chitosan properties in wastewater treatment because they determine the number of accessible sites and the porous structure required for pollutants to attach to [47, 48]. Chitosan powders or flakes are known to be non-porous materials with a low surface area (lower than 10 m^2 g^{-1}) [49]. To increase the surface area of chitosan and improve its potential applications, chemical and physical modifications have been made to it [50, 51]. Phongying et al. [52] derived chitosan from chitin directly and made chitosan nanoscaffolds to increase surface area, particle size, and pore volume. They confirmed that their chitosan scaffolds had a surface area of 55.75 m^2 g^{-1}, which was roughly seven times greater than commercial chitosan flakes (7.70 m^2 g^{-1}). In addition, the chitosan nanoscaffolds had a larger pore volume and pore size. Esquerdo et al. [53] also created chitosan scaffolds and confirmed that the new material had 1135 m^2 g^{-1} specific surface area, 92.2% porosity, and 0.0079 m^3 kg^{-1} pore volume, respectively. These pore volume and surface area sizes are higher when compared to pure chitosan, such as chitosan powders (surface area of 4.2 m^2 g^{-1} and pore volume of $9.5\times10{-}6$ m^3 kg^{-1}), chitosan flakes (surface area range of 4–6 m^2 g^{-1}), chitosan beads (surface area range of 30–40 m^2 g^{-1}), chitosan hydrogel beads (porosity of 85%) [122], and chitosan–graphene mesostructures (surface area of 603.2 m^2 g^{-1}).

Table 2 List of common modifying chemical used for modification of chitosan composite

Modifying chemical	Chitosan composite (form)	Pollutant	Adsorption capacity	References
Epichlorohydrin	Chitosan-Epichlorohydrin (Beads)	Cr	371 mg/g	[56]
		Cu (II)	80.7 mg/g	[57]
Glutaraldehyde	Chitosan-Glutaraldehyde (Beads)	Cr	n.a	[58]
		Cu (II)	31.2 mg/g	[26, 57, 59]
Ethyl acrylate	Chitosan- Ethyl acrylate (Powder)	Basic Blue 41, Basic Red 18	217.4 mg/g, 158.7 mg/g	[9]
Glutaraldehyde/Nylon	Chitosan-Glutaraldehyde/Nylon (Membrane)	Cu (II)	74.7 mg/g	[60]

These studies show that modifying chitosan results in an increase in surface area and, as a result, in porosity and pore volume.

Furthermore, the particle size of the adsorbents has a significant impact on the final solute concentration, and thus on the adsorption process' overall performance. Larger particle sizes have a lower specific surface area, which reduces uptake. As the surface area of the adsorbent increases, new active sites are formed, allowing more solute molecules to bind [54]. Piccin et al. [55] looked into the effects of chitosan particle size, surface area, and pore volume on FD&C Red 40 adsorption. Particle sizes were 0.10, 0.18, and 0.26 mm, with surface areas of 4.2, 3.4, and 1.6 $m^2\ g^{-1}$, respectively. According to the findings, increasing the surface area and decreasing the particle size doubled the adsorption capacity.

5 Modification of Chitosan Composites

The polymers are interesting compounds because they contain specific functional groups. Some biopolymers like chitosan have amino groups and hydroxyl groups which can be altered in variety of ways, both physically or chemically. Chitosan composites have been synthesized by incorporating with other materials. Table 2 lists some of the most common chemicals used to modify chitosan for pollutant removal.

6 Chitosan Magnetite Composites

Magnetic chitosan composites with superparamagnetic properties have been investigated for wastewater treatment as it improves the adsorption through ionic interactions.

According to Zou et al. [61], the oxygen groups of Fe3O4 particles will react with nitrogen group of chitosan and become stabilized on the final structures. It is known that the incorporation of magnetic nanoparticles into the porous polymer matrix can improve the adsorption capacity by means of ionic interactions [62]. Also, magnetization of these particles is quickly changed by an external magnetic field, allowing magnetic adsorbents to be quickly manipulated with magnets and thus easily separated at the end of the process [11, 63–67].

Yang et al. [68] synthesized magnetic chitosan composites by one-single step for the capture of Cr(IV) ions. This composite composed of Fe_3O_4 as cores and chitosan as ion exchange groups, with adsorption capacity of 21.04 mg/g. Yan et al. [69] prepared magnetic chitosan composite containing poly (acrylic acid) (PAA) via simple chemical co-precipitation method for adsorption of Cu(II) from wastewater. The adsorbent had a capacity of 78.0 mg g^{-1} and be regenerated up to six cycles. Mi et al. [70] reported the preparation of magnetic chitosan beads via incorporating N,O-carboxymethyl chitosan-coated magnetic nanoparticles (NOCC-MNPs) with chitosan-citrate gel beads (CCGBs) for removal Cu(II) from wastewater. The combined chelation effects of the electron-donating functional groups in the CCGBs and NOCC-MNPs increased Cu(II) adsorption capacity (35.98 mg g^{-1}) and the magnetized adsorbent made regeneration easy when magnetic field is applied [71].

6.1 Chitosan–Hydroxyapatite Composites

Chitosan and hydroxyapatite (HAp) are reported to be prospective biomaterials for adsorption application. However, they cannot be used directly in acidic medium because of their low chemical stability properties. Park et al. [72] prepared HAp/chitosan composites by a co-precipitation method to improve its mechanical strength and used for adsorption of heavy metals (Pb^{2+} and Cd^{2+}). Adsorption kinetic follows pseudo-second-order kinetic model which is via chemisorption and adsorption isotherm fits to Langmuir model. Authors also compared the adsorption performance between HAp/chitosan composites and chitosan where HAp/chitosan composites showed higher efficiency in removing the respective pollutants than unmodified chitosan [72].

Pereira et al. [73] investigated the potential of amino hydroxyapatite /chitosan hybrids on the removal of diclofenac sodium (DS). The composite was noted to have fast adsorption rate with the adsorption capacity of 125 mg g^{-1} at 15 min. High adsorption capacity in a short time demonstrated that the amino hydroxyapatite /chitosan is a promising biosorbents for the treatment of DS-bearing wastewater [73].

6.2 Chitosan–Montmorillonite Composites

Chitosan-montmorillonite composites have been prepared by the replacement of sodium ions in montmorillonite layers with biopolymeric chitosan for the removal of dyes [74]. The interaction between these two materials can be explained by the electrostatic interaction between the cationic charges of chitosan and the anionic charges of montmorillonite [74]. In this study, sorption performance of the adsorbent was compared with different type of dyes which were cationic, anionic and disperse. Results showed that adsorbent exhibited better adsorption performance for cationic dye (99.3%), followed by anionic dye (67.4%) and disperse dye (68.6%).

A study conducted by Nesic et al. [75] investigated the preparation of chitosan-montmorillonite in a form of membrane for the removal of Bezactiv Orange V–3R. With increasing amounts of montmorillonite (MMT) from 10 to 50% in the membranes, the adsorption capacity of chitosan-montmorillonite membranes increases from 106.8 mg g^{-1} to 740.7 mg g^{-1}, respectively. Since both chitosan and MMT are widely available, these membranes are considerably inexpensive and can be prepared on-site, wherever they are required [75].

6.3 Chitosan–multi-Wall Carbon Nanotube Composites

Chitosan-carbon nanotube (CNT) composites was produced by covalently grafting chitosan with CNT. Guo et al. [76] synthesized chitosan/multi-walled carbon nanotube (CS/MWCNT) for adsorption of phenol. The adsorption capacity of the novel CS/MWCNT for phenol (86.96 mg/g) was higher than the unmodified chitosan (61.69 mg/g) [76]. Another study investigated by Chatterjee et al. [77] on the adsorption performance of chitosan (CS) hydrogel beads incorporated with MWCNT for the removal of congo red (CR). The maximum adsorption capacity obtained was 450 mg g^{-1} and the adsorption data well fitted to Langmuir model.

Also, the adsorption of methyl orange was tested on a composite of multi-walled carbon nanotubes (MWCNT) modified with chitosan (CS) and poly-2–hydroxyethylmethacrylate (pHEMA) [78]. The modified pHEMA–CS-f-MWCNT composite had a better mechanical strength properties with high adsorption capacity of 416.7 mg g − 1 on the removal of methyl orange [78].

6.4 Chitosan/Polyurethane Composites

The synthesis of composites based on chitosan and polyurethane was reported by Li et al. [79] in the removal of malachite green from aqueous solution. Polyurethane has been used in many applications such as insulator in walls and roofs, flexible foam in upholster furniture, medical devices and foot wears [80]. Previous studies have been reported that polyurethane was used to immobilize various kinds of adsorbents such as activated carbon, zeolites and biomass [81, 82] by providing high surface area and open porous structure. It was noted that chitosan/polyurethane composites possess higher adsorption capacity than neat polyurethane in the adsorption of acid dyes [83]. Low adsorption performance by neat polyurethane indicated that the amine groups in its structure might not really reactive to serve as an active site for dye molecules.

6.5 Chitosan/Bentonite Composites

Bentonite contains a high proportion of swelling clays that composed of montmorillonite which has potential to be used in wastewater treatment application [84, 85]. Nonetheless, the characteristic of pure bentonite such as low dispersion and adsorption capacity has led to the modification of bentonite with other materials such as polyethylene glycol, polystyrene, polysiloxane, carboxymethyl cellulose, chitin, and chitosan [86–89]. Normally, the methods used for modification of bentonite are by intercalation polymerization and polymer intercalation either using melt method or solution method [90].

Yang et al. [91] synthesized chitosan/bentonite composite and used it for removal of Cr(VI) in a batch system. It was reported that the adsorption performance of the chitosan/bentonite composite has been reported to be superior to either chitosan or bentonite alone. Meanwhile, Ngah et al. [92] prepared crosslinked chitosan/bentonite composites with the addition of a crosslinker (epichlorohydrin) to adsorb tartrazine. The crosslinker was used to form a bridge that connect between chitosan and bentonite, thus become more stable including in acidic medium. The sythesised chitosan/bentonite composites was reported had better adsorption capacity than unmodified chitosan [93].

6.6 Chitosan/Zeolite Composites

Zeolites has been used widely in wastewater treatment. Clinoptilolite is the most common and inexpensive zeolite mineral, consisting of a microporous arrangement of alumina and silica with the chemical formula $(Na, K, Ca)_4Al_6Si_{30}O_{72}.24H_2O$ [94]. It is well known that zeolite that has been modified with various functional groups has a higher adsorption capacity.

Yang et al. [94] prepared chitosan modified zeolite composite to remove U(VI) in a batch process. The authors found that chitosan modified zeolite showed enhanced adsorption capacity for U(VI) and faster adsorption kinetics than unmodified zeolite. The functional groups from respective materials, $-NH_2$, $NH_3{}^+$ (chitosan), and SiO (zeolite) act as binding sites for U(VI) sorption. Chitosan was used as a support for clinoptilolite zeolites to create a chitosan/zeolite (CZ) composite film [95]. The use

Table 3 The use of chitosan/alumina composites for removal of different type of pollutants in wastewater treatment applications

Pollutant	Adsorption performance	Kinetics	Isotherm	Thermodynamic	References
Methyl orange	35.3 mg/g, 35 °C	Pseudo second-order	Langmuir	Exothermic	[102]
As(III)	56.5 mg/g	n.a	Freundlich	n.a	[100]
Cr(VI)	10.0 mg/g, 50 °C	Pseudo second-order	Dubinin–Radushkevich	Endothermic	[99]
Cu(II)	315.46 mg/g, pH 6	n.a	Langmuir	n.a	[103]
Ni(II)	78.1 mg/g	n.a	Langmuir	n.a	[101]

of chitosan allows for the creation of films with high zeolite content, resulting in enhanced adsorption properties.

6.7 *Chitosan/Alumina Composites*

The use of alumina as an adsorbent has been widely well known [96, 97] due to its amphoteric character properties. Depending on the pH, their acid–base dissociation results in positive ($-OH_2^+$) or negative ($-O$) charges on the surface [98]. Gandhi et al. [99] used alumina to create a new chitosan-based composite for chromium removal from aqueous solutions. With a short adsorption reaction time of 30 min, the synthesised alumina/chitosan composite has an increased chromium a sorption capacity from 3.7 mg g^{-1} (by alumina) and 0.67 mg/g (by chitosan) to 8.62 mg/g. Other studies have been done using chitosan/alumina composites in wastewater treatment applications for removal of methyl orange, As(III), As(V), Cr(VI), Cu(II) and Ni(II), as presented in Table 3.

7 Conclusion

Chitosan is considered to be one of the most promising and applicable materials in adsorption applications owing to its characteristics of abundance, renewable nature, bio–degradability, versatility and ease of structural modification. Nonetheless, their massive applications are hindered due to some limitations such as low adsorption capacity and unstable in acidic medium. Therefore, to improve the adsorption properties of chitosan, surface chemistry modification has been employed. The modification of chitosan can be done by the addition modifying agents or incorporation with other polymers. Findings revealed that the adsorption capacity of chitosan composites has been enhanced when compared to pure chitosan. Nonetheless, the use of chitosan composites in wastewater treatment is currently restricted to laboratory studies. Chitosan has a lot of potential as a water treatment material and its commercialization is seen as the next big step forward.

References

1. Nwe, N., Furuike, T., & Tamura, H.(2011). Chitosan from aquatic and terrestrial organisms and microorganisms: production, properties and applications. Biodegradable materials: production, properties and applications (pp. 29–50). Hauppauge, NY: Nova Science.
2. Singla, A., & Chawla, M. (2001). Chitosan: some pharmaceutical and biological aspects-an update. *Journal of Pharmacy and Pharmacology, 53*(8), 1047–1067.
3. Kamal, T., Ul-Islam, M., Khan, S. B., & Asiri, A. M. (2015). Adsorption and photocatalyst assisted dye removal and bactericidal performance of ZnO/chitosan coating layer. *International Journal of Biological Macromolecules, 81*, 584–590.
4. Khan, S. B., Alamry, K. A., Bifari, E. N., Asiri, A. M., Yasir, M., Gzara, L., & Ahmad, R. Z. (2015). Assessment of antibacterial cellulose nanocomposites for water permeability and salt rejection. *Journal of Industrial and Engineering Chemistry, 24*, 266–275.
5. Dutta, P., Tripathi, S., Mehrotra, G., & Dutta, J. (2009). Perspectives for chitosan based antimicrobial films in food applications. *Food Chemistry, 114*(4), 1173–1182.
6. Pillai, C. K., Paul, W., & Sharma, C. P. (2009). Chitin and chitosan polymers: chemistry, solubility and fiber formation. *Progress in Polymer Science, 34*(7), 641–678.
7. Jardine, A., & Sayed, S. (2018). Valorisation of chitinous biomass for antimicrobial applications. *Pure and Applied Chemistry, 90*(2), 293–304.
8. Pal, P., Pal, A., Nakashima, K., & Yadav, B. K. (2021). Applications of chitosan in environmental remediation: A review. *Chemosphere, 266*, 128934.
9. Sadeghi-Kiakhani, M., Arami, M., & Gharanjig, K. (2013). Preparation of chitosan-ethyl acrylate as a biopolymer adsorbent for basic dyes removal from colored solutions. *Journal of Environmental Chemical Engineering, 1*(3), 406–415.
10. Adarsh, K., & Madhu, G. (2014). A comparative study on metal adsorption properties of different forms of chitosan. *International Journal Of Innovation Research In Science. Engineering and Technology, 3*, 9609–9617.
11. Elwakeel, K. Z. (2010). Removal of Cr (VI) from alkaline aqueous solutions using chemically modified magnetic chitosan resins. *Desalination, 250*(1), 105–112.
12. Ilayaraja, M., Sharmilaparveen, S., & Sayeekannan, R. (2014). Synthesis and adsorption properties of chitosan cross linked with phenol–formaldehyde resin for the removal of heavy metals and dyes from water. *Applied Chemistry, 7*, 16–26.
13. Chen, C.-C., & Chung, Y.-C. (2006). Arsenic removal using a biopolymer chitosan sorbent. *Journal of Environmental Science and Health, Part A, 41*(4), 645–658.
14. Ngah, W. W., Ab Ghani, S., & Kamari, A. (2005). Adsorption behaviour of Fe (II) and Fe (III) ions in aqueous solution on chitosan and cross-linked chitosan beads. *Bioresource Technology, 96*(4), 443–450.
15. Darras, V., Nelea, M., Winnik, F. M., & Buschmann, M. D. (2010). Chitosan modified with gadolinium diethylenetriaminepentaacetic acid for magnetic resonance imaging of DNA/chitosan nanoparticles. *Carbohydrate Polymers, 80*(4), 1137–1146.
16. Dehaghi, S. M., Rahmanifar, B., Moradi, A. M., & Azar, P. A. (2014). Removal of permethrin pesticide from water by chitosan–zinc oxide nanoparticles composite as an adsorbent. *Journal of Saudi Chemical Society, 18*(4), 348–355.
17. Qian, L., & Zhang, H. (2010). Green synthesis of chitosan-based nanofibers and their applications. *Green Chemistry, 12*(7), 1207–1214.

18. Iqbal, J., Wattoo, F. H., Wattoo, M. H. S., Malik, R., Tirmizi, S. A., Imran, M., & Ghangro, A. B. (2011). Adsorption of acid yellow dye on flakes of chitosan prepared from fishery wastes. *Arabian Journal of Chemistry, 4*(4), 389–395.
19. Verbych, S., Bryk, M., Chornokur, G., & Fuhr, B. (2005). Removal of copper (II) from aqueous solutions by chitosan adsorption. *Separation Science and Technology, 40*(8), 1749–1759.
20. Aydın, Y. A., & Aksoy, N. D. (2009). Adsorption of chromium on chitosan: Optimization, kinetics and thermodynamics. *Chemical Engineering Journal, 151*(1–3), 188–194.
21. Ahmad, A., Sumathi, S., & Hameed, B. (2005). Adsorption of residue oil from palm oil mill effluent using powder and flake chitosan: Equilibrium and kinetic studies. *Water Research, 39*(12), 2483–2494.
22. Hassan, M. A., Li, T. P., & Noor, Z. Z. (2009). Coagulation and flocculation treatment of wastewater in textile industry using chitosan. *Journal of Chemical and Natural Resources Engineering, 4*(1), 43–53.
23. Jagaba, A., Kutty, S., Hayder, G., Latiff, A., Aziz, N., Umaru, I., Ghaleb, A., Abubakar, S., Lawal, I., & Nasara, M. (2020). Sustainable use of natural and chemical coagulants for contaminants removal from palm oil mill effluent: A comparative analysis. *Ain Shams Engineering Journal, 11*(4), 951–960.
24. Rao, L. N. (2015). Coagulation and flocculation of industrial wastewater by chitosan. *International Journal of Engineering and Applied Sciences, 2*(7), 257870.
25. Qu, B., & Luo, Y. (2020). Chitosan-based hydrogel beads: Preparations, modifications and applications in food and agriculture sectors–A review. *International Journal of Biological Macromolecules, 152*, 437–448.
26. Ngah, W. W., & Fatinathan, S. (2008). Adsorption of Cu (II) ions in aqueous solution using chitosan beads, chitosan–GLA beads and chitosan–alginate beads. *Chemical Engineering Journal, 143*(1–3), 62–72.
27. Muedas-Taipe, G., Mejía, I. M. M., Santillán, F. A., Velásquez, C. J., & Asencios, Y. J. (2020). Removal of azo dyes in aqueous solutions using magnetized and chemically modified chitosan beads. *Materials Chemistry and Physics, 256*, 123595.
28. Trifković, K. T., Milašinović, N. Z., Djordjević, V. B., Krušić, M. T. K., Knežević-Jugović, Z. D., Nedović, V. A., & Bugarski, B. M. (2014). Chitosan microbeads for encapsulation of thyme (Thymus serpyllum L.) polyphenols. *Carbohydrate Polymers, 111*, 901–907
29. Gómez-Mascaraque, L. G., Sanchez, G., & López-Rubio, A. (2016). Impact of molecular weight on the formation of electrosprayed chitosan microcapsules as delivery vehicles for bioactive compounds. *Carbohydrate Polymers, 150*, 121–130.
30. Milašinović, N., Čalija, B., Vidović, B., Sakač, M. C., Vujić, Z., & Knežević-Jugović, Z. (2016). Sustained release of α-lipoic acid from chitosan microbeads synthetized by inverse emulsion method. *Journal of the Taiwan Institute of Chemical Engineers, 60*, 106–112.
31. Oladipo, A. A., & Gazi, M. (2017). Targeted boron removal from highly-saline and boron-spiked seawater using magnetic nanobeads: Chemometric optimisation and modelling studies. *Chemical Engineering Research and Design, 121*, 329–338.
32. Xu, L., Huang, Y.-A., Zhu, Q.-J., & Ye, C. (2015). Chitosan in molecularly-imprinted polymers: Current and future prospects. *International Journal of Molecular Sciences, 16*(8), 18328–18347.
33. Zhang, M., Zhang, Y., & Helleur, R. (2015). Selective adsorption of Ag+ by ion-imprinted O-carboxymethyl chitosan beads grafted with thiourea–glutaraldehyde. *Chemical Engineering Journal, 264*, 56–65.
34. Younis, M. R., Bajwa, S. Z., Lieberzeit, P. A., Khan, W. S., Mujahid, A., Ihsan, A., & Rehman, A. (2016). Molecularly imprinted porous beads for the selective removal of copper ions. *Journal of Separation Science, 39*(4), 793–798.
35. Qin, C., Zhou, J., Zhang, Z., Chen, W., Hu, Q., & Wang, Y. (2019). Convenient one-step approach based on stimuli-responsive sol-gel transition properties to directly build chitosan-alginate core-shell beads. *Food Hydrocolloids, 87*, 253–259.

36. Song, M.-H., Kim, S., Reddy, D. H. K., Wei, W., Bediako, J. K., Park, S., & Yun, Y.-S. (2017). Development of polyethyleneimine-loaded core-shell chitosan hollow beads and their application for platinum recovery in sequential metal scavenging fill-and-draw process. *Journal of Hazardous Materials, 324*, 724–731.
37. Wang, K., Ma, H., Pu, S., Yan, C., Wang, M., Yu, J., Wang, X., Chu, W., & Zinchenko, A. (2019). Hybrid porous magnetic bentonite-chitosan beads for selective removal of radioactive cesium in water. *Journal of Hazardous Materials, 362*, 160–169.
38. Choi, S.-W., Zhang, Y., Yeh, Y.-C., Wooten, A. L., & Xia, Y. (2012). Biodegradable porous beads and their potential applications in regenerative medicine. *Journal of Materials Chemistry, 22*(23), 11442–11451.
39. Pu, S., Ma, H., Zinchenko, A., & Chu, W. (2017). Novel highly porous magnetic hydrogel beads composed of chitosan and sodium citrate: An effective adsorbent for the removal of heavy metals from aqueous solutions. *Environmental Science and Pollution Research, 24*(19), 16520–16530.
40. de Moraes, M. A., Cocenza, D. S., da Cruz Vasconcellos, F., Fraceto, L. F., & Beppu, M. M. (2013). Chitosan and alginate biopolymer membranes for remediation of contaminated water with herbicides. *Journal of Environmental Management, 131*, 222–227.
41. der Merwe, V., & Christoffel, H. (2006). Chitosan membranes for the removal of zinc from simulated wastewater. North-West University.
42. Li, C., Lou, T., Yan, X., Long, Y.-Z., Cui, G., & Wang, X. (2018). Fabrication of pure chitosan nanofibrous membranes as effective absorbent for dye removal. *International Journal of Biological Macromolecules, 106*, 768–774.
43. Zhao, L.-M., Shi, L.-E., Zhang, Z.-L., Chen, J.-M., Shi, D.-D., Yang, J., & Tang, Z.-X. (2011). Preparation and application of chitosan nanoparticles and nanofibers. *Brazilian Journal of Chemical Engineering, 28*, 353–362.
44. Zhang, H., Tachikawa, H., Gao, X.-D., & Nakanishi, H. (2014). Applied usage of yeast spores as chitosan beads. *Applied and Environmental Microbiology, 80*(16), 5098–5105.
45. de Pinho Neves, A. L., Milioli, C. C., Müller, L., Riella, H. G., Kuhnen, N. C., & Stulzer, H. K. (2014). Factorial design as tool in chitosan nanoparticles development by ionic gelation technique. *Colloids and Surfaces A: Physicochemical and Engineering Aspects, 445*, 34–39.
46. Orellano, M. S., Longo, G. S., Porporatto, C., Correa, N. M., & Falcone, R. D. (2020). Role of micellar interface in the synthesis of chitosan nanoparticles formulated by reverse micellar method. *Colloids and Surfaces A: Physicochemical and Engineering Aspects, 599*, 124876.
47. Vieira, M. L., Pinheiro, C. P., Silva, K. A., Lutke, S. F., Cadaval, T. R. S. A., Dotto, G., & Pinto, L. A. D. A. (2019). Chitosan and cyanoguanidine-crosslinked chitosan coated glass beads and its application in fixed bed adsorption. *Chemical Engineering Communications, 206*(11), 1474–1486.
48. Jayasuriya, A. (2017). Production of micro-and nanoscale chitosan particles for biomedical applications, in Chitosan Based Biomaterials (vol. 1, pp. 185–209). Elsevier.
49. de Almeida Pinto, L. A., Dotto, G. L., & Campana-Filho, S. P. (2017). Frontiers in biomaterials: Chitosan based materials and Its applications (Vol. 3). Bentham Science Publishers.
50. da Silva Alves, D. C., Healy, B., Pinto, L. A., Cadaval, T. R., & Breslin, C. B. (2021) Recent developments in chitosan-based adsorbents for the removal of pollutants from aqueous environments. *Molecules, 26*(3), 594.
51. Jiménez-Gómez, C. P., & Cecilia, J. A. (2020). Chitosan: A Natural biopolymer with a wide and varied range of applications. *Molecules, 25*(17), 3981.
52. Phongying, S., Aiba, S.-I., & Chirachanchai, S. (2007). Direct chitosan nanoscaffold formation via chitin whiskers. *Polymer, 48*(1), 393–400.
53. Esquerdo, V. M., Cadaval, T., Jr., Dotto, G., & Pinto, L. (2014). Chitosan scaffold as an alternative adsorbent for the removal of hazardous food dyes from aqueous solutions. *Journal of Colloid and Interface Science, 424*, 7–15.
54. Karimi, S., Yaraki, M. T., & Karri, R. R. (2019). A comprehensive review of the adsorption mechanisms and factors influencing the adsorption process from the perspective of bioethanol dehydration. *Renewable and Sustainable Energy Reviews, 107*, 535–553.

55. Piccin, J. S., Dotto, G. L., Vieira, M. L., & Pinto, L. A. (2011). Kinetics and mechanism of the food dye FD&C Red 40 adsorption onto chitosan. *Journal of Chemical & Engineering Data, 56*(10), 3759–3765.
56. Jaros, K., Kaminski, W., Albinska, J., & Nowak, U. (2005). Removal of heavy metal ions: Copper, zinc and chromium from water on chitosan beads. *Environment Protection Engineering, 31*(3/4), 153.
57. Ngah, W. W., Endud, C., & Mayanar, R. (2002). Removal of copper (II) ions from aqueous solution onto chitosan and cross-linked chitosan beads. *Reactive and Functional Polymers, 50*(2), 181–190.
58. Sugashini, S. and Gopalakrishnan, S. (2012). Studies on the performance of protonated cross linked chitosan beads (PCCB) for chromium removal. *Research Journal of Chemical Sciences*. ISSN. 2012; 2231: 606X.
59. Patrulea, V., Negrulescu, A., Mincea, M. M., Pitulice, L. D., Spiridon, O. B., & Ostafe, V. (2013). Optimization of the removal of copper (II) ions from aqueous solution on chitosan and cross-linked chitosan beads. *BioResources, 8*(1), 1147–1165.
60. Prakash, N., Sudha, P., & Renganathan, N. (2012). Copper and cadmium removal from synthetic industrial wastewater using chitosan and nylon 6. *Environmental Science and Pollution Research, 19*(7), 2930–2941.
61. Zou, W., Geng, H., Lin, M., & Xiong, X. (2012). Facile one-pot preparation of superparamagnetic chitosan sphere and its derived hollow sphere. *Journal of Applied Polymer Science, 123*(6), 3587–3594.
62. Luo, X., & Zhang, L. (2009). High effective adsorption of organic dyes on magnetic cellulose beads entrapping activated carbon. *Journal of Hazardous Materials, 171*(1–3), 340–347.
63. Elwakeel, K. Z., Atia, A. A., & Donia, A. M. (2009). Removal of Mo (VI) as oxoanions from aqueous solutions using chemically modified magnetic chitosan resins. *Hydrometallurgy, 97*(1–2), 21–28.
64. Hritcu, D., Dodi, G., & Popa, M. I. (2012). Heavy metal ions adsorption on chitosan-magnetite microspheres. *International Review of Chemical Engineering, 4*(3), 364–368.
65. Liu, X., Hu, Q., Fang, Z., Zhang, X., & Zhang, B. (2009). Magnetic chitosan nanocomposites: A useful recyclable tool for heavy metal ion removal. *Langmuir, 25*(1), 3–8.
66. Yuwei, C., & Jianlong, W. (2011). Preparation and characterization of magnetic chitosan nanoparticles and its application for Cu (II) removal. *Chemical Engineering Journal, 168*(1), 286–292.
67. Ren, Y., Zhang, M., & Zhao, D. (2008). Synthesis and properties of magnetic Cu (II) ion imprinted composite adsorbent for selective removal of copper. *Desalination, 228*(1–3), 135–149.
68. Yang, W.-C., Tang, Q.-Z., Dong, S.-Y., Chai, L.-Y., & Wang, H.-Y. (2016). Single-step synthesis of magnetic chitosan composites and application for chromate (Cr (VI)) removal. *Journal of Central South University, 23*(2), 317–323.
69. Yan, H., Yang, L., Yang, Z., Yang, H., Li, A., & Cheng, R. (2012). Preparation of chitosan/poly (acrylic acid) magnetic composite microspheres and applications in the removal of copper (II) ions from aqueous solutions. *Journal of Hazardous Materials, 229*, 371–380.
70. Mi, F.-L., Wu, S.-J., & Chen, Y.-C. (2015). Combination of carboxymethyl chitosan-coated magnetic nanoparticles and chitosan-citrate complex gel beads as a novel magnetic adsorbent. *Carbohydrate Polymers, 131*, 255–263.
71. Namdeo, M., & Bajpai, S. (2008). Chitosan–magnetite nanocomposites (CMNs) as magnetic carrier particles for removal of Fe (III) from aqueous solutions. *Colloids and Surfaces A: Physicochemical and Engineering Aspects, 320*(1–3), 161–168.
72. Park, S., Gomez-Flores, A., Chung, Y. S., & Kim, H. (2015). Removal of cadmium and lead from aqueous solution by hydroxyapatite/chitosan hybrid fibrous sorbent: kinetics and equilibrium studies. *Journal of Chemistry*.
73. Pereira, M., França, D., Araújo, R. C., Silva Filho, E. C., Rigaud, B., Fonseca, M., & Jaber, M. (2020). Amino hydroxyapatite/chitosan hybrids reticulated with glutaraldehyde at different pH values and their use for diclofenac removal. *Carbohydrate Polymers, 236*, 116036.

74. Lertsutthiwong, P., Noomun, K., Khunthon, S., & Limpanart, S. (2012). Influence of chitosan characteristics on the properties of biopolymeric chitosan–montmorillonite. *Progress in Natural Science: Materials International, 22*(5), 502–508.
75. Nesic, A. R., Velickovic, S. J., & Antonovic, D. G. (2012). Characterization of chitosan/montmorillonite membranes as adsorbents for Bezactiv Orange V-3R dye. *Journal of Hazardous Materials, 209*, 256–263.
76. Guo, M., Wang, J., Wang, C., Strong, P., Jiang, P., Ok, Y. S., & Wang, H. (2019). Carbon nanotube-grafted chitosan and its adsorption capacity for phenol in aqueous solution. *Science of the Total Environment, 682*, 340–347.
77. Chatterjee, S., Lee, M. W., & Woo, S. H. (2010). Adsorption of congo red by chitosan hydrogel beads impregnated with carbon nanotubes. *Bioresource Technology, 101*(6), 1800–1806.
78. Mahmoodian, H., Moradi, O., Shariatzadeha, B., Salehf, T. A., Tyagi, I., Maity, A., Asif, M., & Gupta, V. K. (2015). Enhanced removal of methyl orange from aqueous solutions by poly HEMA–chitosan-MWCNT nano-composite. *Journal of Molecular Liquids, 202*, 189–198.
79. Li, X., Li, J., Cai, L., Li, T., Liu, X., & Li, J. (2016). Malachite green adsorption behavior of polyurethane/chitosan composite foam. *Cellular Polymers, 35*(1), 1–18.
80. Zia, K. M., Bhatti, H. N., & Bhatti, I. A. (2007). Methods for polyurethane and polyurethane composites, recycling and recovery: A review. *Reactive and Functional Polymers, 67*(8), 675–692.
81. Jang, S. H., Min, B. G., Jeong, Y. G., Lyoo, W. S., & Lee, S. C. (2008). Removal of lead ions in aqueous solution by hydroxyapatite/polyurethane composite foams. *Journal of Hazardous Materials, 152*(3), 1285–1292.
82. Pinto, M. L., Pires, J., Carvalho, A. P., de Carvalho, M. B., & Bordado, J. C. (2006). Synthesis and regeneration of polyurethane/adsorbent composites and their characterization by adsorption methods. *Microporous and Mesoporous Materials, 89*(1–3), 260–269.
83. Ngah, W. W., Teong, L., & Hanafiah, M. M. (2011). Adsorption of dyes and heavy metal ions by chitosan composites: A review. *Carbohydrate Polymers, 83*(4), 1446–1456.
84. Holzer, L., Münch, B., Rizzi, M., Wepf, R., Marschall, P., & Graule, T. (2010). 3D-microstructure analysis of hydrated bentonite with cryo-stabilized pore water. *Applied Clay Science, 47*(3–4), 330–342.
85. Li, Q., Yue, Q.-Y., Sun, H.-J., Su, Y., & Gao, B.-Y. (2010). A comparative study on the properties, mechanisms and process designs for the adsorption of non-ionic or anionic dyes onto cationic-polymer/bentonite. *Journal of Environmental Management, 91*(7), 1601–1611.
86. Seilkhanova, G. A., Imangaliyeva, A. N., Mastai, Y., & Rakhym, A. B. (2019). Bentonite polymer composite for water purification. *Bulletin of Materials Science, 42*(2), 1–8.
87. Alekseeva, O. V., Rodionova, A. N., Bagrovskaya, N. A., Noskov, A. V., & Agafonov, A. V. (2019). Bentonite filler effect on structure and properties of polystyrene-based composites. *Iranian Polymer Journal, 28*(2), 123–133.
88. Leshaf, A., Cherif, H. Z., & Benmansour, K. (2019). Adsorption of Acidol Red 2BE-NW Dye from aqueous solutions on carboxymethyl cellulose/organo-bentonite composite: Characterization, kinetic and thermodynamic studies. *Journal of Polymers and the Environment, 27*(5), 1054–1064.
89. Mukhopadhyay, R., Bhaduri, D., Sarkar, B., Rusmin, R., Hou, D., Khanam, R., Sarkar, S., Biswas, J. K., Vithanage, M., & Bhatnagar, A. (2020). Clay–polymer nanocomposites: Progress and challenges for use in sustainable water treatment. *Journal of Hazardous Materials, 383*, 121125.
90. Shi, X., & Gan, Z. (2007). Preparation and characterization of poly (propylene carbonate)/montmorillonite nanocomposites by solution intercalation. *European Polymer Journal, 43*(12), 4852–4858.
91. Yang, J., Huang, B., & Lin, M. (2020). Adsorption of hexavalent chromium from aqueous solution by a chitosan/bentonite composite: Isotherm, kinetics, and thermodynamics studies. *Journal of Chemical & Engineering Data, 65*(5), 2751–2763.
92. Ngah, W. S. W., Ariff, N. F. M., & Hanafiah, M. A. K. M. (2010). Preparation, characterization, and environmental application of crosslinked chitosan-coated bentonite for tartrazine adsorption from aqueous solutions. *Water, Air, and Soil Pollution, 206*(1), 225–236.

93. Crini, G., & Badot, P.-M. (2008). Application of chitosan, a natural aminopolysaccharide, for dye removal from aqueous solutions by adsorption processes using batch studies: A review of recent literature. *Progress in Polymer Science, 33*(4), 399–447.
94. Yang, T., Zhang, W., Liu, H., & Guo, Y. (2020). Enhanced removal of U (VI) from aqueous solution by chitosan-modified zeolite. *Journal of Radioanalytical and Nuclear Chemistry, 323*(2), 1003–1012.
95. do Nascimento Sousa, S. D., Santiago, R. G., Maia, D. A. S., de Oliveira Silva, E., Vieira, R. S., & Bastos-Neto, M. (2020). Ethylene adsorption on chitosan/zeolite composite films for packaging applications. *Food Packaging and Shelf Life, 26*,100584.
96. Ahmed, W., Rego, C., Cherry, R., Afzal, A., Ali, N., & Hassan, I. (2000). CVD diamond: Controlling structure and morphology. *Vacuum, 56*(3), 153–158.
97. Khattak, A., Afzal, M., Saleem, M., Yasmeen, G., & Ahmad, R. (2000). Surface modification of alumina by metal doping. *Colloids and Surfaces A: Physicochemical and Engineering Aspects, 162*(1–3), 99–106.
98. Szekeres, M., Tombácz, E., Ferencz, K., & Dékány, I. (1998). Adsorption of salicylate on alumina surfaces. *Colloids and Surfaces A: Physicochemical and Engineering Aspects, 141*(3), 319–325.
99. Gandhi, M. R., Viswanathan, N., & Meenakshi, S. (2010). Preparation and application of alumina/chitosan biocomposite. *International Journal of Biological Macromolecules, 47*(2), 146–154.
100. Boddu, V. M., Abburi, K., Talbott, J. L., Smith, E. D., & Haasch, R. (2008). Removal of arsenic (III) and arsenic (V) from aqueous medium using chitosan-coated biosorbent. *Water Research, 42*(3), 633–642.
101. Boddu, V. M., Abburi, K., Randolph, A. J., & Smith, E. D. (2008). Removal of copper (II) and nickel (II) ions from aqueous solutions by a composite chitosan biosorbent. *Separation Science and Technology., 43*(6), 1365–1381.
102. Zhang, J., Zhou, Q., & Ou, L. (2012). Kinetic, isotherm, and thermodynamic studies of the adsorption of methyl orange from aqueous solution by chitosan/alumina composite. *Journal of Chemical & Engineering Data, 57*(2), 412–419.
103. Guijuan, J., Weiwei, B., Guimei, G., Baichao, A., Haifeng, Z., & Shucai, G. (2012). Removal of Cu (II) from aqueous solution using a novel crosslinked alumina-chitosan hybrid adsorbent. *Chinese Journal of Chemical Engineering, 20*(4), 641–648.

Development of Nipah Palm Fibre Extraction Process as Reinforcing Agent in Unsaturated Polyester Composite

Syed Tarmizi Syed Shazali, Tracy Dickie, and Noor Hisyam Noor Mohamed

1 Introduction

Nipah palm or scientifically known as *Nyipa fruticans* is a palm that grows in an aquatic environment as shown in Fig. 1. Nipah is the most abundant palm in the mangrove forest of the South, Southeast Asia and the Oceania. Nipah palm grows in the mangrove forest, it is also known as mangrove palm. The growth area of nipah palm is where the fresh water meets the salt water of the mangrove forest. The muddy and brackish water of this area are the natural habitat of nipah palm. Nipah palm are rarely seen on the seashores, but more to the estuarine area. This palm is trunkless and has its fronds grow from the roots. The trunk submerged in the water during high tide and exposed to the air during low tide. The submerged fronds are bulky and spongy compared to the upper fronds. The spongy nature of the lower fronds is assumed to provide the palm the buoyancy during high tide. The upper frond is rigid and study, supporting the towering. They can grow up to 10 m tall [1, 2] (Fig. 1).

Historically, this palm provides a wide diversity of use towards the indigenous people living near the forest [1]. Nipah palm is abundant along the riverbanks of Sarawak River. It is locally known as *pokok apong* by the Sarawakians. It is well-known for its sugar sap; it has been collected by the population nearby and has been used as sweetener in local delicacies known as *gula apong* by locals as shown in Fig. 3a. This sugar sap is also the source of income for them. Besides that, the dried and fresh young leaves of nipah palm are traditionally used to make baskets, roof thatches, food wrappers, cigarette wrapper (Fig. 3b) and many more.

There are two forms of nipah found in Malaysia, *nipah gala* and *nipah padi*. Nipah palm found in Malaysia and Papua New Guinea is bigger than the one found in Philippines. Nipah palm largest population is found in Indonesia (700,000 ha), Papua New Guinea (500,000 ha) and Philippines (8,000 ha) [4]. Mangrove forest

S. T. Syed Shazali (✉) · T. Dickie · N. H. Noor Mohamed
Faculty of Engineering, Universiti Malaysia Sarawak, 94300 Kota Samarahan, Sarawak, Malaysia
e-mail: starmizi@unimas.my

Sapuan S. M. and I. Ahmad (eds.), *Composites from the Aquatic Environment*,
Composites Science and Technology, https://doi.org/10.1007/978-981-19-5327-9_8

Fig. 1 Nipah palm along *Sungai Kuap*, Kuching (red arrow is showing the lower frond)

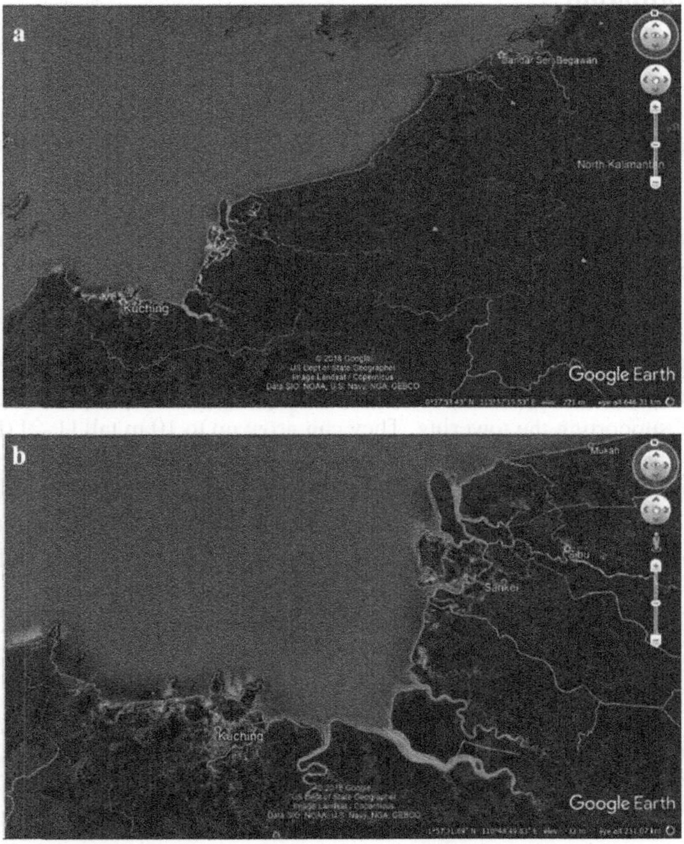

Fig. 2 Mangrove forest distribution. **a** Sarawak, **b** Kuching, Samarahan, Sibu and Sarikei [3]

Fig. 3 **a** Sugar sap, **b** cigarette wrapper

occupying slightly more than 0.09 million hectares in Sarawak [3]. Figure 2 show the distribution of mangrove forest in Sarawak as marked in red. Mangrove forest is concentrated within Kuching, Samarahan, Sarikei and Sibu division as shown Fig. 2b. The actual data for nipah palm population within the mangrove forest is not available. It has yet been explored by the Forest Department of Sarawak due to no requirement for the actual population. Sarawak has nine major type of forest, which is rich and diverse with flora and fauna species. This explained the lack of work done on nipah palm in Sarawak, as priority is given to other flora and fauna in the deep rainforest.

Nipah palm is the most utilised mangrove species. It is the most valued palm for the population living at the growth area. Nipah palm is known as *pokok apong* in Sarawak and *attap palm* in Singapore. The commercial value of nipah palm in Sarawak is still lacking compared to its neighbouring country. Indonesia and Philippines are conducting studies to diversify the usage of nipah palm product. For instance, Indonesia is exploring renewable energy derived from plant to meet their growing demand on energy. Recent work reported on nipah palm fibre is to extract fibre from nipah flower stalk to produce composite based on recycled polypropylene matrix [5]. Nipah palm sap has been investigated for its potential as bioethanol production [6]. Ethanol produced from nipah sap was reported to be better than sugarcane, cassava, coconut and potato [7]. Philippines is the third largest area with nipah plantation in Southeast Asia [8]. It is reported that a group of enterprising farmers in Philippines has come out with natural sweetener from nipah sap besides its common usage such as roofing material, vinegar (*sukang paombong*) and wine (*laksoy*). Another application of nipah as reported in the literature is as MDF board, made with coconut [9].

Natural fibre composite has been utilized in automotive industry. The trend has shown tremendous increased by most major vehicle manufacturers [10]. The urge to reduce fuel consumption of motor vehicle and improving its energy efficiency can be achieved by reducing the weight of the vehicle, natural fibre composite fits in the criteria as lightweight and strong material to achieve this. A few research works were investigating natural fibre-based packaging materials suitable for food and pharmaceutical application. Another emerging area is focusing the used of bio-based polymers and composites is the electronic application [11].

This investigation is to extract nipah palm fibre as reinforcing agent in polyester composite. The characterization of fibre in terms of its physical properties, chemical composition, thermal behaviour and its morphology were investigated. This investigation will provide new knowledge on the extraction method, composite fabrication, characterization of nipah fibre, mechanical test, namely, tensile test and water absorption test.

2 Materials and Method

2.1 Preparation of Raw Materials

Nipah fronds were collected from Sungai Kuap, Samarahan Division, Sarawak. The source of nipah palm fibres is from the fronds. Nipah fronds are found to have two distinctive parts, the lower fronds and upper fronds. The lower fronds are the submerged fronds during high tide and it has spongy structure inside, surrounded by hard and waxy dark brown bark. The upper fronds contain more rigid bundles of fibres as compared to the lower fronds as they are less spongy.

2.1.1 Fibre Extraction

Nipah fibre palm was extracted using wet extraction method or also known as water retting process. A whole lot of nipah fronds were prepared for retting process in a fresh water fish pond. Fresh water pond was chosen to allow biodegradation of the non-cellulosic component of the fronds. Nipah fronds were cut to about 50 cm length, placed into a plastic bag, weighted down with big stone and dropped into the fish pond. At 12 weeks as in Fig. 4a, b, the fronds emitted an extremely strong decompose odour. When the fronds were opened and cleaned under running water, the fibres were separated easily from the fibre bundles as shown in Fig. 4c. The extraction process of the fibre was done manually under running tap water. The source of the tap water is directly from the water catchment area at Mount. Sadong, Serian Division, Sarawak. The water-retted fibres were easier to be pulled out from its fibre bundles as the viscid substance has loosened up. Figure 4d shows the extraction process of fibres consisting of entangled Fibre A and Fibre B. Fibre bundles were easily separated once the water retting process is optimum. Fibres were pulled out one by one and cleaned at the same time. The extracted Fibre A are as shown Fig. 4e, while Fibre B are the remaining of the fibre bundle once Fibre A had been pulled out. It was placed in a basin that contains water as it was easier to remove the dirt. Figure 4f shows the extracted Fibre B. The detail of the extraction process of nipah palm fibre is published in previous work [12].

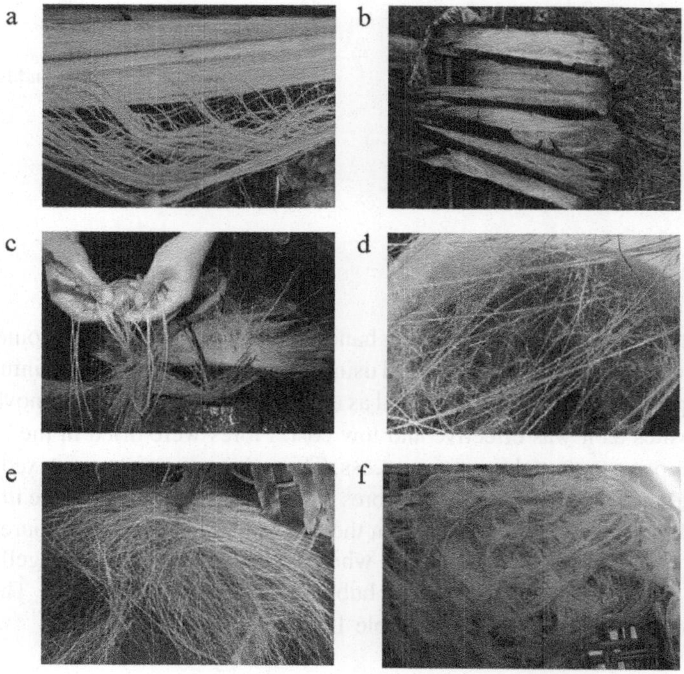

Fig. 4 Water retting process **a** 12 week—upper frond, **b** 12 week—lower frond, **c** extracting process, **d** extracting Fibre A and Fibre B, **e** freshly extracted Fibre A, **f** freshly extracted Fibre B

2.1.2 Cleaning and Drying Process

The cleaning process of the fibres after the extraction process was carried out by soaking the fibres in detergent (Sunlight Dishwashing Liquid) for one hour at 70 °C. The fibres were then cleaned under running pipe water for about 15 min and dried under the hot sun and proceeded to oven drying at 80 °C for 48 h. Distilled water soaking was not conducted in this work. Preliminary soaking test of the frond was conducted using tape water (chlorinated), however due to extremely slow decaying process, the test was discontinued.

2.2 Sample Preparation

2.2.1 Single Fibre Test

The single fibre test was performed on nipah palm fibre to determine its axial tensile modulus, ultimate strength and failure strain according to ASTM D3379-75. The gauge length was set at 50 mm and the test speed was set at 0.5 mm/min.

Table 1 Composite loading and treatment condition

	Composite loading condition
5 W/0 N	5 wt% fibre with untreated condition
5 W/4 N	5 wt% fibre with 4% NaOH
5 W/10 N	5 wt% fibre with 10% NaOH
15 W/8 N	15 wt% fibre with 8% NaOH

2.2.2 Composite Preparation

The composites were fabricated using hand lay-up and compression moulding technique. The composites were fabricated using a 230 mm × 230 mm aluminium mould. Vaseline pure petroleum jelly was used as a release agent to ease the removal process of composites as it was effective and low cost. Fibres were dried in the oven at 70 °C for 1 h prior to the fabrication process. The polyester resin was mixed with two to three drops of hardener while the fibres were in the final drying stage in the oven. The dried fibres were then arranged on the mould and the resin was poured into the mould cavity. The mould was covered when the resin had reached the gelling stage, this is to reduce the formation of air bubbles in the composite plate. The loading condition of nipah composite is in Table 1.

2.2.3 Alkali Treatment

Fibre A were treated with 6, 8, 10 and 15% natrium hydroxide, while fibre B were treated with 2 and 4% natrium hydroxide. Composites for fibre A were fabricated with 6, 8 and 10% natrium hydroxide treatment while only for 2% natrium hydroxide treatment for fibre B.

2.3 *Characterisation*

2.3.1 Diameter and Density Measurement

The diameter of the fibre was measured from observation under the optical microscope. The measurement was taken at 100 different locations along each fibre, and four fibre samples were measured for each type of fibre. The average cross-sectional area (A) was calculated from two fibre perpendicular diameters, d_1 and d_2, $A = \pi d_1 d_2/4$. The density of Fibre A and B were measured using water pycnometer procedure according to ASTM D 854.

2.3.2 Chemical Analysis

The chemical composition evaluation of the fibres was carried out at Forest Research Institute of Malaysia (FRIM) and ENVIC Laboratory Sdn. Bhd. The standards are TAPPI T 203CM-99 (α-cellulose), TAPPI T 222 om-02 (lignin) and aqueous alkali extraction (hemicellulose).

2.3.3 Fourier Transforms Infrared Spectroscopy (FTIR)

FTIR was carried out using Nicolet iS10. All the spectra were recorded in the wavenumber range between 400 and 4000 cm^{-1}.

2.3.4 Thermogravimetric Analysis (TGA)

TGA was carried out using Mettle Toledo with the nitrogen was used as the carrier gas, and the heating rate of 10 °C/min heated ranging from 30 to 700 °C.

2.3.5 X-Ray Diffraction (XRD)

XRD (P8Advan-Bruker) was used to analyse the crystallinity before and after alkali treatment. Measurement was taken from 2θ of 10–90 °C using Cu Kα X-ray source.

2.3.6 Morphology

Scanning Electron Microscopy (SEM) was used to study fibre morphology. Hitachi TM3030 at an acceleration voltage of 15 kV, samples were gold coated to improve the surface conductivity.

2.3.7 Mechanical Test

Tensile test for nipah composites was performed using the Testometric (25 kN) according to ASTM D638-03. The crosshead speed was set at 2 mm/min, with the gauge length of 50 mm.

2.3.8 Water Absorption

Water absorption test was conducted based on ASTM D570-98. A long-term immersion in distilled water up to 28 days at normal room temperature.

3 Results and Discussions

Water retting process for nipah fronds was conducted at a fresh water fish pond, which took about 12 weeks to be completed. This was when the fibres could be easily extracted without too much force and most unlikely to introduce premature defects to the fibres. As reported in the literature, water retting process is a time-consuming process [13]. At this particular point of 12 weeks, the fibre separation was very easy, because each fibre strand could be pulled, leaving behind the non-cellulosic substance. The whole fronds emitted strong decomposed odours, as compared to the previous checking points at week 8 and week 10. The wet extraction process produced two types of fibres, namely Fibre A and Fibre B, only Fibre A is discussed in this chapter. It was observed that lower fronds of nipah palm were easier to be extracted due to the larger number of spongy media as compared to the upper fronds. The lower fronds were normally bigger in diameter compared to the upper fronds and were observed to be softer once they reached the 12 weeks water retting process. They were easily split up manually requiring less force. The less exposed part of the fronds was at times difficult to be extracted even though the 12 weeks of water retting process were completed. This could be due to the age of the palm itself. The effects of harvest time were studied on hemp fibres [14]. During the collection process, the fronds were not collected from one single palm tree but also from other nipah palms. Another reason could be the effect of fronds diameter, where the lower fronds could reach 200–300 mm in diameter while the upper fronds were around 100 mm. The fibres extracted from this process are shown in Fig. 4e, f. Some of the nipah fronds were left in the pond for more than 12 weeks, to observe the over retting condition of the fibres. The fibres were brittle and broke into tiny pieces due to the excessive water retting process. Fibre separation was difficult as the tiny pieces of Fibre A entangled within the Fibre B mesh. This study concluded that 12 weeks is the optimum period for nipah fibre to be easily extracted.

The nipah fibre density and the diameter are shown in Table 2. The chemical analysis of nipah fibre is shown in Table 3. As stated in the literature, the nipah palm chemical composition is said to be very similar to oil palm [15, 16]. Nipah and oil palm are from the same palmae family, therefore, it is expected that there are some similarities in terms of its chemical composition.

Table 2 Nipah palm fibre physical properties (average)

Physical properties	
Density (g/cm^3)	1.00
Diameter (mm)	0.53 ± 0.06

Table 3 Chemical composition of nipah fibre

Natural fibre	α-Cellulose (% w/w)	Hemicellulose (% w/w)	Lignin (% w/w)	References
Nipah fibre	27.0	23.0	21.0	–
Oil palm (frond)	39.5	29.8	21.2 (Klason)	[15]
Oil palm (trunk)	30.6	33.2	24.7 (Klason)	

3.1 Effect of Alkali Treatment Towards Nipah Fibre

From Fig. 5, it is clear that alkali treatment has effected significant physical changes towards nipah palm fibre. The effect of 6, 8, 10 and 12% alkali treatment had changed the physical appearance of the fibres. The colour of the fibres had turned from a light shade of brown to dark brown due to the alkali treatment. All the treated fibres showed fibre bending and twisting. The higher the alkali concentration, the greater the effect of fibre twisting. The 12% alkali treated nipah fibres were badly twisted and fibre breakage was noticeable. This indicates that nipah fibres undergone destruction due to high alkali concentration. This could be due to excessive stripping of undesirable outer layer of the fibres such as hemicellulose and lignin. Hemicelluloses are amorphous and hydrophilic and soluble in alkali solution. Lignin is known to provide structural strength towards the fibre. Lignin is amorphous and hydrophobic in nature and soluble in alkali solution [17]. Both hemicellulose and lignin can be removed by alkali solution. Once removed, the bulk lignin will be disrupted, causing disaggregation of micro fibril of the fibres [18]. This is concluded by some shrinkage of the fibres in Fig. 6. The shrinkage values are almost similar for 6% alkali, 10% alkali and 12% alkali, which are between 54 and 55%. Fibre treated with 8% alkali shows shrinkage at 59%. The value of area reduction showed fluctuating pattern, it did not show a continual decrease pattern as the alkali concentration increase. This result is in agreement with the finding by [19] on hemp fibre. These reductions were due to fibre degradation and delignification of the treated fibres. Sugarcane fibre bundles were reported to show better lignin and hemicellulose removal at 5% alkali [20, 21], while Napier grass experienced 12–45% diameter reduction after alkali treatment [17]. Screw pine (*Pandanus Odoratissimus*) fibre reported the highest cross-sectional area reduction of 42.1% for alkali concentration of 15% [22]. Kenaf fibre cross-sectional area was reported to show rapid decrement pattern when subjected to various immersion time, alkali concentration and temperature [23].

The thermogravimetric analysis result for untreated and alkali treated nipah fibres are shown in Fig. 7. The results are tabulated in Table 4. The residue values were similar to the value of 32.8% reported for lignin. More residues indicating that the alkali treatment has indeed increased the temperature stability of the treated fibres even at a higher temperature.

It was observed from Fig. 7 that the untreated line shows two decomposition steps, 300 and 470 °C and the 6% alkali treated fibres also shows two decomposition steps at 250 and 550 °C. The two steps exist may be due to partial removal of

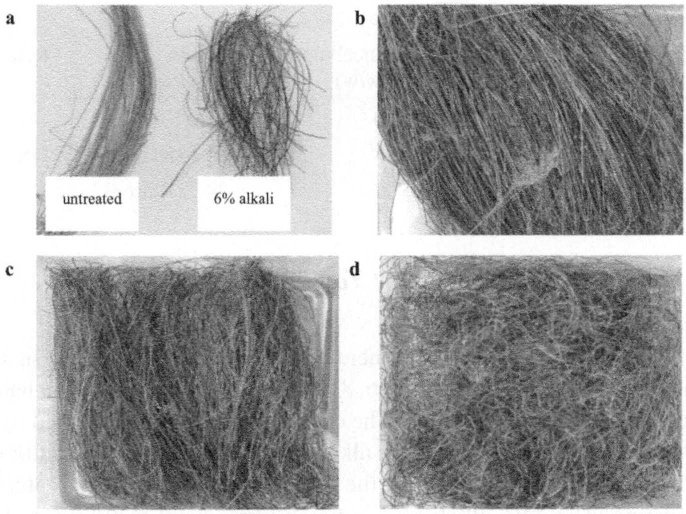

Fig. 5 **a** Untreated and 6% alkali treated fibres, **b** 8% alkali treated fibre, **c** 10% alkali treated fibre, **d** 12% alkali treated fibre

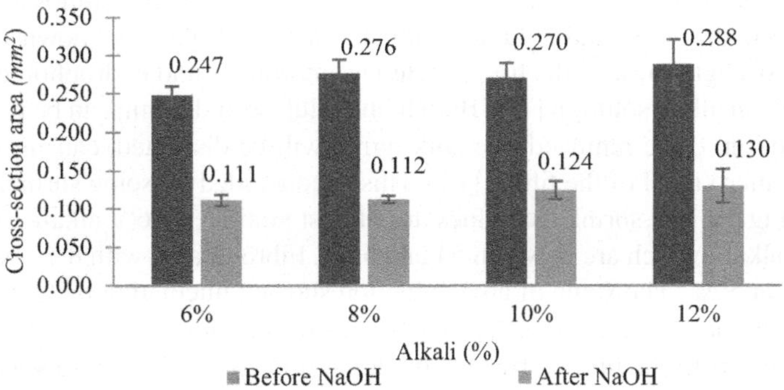

Fig. 6 Fibre cross-section area before and after alkali treatment

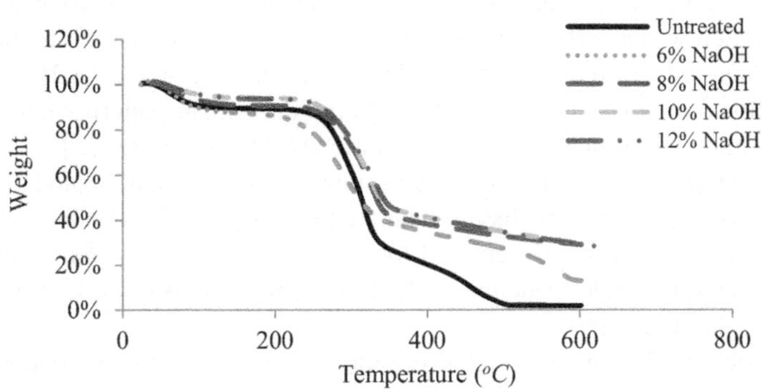

Fig. 7 TGA analysis

Table 4 Residue results of based on treatment condition

Treatment condition	Residue	
	400 °C (%)	600 °C (%)
Untreated	20	3
6%	35	12
8%	38	30
10%	45	30
12%	45	30

hemicellulose, lignin and waxes. However, at 8, 10 and 12% alkali treatment, it shows one decomposition step, which was the decomposition of α-cellulose and lignin. This indicates that a higher alkali percentage completely removes hemicellulose, lignin and waxes [24]. Similar observation was made in other natural fibres [25–27]. In order to probe further, spectral analysis was carried for the untreated and alkali treated fibres.

The FTIR spectra of nipah palm fibre are as shown in Fig. 8. Untreated, 6, 8, 10 and 12% alkali treated FTIR spectra show peaks dominated at 3335.03 cm^{-1}. These spectra could be attributed by O–H stretching vibration. The peak for treated nipah is broader and more intense indicating more –OH group existed due to the treatment. The peak at 1726.26 cm^{-1} for treated nipah fibre slowly disappears after the treatment with an increasing alkali aqueous solution. This is due to the removal of acid, lignin and other natural fibre constituents [28, 29].

Hydrolysis occurs after the alkali treatment. This breaks down the ester bond, which explains the disappearance of 1726.26 cm^{-1} peak. The disappearance of the 1726.26 and 1665.81 cm^{-1} peaks after alkali treatment indicates that either the carboxylic acid and acetyl groups were destroyed by the alkali treatment or the macromolecules containing these functional groups were selectively dissolved from the fibre bundles under strong alkali condition. The peaks observed the 1000 and 1500 cm^{-1} range show the presence of hemicellulose [24]. Based on peaks for nipah fibre between untreated and 6, 8, 10 and 12% alkali treated, the intensity of the peaks

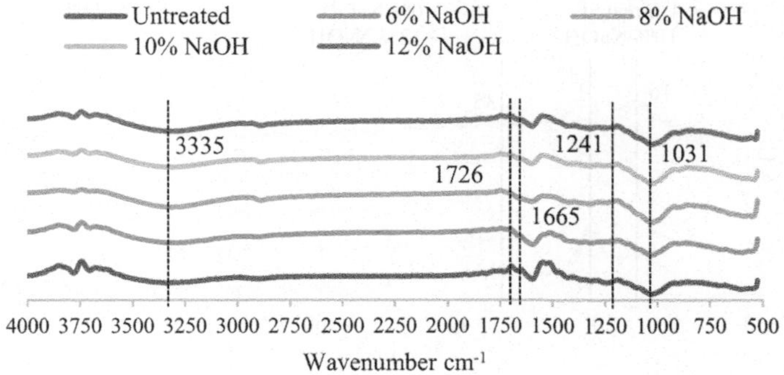

Fig. 8 FTIR

is slightly reduced, which indicates a slight removal of hemicellulose compounds, while the peaks at 1031.20 and 1030.90 cm^{-1} of the untreated and treated fibre indicate the stretching vibrations of C–O. The peak at 1241.48 cm^{-1} can also readily assigned to the C–O stretching mode of acetyl groups in lignin [30]. All the major peaks of nipah palm fibre are similar with the findings found by [31].

The X-ray diffraction graph obtained for nipah fibres is shown in Fig. 9. The major peaks observed for all fibre samples were at 2θ diffraction angles of 16°, 22° and 35°. The overall pattern shows amorphous structure of the nipah fibres. The sharpest peak was observed for 8% alkali, indicating highest crystallinity value of 50%, due to the removal of amorphous structure [32]. This result is supported by the single fibre tensile strength, where the 8% alkali depicted the highest tensile strength as shown in Fig. 10. The alkali treatment has removed fractions of amorphous in the fibre, leaving behind the crystalline structures in the fibres. This explains the sharpest peak at 8% alkali. The crystallinity index calculation was based on Segal Empirical method [33]. The degree of crystallinity of untreated fibre was 39.7 and the 8% alkali treated fibre was 50%. The overall result is in Table 5. This shows that the percentage of crystallinity index of treated 8% alkali fibre was 11% higher than the untreated nipah fibre. The increase of the percentage of crystallinity contributes to the enhancement of the tensile properties of the 8% alkali nipah fibre due to the restructuring of cellulose [34].

The single fibre tensile strength as depicted in Fig. 10a, shows the overall tensile strength between untreated and treated single fibres. All the symbols show individual result from each of the test. This shows the variation of the result instead of the average result. The result clearly showed that 8% alkali treated fibres had a higher tensile strength compared to the other treated fibres. The alkali treatment caused the removal of non-cellulosic content in the fibre. This removal allowed the cellulose to position itself when subjected to loading and facilitate better load transfer [35]. The cellulose chains were no longer in constraint state, therefore the fibrils were able to position itself towards closer packing arrangement to improve fibre strength and tensile properties [36]. However, at higher alkali concentration, the tensile strength drops. This was due to fibre damage caused by extreme reaction of the alkali towards

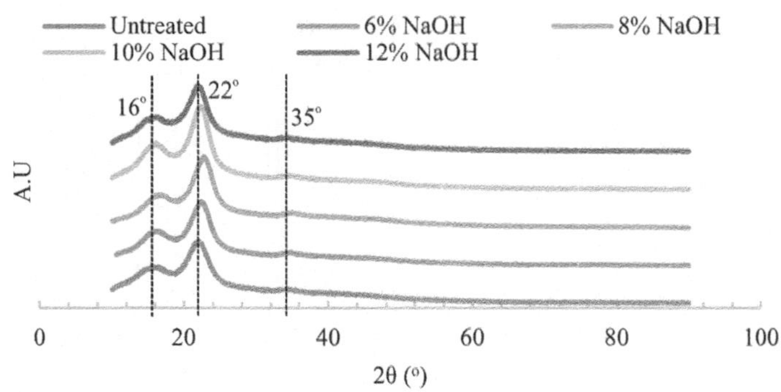

Fig. 9 X-ray diffraction

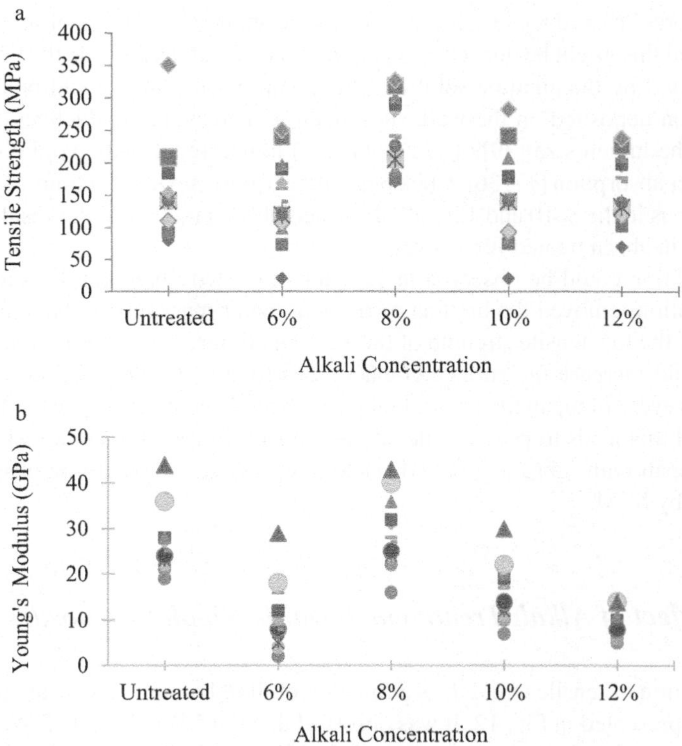

Fig. 10 **a** Single fibre tensile strength, **b** Young's modulus

Table 5 Crystallinity index based on treatment condition

Treatment condition	Crystallinity index (%)
Untreated	39.7
6% alkali	42.3
8%	50.0
10%	42.0
12%	43.9

the fibres. This could also due to the existence of deep pores at higher alkali concentration. The increase number of deep pores leads to the decreasing of tensile strength [37].

As for the fracture surface of the single fibres shown in Fig. 11, they were almost similar, literary not flat surface, which are basic criteria for fibrous material. The vessel of the fibre for untreated and 6% alkali was round at the centre and the fibre lumen was elliptical for nipah fibre, and this can also be observed in bagasse fibre [38]. The presence of vessel and lumens at the fibre surface explained the light weight of the fibre. The vessel had a very smooth internal surface. The elliptical lumen for treated fibres showed some compression. However, no change was observed for the vessel, and this might be due to the location of the vessel deep inside the fibre, which was affected by the alkaline solution. The compressed lumen could be due to the sodium ion deposited on the wall, subsequently increasing the thickness and thus, reduces the lumen size [39]. The compressed lumen reduces the void content and fibre water absorption [17, 36]. Click or tap here to enter text.. The compressed lumen was obvious in the 8, 10 and 12% alkali treated fibres, and the similar result was also observed in abaca treated fibres [30].

Fibrillation could be observed in 12% alkali treated fibre, and the strong alkali concentration removed the binding material at the primary nipah fibre bundle. This explained the low tensile strength of the 12% alkali fibres. The fibrillation in treated fibres could increase the surface contact areas for matrix interlocking as reported [22]. However, for nipah fibre, fibrillation was only observed clearly in the 12% alkali fibres and this leads to poor tensile strength result. Fibrillation could exist in other treated nipah palm fibre, but its existence may not be obvious to be observed and captured by SEM.

3.2 Effect of Alkali Treatment Towards Nipah Composites

The variation of tensile strength as a function of fibre loading and the alkali concentration is presented in Fig. 12. It was observed that the 5 W/0 N and 15 W/8 N alkali composites showed tensile strength at 29.36 MPa and 28.29 MPa respectively.

Similar pattern was observed for the Young's modulus value for the composites, and the highest was contributed by the 5 W/10 N composites at 471 MPa. There was an improvement of 42% of tensile strength, 35% of Young's modulus value and 70% of strain at break between 5 W/0 N composites and 5 W/10 N composites.

It was noted that the 10% alkali treated fibre composites at a lower loading condition produced the best tensile strength, whereas, the 8% alkali treated fibre composites produced the best result at the highest loading condition. The single fibre test result showed 8% alkali had the highest tensile strength value and it was expected that the nipah fibre reinforced polyester composites would yield the same result. However, the result showed the opposite. It was believed that the 5 W/10 N composites had better interlocking and load transfer between fibres and polyester matrix. At higher loading the fibre direction might not be favourable to the load direction and packed fibre agglomeration may result to improper wetting condition. It was reported that alkali treatment can have different effect towards the mechanical properties of the fibres and its composites when they were used as reinforcement agent [24]. This could be the reason of fibre strength variation. Since the single fibre test of the 8% alkali showed the highest value, its composites strength showed a significant increase from 5 to 15 wt% loading. The 10% alkali nipah fibre reinforced polyester composites showed a reduction in strength as the loading gets higher. It was believed that a rougher fibre with a lower loading leads to a greater strength, but as the amount of fibre increases, it leads to improper wetting condition. From this result, it should be

Fig. 11 Fibre **a** untreated, **b** 6% alkali, **c** 8% alkali, **d** 10% alkali, **e** 12% alkali

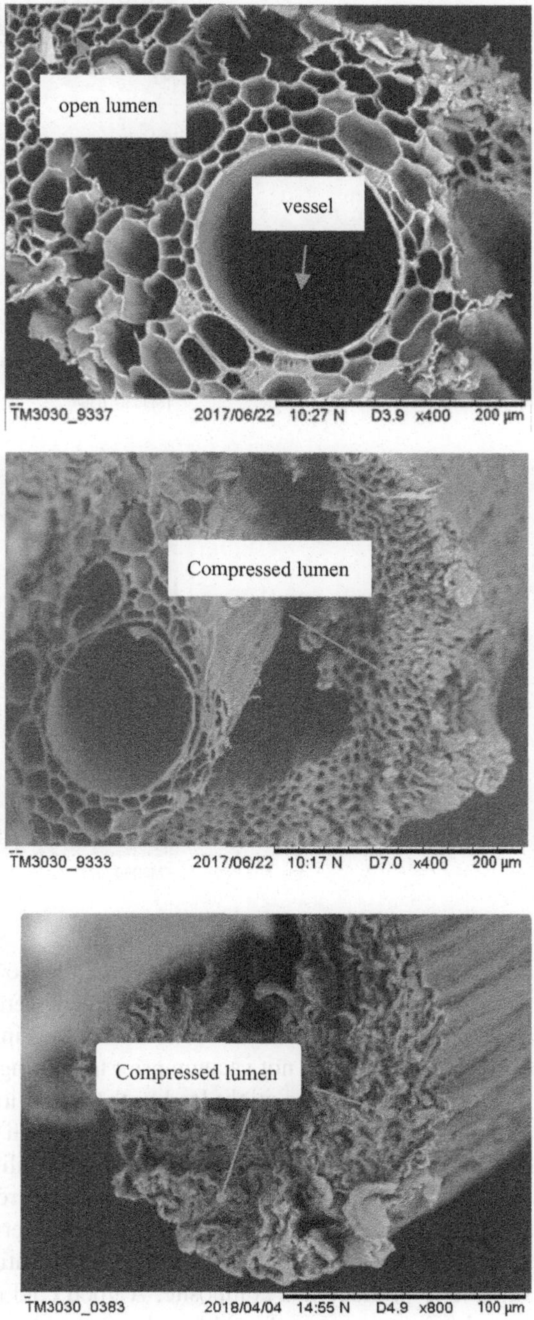

Fig. 11 (continued)

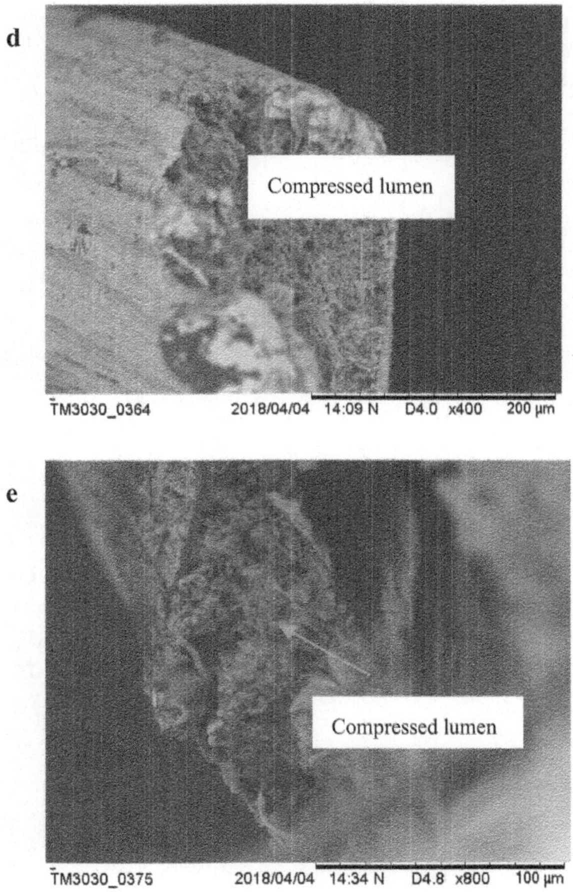

noted that the incorporation of nipah fibre failed to reinforce polyester composites as compared to the neat polyester strength. There were a few possible factors contribute to this result. First, the existence of air bubbles in the fabricated composites as the polyester mixture was not vacuum prior to pouring into the mould. Second, the brittleness of the polyester might lead to the formation of micro cracks during cutting process. The samples were hand saw, even though great care was taken into account during the cutting process. Third, the compatibility between nipah palm fibre and the polyester matrix may not ideal to provide the reinforcement to the composites as can be seen in Fig. 13a, fibre breakage was observed at some point of the fracture surface as shown in Fig. 13b. Fourth, the fabrication method, hand lay-up may not suitable to produce this composite. A good cold or hot press machine that is able to maintain its pressure for 24 h and equip with a vacuum pump to suck out all the air bubbles during the pressing process might produce a good plate of composite. However, the alkali treatment had successfully increased the tensile strength of the composites compared to the untreated nipah fibre composites.

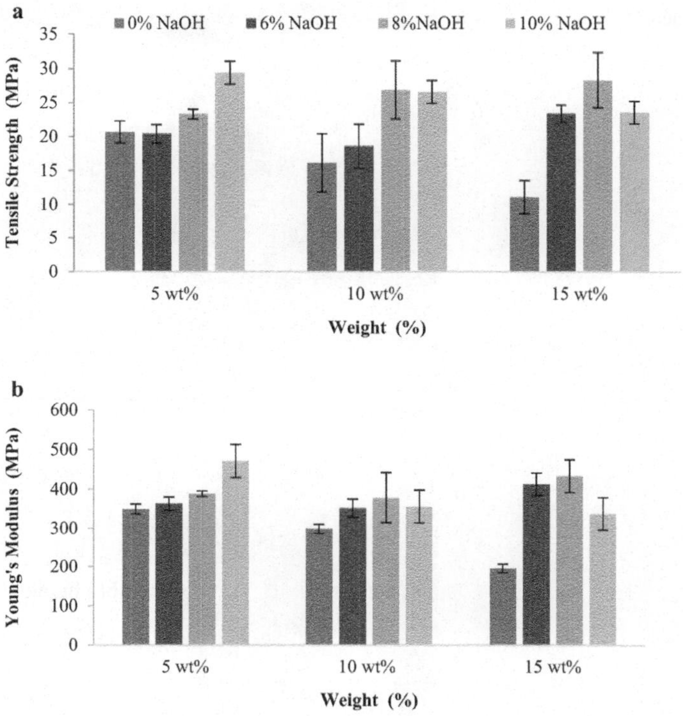

Fig. 12 **a** Tensile strength, **b** Young's modulus

This observation was also observed in other natural fibre polyester composites [40, 41]. It was reported that coir fibre has to be loaded up to 45% or higher in order to obtain a significant reinforcing effect, and lack of reinforcing effect was due to low modulus of elasticity as compared to neat polyester [42]. Piassava fibres reinforced epoxy composites was reported to show similar findings as the composites failed to reinforce up to 30% of volume fraction [43]. The napier grass fibre was reported to reinforce the polyester composites at a value of 15.64 MPa for short fibre. The tensile strength of the pure polyester is 10.1 MPa [44]. The strength value obtained in nipah fibre composites was much higher than the value reported by napier grass. However, the alkali treatment had indeed improved the wetting between nipah fibres and polyester matrix. The removal of surface impurities, non-cellulosic material and waxes resulted in cleaner and rougher fibre surface and improves the composite strength. This observation was observed somewhere else [22, 35].

The water absorption curve tested in distilled water at normal room temperature (28 °C) is presented in Fig. 14. The results showed an increase in total absorption for up to 28 days, and as can be observed, the weight increased as a function of its fibre loading. This was expected as natural fibre like nipah fibre is hydrophilic in nature. The water absorption of nipah fibre composites followed the Fickian's behaviour, and this was also observed in kenaf fibre reinforced polyester composites as reported

Fig. 13 Fracture morphology

a

b

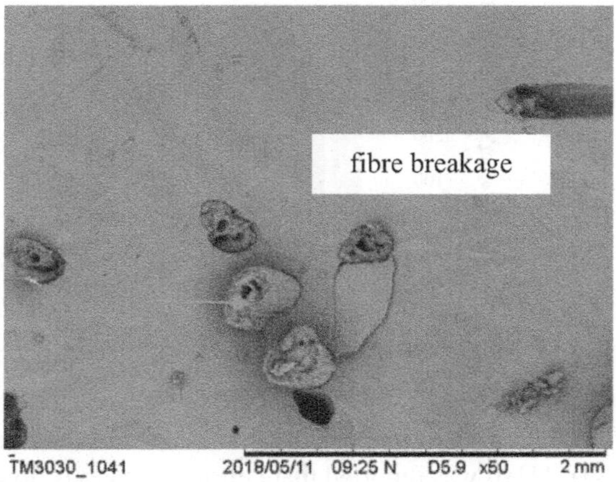

[45]. The highest fibre loading conducted was at 20 wt% of the untreated fibres. There was an increase of 7% weight gain between the 15 and 20 wt% fibre loading.

It should be noted that the untreated fibres had a maximum peak at day 22 while the treated fibres had its maximum peak at day 18. The surface treatment has shortened the peak period for treated fibres. The treated fibres regardless of its loading condition, its weight gain curve falls below the untreated fibres. The alkali treatment had significantly reduced the water absorption rate in treated fibres. It had created a protective layer on the fibre surface that was resistant to water absorption. The compressed lumen of treated fibres also reduced the water absorption capacity. It was also reported in the literature that alkali treated fibres showed a compressed lumen structure, and this caused the reduction of void content and less water absorption capacity of the treated fibres [17, 36, 39].

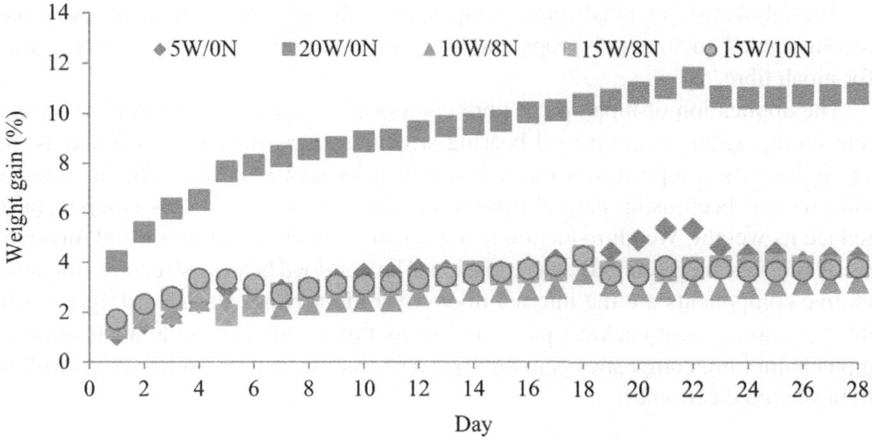

Fig. 14 Result of water absorption test

4 Conclusion

The main objective of this research work is to investigate the physical, thermal, chemical and morphological properties of nipah fibre. This research work provides a new knowledge on nipah palm fibre characterisation and its composites properties.

The alkali treatment conducted on nipah fibre had shown better interfacial properties on its composite properties. Similar improvement reported by other natural fibre treated with alkali is also observed in nipah palm fibre. This provides a full potential for nipah palm fibre as reinforcing agent in composites. The 8% alkali treatment showed the highest strength on single fibre, while the 10% alkali treatment displayed the highest strength of its composites.

Microstructural changes of the treated fibres showed improvement due to the alkali treatment, where the roughened surface provides a good site for fibre interlocking and improving the strength of its composites. The collapsed lumen of treated fibres had improved its water absorption capacity. The water resistance layer provided by the alkali treatment helped in resisting water absorption of the composites.

There are a few future recommendations for future research such as to conduct the various fibre loading condition such as long and continuous composites. Long and continuous was not investigated in this research work, due to the difficulty to align the mould. Proper equipment to align the fibre during fabrication is needed to successfully produce long and align nipah fibre composites.

Impact and flexural strength for nipah composites are recommended to be investigated. It is also recommended to use vacuum bagging method to fabricate the composites. This is to eliminate issues with air bubbles. The air bubbles are due to the reaction between unsaturated polyester resins with MEKP hardener.

The fabrication of nipah fibre composites with other polymer matrices is also recommended such as polypropylene. This will open more research opportunities for nipah fibre.

The application of nipah palm fibre composite is recommended to be used as a reinforcing agent in a non-load bearing structure, or as filler in composite as the strength of its composites is much lesser than its neat polyester. The automotive industry has been using natural fibre reinforced composites in cars component to reduce its weight. Weigh reduction is important for fuel consumption and this indirectly reduces the greenhouse gas emission. The non-load bearing structures for automotive components are the internal door cover, door panel, dashboard, tyre cover, interior carpets, seat backrest panel and many more. Other possible application of nipah palm fibre composites is in the food and pharmaceutical packaging as well as in bio-based electronics.

References

1. Teo, S., Ang, W. F., Lok, A. F. S. L., Kurukulasuriya, B. R., & Tan, H. T. W. (2010). The status and distribution of the Nipah palm Nypa fruticans Wurmb. (Arecaceae), in Singapore. *Nature in Singapore, 3*(February), 45–52.
2. Tsuji, K., et al. (2011). Biological and ethnobotanical characteristics of Nipa Palm (Nypa fructicans wurmb.): A review. *Sains Malaysiana, 40*(12), 1407–1412.
3. Sarawak Forestry Corporation (SFC). (2018). *Mangrove forest in Sarawak.*
4. (2019). Nypa fruticans (nipa palm). Retrieved September 30, 2021, from http://www.cabi.org/isc/datasheet/36772
5. Cao, L. N. H., Nguyễn, T. B. T., Lương, H. V. T., Mai, V. P. T., & Trần, N. P. L. (2021). Nghiên cứu vật liệu composite thân thiện môi trường từ sợi cuống dừa nước và nhựa polyethylene tỷ trọng cao tái chế. *Can Tho University Journal of Science, 57*(6), 42–52. https://doi.org/10.22144/ctu.jvn.2021.171
6. Mohammed, L., Ansari, M. N. M., Pua, G., Jawaid, M., & Islam, M. S. (2015). A review on natural fiber reinforced polymer composite and its applications. *International Journal of Polymer Science, 2015*. https://doi.org/10.1155/2015/243947
7. Hamilton, L. S., & Murphy, D. H. (1988). Use and management of Nipa palm (Nypa fruticans, arecaceae): A review. *Economic Botany, 42*(2), 206–213. https://doi.org/10.1007/BF02858921
8. Neves, S., Muylaert, F., Rodrigues, G., Leite, R., & Picanço, M. (2015). Tensile strength of polyester composites reinforced with thinner buriti fibers. *805,* 466–471. https://doi.org/10.4028/www.scientific.net/MSF.805.466
9. Kruse, K. (2001). *Properties of nipa- and coconut fibers and production and properties of particle- and MDF-boards made from nipa and coconut* (pp. 1–41).
10. Faruk, O., Bledzki, A. K., Fink, H. P., & Sain, M. (2014). Progress report on natural fiber reinforced composites. *Macromolecular Materials and Engineering (Wiley-VCH Verlag), 299*(1), 9–26. https://doi.org/10.1002/mame.201300008
11. Salit, M. S., Jawaid, M., bin Yusoff, N., & Hoque, M. E. (2015). *Manufacturing of natural fibre reinforced polymer composites* (pp. 1–383). https://doi.org/10.1007/978-3-319-07944-8
12. Dickie, T., Syazali, S. T. B. S., & Noor Mohamed, N. H. B. (2020). Development of nipah palm fibre extraction process. *Materials Science Forum, 997,* 57–65. https://doi.org/10.4028/www.scientific.net/MSF.997.57
13. Tamburini, E., León, A. G., Perito, B., di Candilo, M., & Mastromei, G. (2004). Exploitation of bacterial pectinolytic strains for improvement of hemp water retting. *Euphytica, 140*(1–2), 47–54. https://doi.org/10.1007/s10681-004-4754-y

14. Liu, M., et al. (2015). Effect of harvest time and field retting duration on the chemical composition, morphology and mechanical properties of hemp fibers. *Industrial Crops and Products, 69*, 29–39. https://doi.org/10.1016/j.indcrop.2015.02.010
15. Masashi Shibata, S. S., Varman, M., Tono, Y., & Miyafuji, H. (2008). Characterization in chemical composition of the oil palm (Elaeis guineensis). *Journal of the Japan Institute of Energy,* (87), 383–388. https://doi.org/10.1046/j.1365-2559.2002.14891.x
16. Tamunaidu, P., & Saka, S. (2011). Chemical characterization of various parts of nipa palm (Nypa fruticans). *Industrial Crops and Products, 34*(3), 1423–1428. https://doi.org/10.1016/j.indcrop.2011.04.020
17. Ridzuan, M. J. M., Abdul Majid, M. S., Afendi, M., Aqmariah Kanafiah, S. N., Zahri, J. M., & Gibson, A. G. (2016). Characterisation of natural cellulosic fibre from Pennisetum purpureum stem as potential reinforcement of polymer composites. *Materials and Design, 89,* 839–847. https://doi.org/10.1016/j.matdes.2015.10.052
18. Yusriah, L., Sapuan, S. M., Zainudin, E. S., & Mariatti, M. (2014). Characterization of physical, mechanical, thermal and morphological properties of agro-waste betel nut (Areca catechu) husk fibre. *Journal of Cleaner Production, 72*, 174–180. https://doi.org/10.1016/j.jclepro.2014.02.025
19. Kabir, M. M., Wang, H., Lau, K. T., & Cardona, F. (2013). Tensile properties of chemically treated hemp fibres as reinforcement for composites. *Composites Part B: Engineering, 53*, 362–368. https://doi.org/10.1016/j.compositesb.2013.05.048
20. Hossain, M. K., et al. (2014). Comparative mechanical and thermal study of chemically treated and untreated single sugarcane fiber bundle. *Industrial Crops and Products, 58*(November 2017), 78–90. https://doi.org/10.1016/j.indcrop.2014.04.002
21. Yue, Y., Han, J., Han, G., Aita, G. M., & Wu, Q. (2015). Cellulose fibers isolated from energycane bagasse using alkaline and sodium chlorite treatments: Structural, chemical and thermal properties. *Industrial Crops and Products, 76*, 355–363. https://doi.org/10.1016/j.indcrop.2015.07.006
22. Abral, H., et al. (2012). Alkali treatment of screw pine (Pandanus odoratissimus) fibers and its effect on unsaturated polyester composites. *Polymer-Plastics Technology and Engineering, 51*(August 2013), 12–18. https://doi.org/10.1080/03602559.2011.593090
23. Hashim, M. Y., Amin, A. M., Mohd, O., & Marwah, F. (2017). *The effect of alkali treatment under various conditions on physical properties of kenaf fiber.*
24. Mwaikambo, L. Y., & Ansell, M. P. (2002). Chemical modification of hemp, sisal, jute, and kapok fibers by alkalization. *Journal of Applied Polymer Science, 84*(12), 2222–2234. https://doi.org/10.1002/app.10460
25. Mwaikambo, L. Y. (2008). Kapok/cotton fabric—polypropylene composites. *Polymer Testing, 19(2000)*, 905–918.
26. Reddy, K. O., Maheswari, C. U., Reddy, D. J. P., & Rajulu, A. V. (2009). Thermal properties of Napier grass fibers. *Materials Letters, 63*(27), 2390–2392. https://doi.org/10.1016/j.matlet.2009.08.035
27. Wang, Z., Li, J., Barford, J. P., Hellgradt, K., & Mckay, G. (2016). A comparison of chemical treatment methods for the preparation of rice husk cellulosic fibers. (1), 67–77.
28. Sain, M., & Panthapulakkal, S. (2006). Bioprocess preparation of wheat straw fibers and their characterization. *Industrial Crops and Products, 23*(1), 1–8. https://doi.org/10.1016/j.indcrop.2005.01.006
29. Valadez-Gonzalez, A., Cervantes-Uc, J. M., Olayo, R., & Herrera-Franco, P. J. (1999). Effect of fiber surface treatment on the fiber-matrix bond strength of natural fiber reinforced composites. *Composites Part B: Engineering, 30*(3), 309–320. https://doi.org/10.1016/S1359-8368(98)00054-7
30. Cai, M., Takagi, H., Nakagaito, A. N., Li, Y., & Waterhouse, G. I. N. (2016). Effect of alkali treatment on interfacial bonding in abaca fiber-reinforced composites. *Composites Part A: Applied Science and Manufacturing, 90*(August), 589–597. https://doi.org/10.1016/j.compositesa.2016.08.025

31. Rasidi, M. S. M., Husseinsyah, S., & Leng, T. P. (2014). Chemical modification of nypa fruticans filled polylactic acid/recycled low-density polyethylene biocomposites. *BioResources, 9*(2), 2033–2050.
32. Júnior, A. E. C., Barreto, A. C. H., Rosa, D. S., Maia, F. J. N., Lomonaco, D., & Mazzetto, S. E. (2015). Thermal and mechanical properties of biocomposites based on a cashew nut shell liquid matrix reinforced with bamboo fibers. *Journal of Composite Materials, 49*(18), 2203–2215. https://doi.org/10.1177/0021998314545182
33. Segal, L., Creely, J. J., MartinJr, A. E., & Conrad, C. M. (1959). An empirical method for estimating the degree of crystallinity of native cellulose using the X-ray diffractometer. *Têxtil Research Journal, 29*, 786–794. https://doi.org/10.1177/004051755902901003
34. Haameem, J. A. M., Abdul Majid, M. S., Afendi, M., Haslan, M., & Fahmi, I. (2016). Alkaline treatment and thermal properties of Napier grass fibres. *International Journal of Automotive and Mechanical Engineering, 13*(1), 3238–3247. https://doi.org/10.15282/ijame.13.1.2016.10.0270
35. Vardhini, K. J. V., Murugan, R., Selvi, C., & Surjit, R. (2016). Optimisation of alkali treatment of banana fibres on lignin removal. *Indian Journal of Fibre and Textile Research, 41*(2), 156–160.
36. Reddy, K. O., Maheswari, C. U., Shukla, M., & Rajulu, A. V. (2012). Chemical composition and structural characterization of Napier grass fibers. *Materials Letters, 67*(1), 35–38. https://doi.org/10.1016/j.matlet.2011.09.027
37. Han, S. O., & Choi, H. Y. (2010). Morphology and surface properties of natural fiber treated with electron beam. *Microscopy: Science, Technology, Applications and Education, 1880*, 1880–1887.
38. Guimarães, J. L., Frollini, E., da Silva, C. G., Wypych, F., & Satyanarayana, K. G. (2009). Characterization of banana, sugarcane bagasse and sponge gourd fibers of Brazil. *Industrial Crops and Products, 30*(3), 407–415. https://doi.org/10.1016/j.indcrop.2009.07.013
39. Cai, M., Takagi, H., Nakagaito, A. N., Kusaka, K., Katoh, M., & Li, Y. (2015). Influence of alkali concentration on morphology and tensile properties of abaca fibers. *Advanced Materials Research, 1110*, 302–305. https://doi.org/10.4028/www.scientific.net/AMR.1110.302
40. Boynard, C. A., & D'Almeida, J. R. M. (2000). Morphological characterization and mechanical behavior of sponge gourd (Luffa cylindrica)-polyester composite materials. *Polymer—Plastics Technology and Engineering, 39*(3), 489–499. https://doi.org/10.1081/PPT-100100042
41. Melkamu, A., Kahsay, M. B., & Tesfay, A. G. (2018). Mechanical and water-absorption properties of sisal fiber (Agave sisalana)-reinforced polyester composite. *Journal of Natural Fibers*, 1–9. https://doi.org/10.1080/15440478.2018.1441088.
42. Monteiro, S. N., Terrones, L. A. H., & D'Almeida, J. R. M. (2008). Mechanical performance of coir fiber/polyester composites. *Polymer Testing, 27*(5), 591–595. https://doi.org/10.1016/j.polymertesting.2008.03.003
43. Nascimento, D. C. O., Lopes, F. P. D., & Monteiro, S. N. (2010). *Tensile behavior of lignocellulosic fiber reinforced polymer composites: Part I piassava/epoxy (a) (b)* (pp. 189–194).
44. Haameem, M., Majid, A., Afendi, M., Marzuki, H. F. A., Fahmi, I., & Gibson, A. G. (2016). Mechanical properties of Napier grass fibre/polyester composites. *Composite Structures, 136*, 1–10. https://doi.org/10.1016/j.compstruct.2015.09.051
45. Mazuki, A. A. M., Akil, H. M., Safiee, S., Ishak, Z. A. M., & Bakar, A. A. (2011). Degradation of dynamic mechanical properties of pultruded kenaf fiber reinforced composites after immersion in various solutions. *Composites Part B: Engineering, 42*(1), 71–76. https://doi.org/10.1016/j.compositesb.2010.08.004

Life Cycle Assessment for Microalgal Biocomposites

Mohd Danish Ahmad, Imran Ahmad, Norhayati Abdullah, Iwamoto Koji, Shaza Eva Mohamad, Ali Yuzir, Shristy Gautam, and Mostafa El-Sheekh

1 Introduction

The circular economy model (CEM) has garnered a lot of attention as a solution to the present difficulties caused by rising consumption and production needs. To restrict the use of raw materials and natural resources, this approach is primarily on the basis of a resource, recovery, and recycling strategy [1]. CEM effectively provides a simultaneous examination of social, economic, and environmental problems, which was previously unavailable [2]. The circular economy has recently depended on the biorefinery idea, which uses biomass and renewable sources of energy to reduce emissions of greenhouse gases and supplements waste management [3].

A biorefinery is critical in the transition to a net-zero waste society [4]. More than 400 million tonnes of plastic are manufactured each year in the world, with nearly a third of it ending up as plastic garbage in landfills, freshwater lakes, rivers, and oceans [5]. Petroleum-based plastics and petrochemical-based polymers, despite their wide range of uses, are non-biodegradable and can create a slew of issues across

M. D. Ahmad
Department of Post-Harvest Engineering and Technology, Aligarh Muslim University, Aligarh 202002, India

I. Ahmad (✉) · N. Abdullah (✉) · I. Koji · S. E. Mohamad · A. Yuzir
Malaysia-Japan International Institute of Technology, Universiti Teknologi Malaysia, 54100 Kuala Lumpur, Malaysia
e-mail: mustafwibinqamar@gmail.com

N. Abdullah
e-mail: norhayati@utm.my

S. Gautam
Department of Molecular Biology and Genetic Engineering, School of Bioengineering and Biosciences, Lovely Professional University, Phagwara 144411, Punjab, India

M. El-Sheekh
Tanta University, Tanta 31527, Egypt

Sapuan S. M. and I. Ahmad (eds.), *Composites from the Aquatic Environment*, Composites Science and Technology, https://doi.org/10.1007/978-981-19-5327-9_10

the ecosystem. A fast expansion in synthetic plastics production has been linked to significant energy consumption, GHG emissions, and the release of toxic compounds. Many academicians and researchers have tried over the years to identify ecologically benign and sustainable materials to replace plastic production. Bioplastics, which may be made from renewable biomass, agricultural wastes, and microorganisms, are seen as viable alternatives to traditional plastics. As compared to petroleum-based traditional polymers, their manufacture requires very less energy [6].

The initial generation of bioplastics that are made from raw materials, necessitate arable land, groundwater and nutrients, putting food production in jeopardy. The use of agricultural residues can help solve this problem. However, these materials are fewer and inadequate for the manufacturing of bioplastics. As a result, bioplastics made from rapid-growing microbes like bacteria and microalgae have gotten a lot of interest as well. Apart from their high potential for CO_2 mitigation, microalgae require less water to cultivate as compared with land crops, and also, unlike food crops, algae does not use arable land as they can be grown in open and closed photobioreactors, implying that they can be used as a versatile resource for bioplastic production whilst also posing less of a threat to food security [7].

Microalgae (and certain cyanobacteria) can produce large amounts of the lipids, proteins, and carbohydrates, which are the most important components of bioplastics, biopolymers, and biobased polyurethane. Microalgal biomass, whether utilised directly as biofuels or as a raw material for secondary process, has recently been identified as a possible source of material for contributing a variety of industries, including bioplastic manufacture [8].

The term "life cycle assessment" refers to a set of techniques for estimating the environmental impacts of products as they are processed throughout their life span, sometimes known as "cradle to grave" approach [9]. LCA is a process of monitoring and control for examining the environmental implications of items, processes, or behaviors across their entire life cycle. This is done for a product by (1) aggregating the essential inputs and outcomes of a series of operations, (2) assessing the potential consequences among the list, and (3) interpreting the final findings in terms of the assessment's goal and scope [10]. LCA is a useful method for discovering environmental "pinch spots," cost-cutting opportunities, and process trade-offs. As a consequence of these advantages, as well as a shift in public perception toward environmental conservation, LCA has become a widely used industrial tool to evaluate and select novel materials and methods.

It is increasingly extending its position in the construction [11], aerospace [12], wind [13], and automotive sectors [14]. The automobile industry is now required to minimize pollution levels and encourage end-of-life reprocessing (EOL), also with the goal of fulfilling recycling objectives (at least 85% in the EU) [15]. The production chain modelling and quantitative methods provided by the criteria for LCA evaluation may help in making informed changes toward these objectives. The LCA approach has been chosen in the composites industry to emphasize the advantages of lighter materials for replacements of standard materials [16]. Light-weight products like aluminum, magnesium, Carbon Fiber Reinforced Polymeric materials (CFRP), or

Glass Fiber Reinforced composites (GFRP) seem to be more durable than ordinary steel over time [17].

The LCA framework is based on ISO (International Organization of Standardization) guidelines (ISO 14040 2006; ISO 14044 2006). There are four different types of LCAs: (1) identify the goal and scope, (2) identify the life cycle inventories, (3) assessing the effect, and (4) interpreting the results [18].

1.1 LCA Goals and Scope

The goal and scope identify the systems that will be investigated, and the environmental assessment domains and any limitations or assumptions that will be used throughout the assessment. An evaluation of alternative technologies for something like a reference unit, that is a functional unit, such as fixed quantity of material or even a particular part, such as an aviation undercarriage support beam, is used for material selection [19].

1.2 Life Cycle Inventory (LCI)

The life cycle inventory identifies and summarizes all of the unit operation flows important to the production system's (LCI). CFRP product technologies include the whole life cycle of a product, from harvesting of raw materials through end-of-life recycling. The LCI (Fig. 1) is comprised of critical stages of the functional unit's life span that are constrained by system restrictions [20].

Individual LCA studies, such as TEA, are analyzed with the purpose of finding the most common technologies utilized in LCA. This section also explains the most common performance indicators generated by LCA and how they are used in decision-making, such as

- A life cycle assessment (LCA) may be used to optimize processes depending on environmental factors.
- A life cycle assessment (LCA) compares several manufacturing methods to arrive at the same end product.
- LCA is used to determine the most ecologically friendly route for a new technology while it is still in the design stage. Bennion et al. [21] studied several thermochemical techniques for creating biofuel from microalgae. In order to identify creative solutions with the least degree of environmental impact, two distinct conversion techniques for manufacturing biofuel from microalgae are being studied.
- LCA may also be used to evaluate existing alternative design paradigms; three alternative process design models were analyzed using LCA to determine greenhouse gas (GHG) emissions and energy production. The most productive process design with the lowest GHG emissions is identified using the LCA data [22].

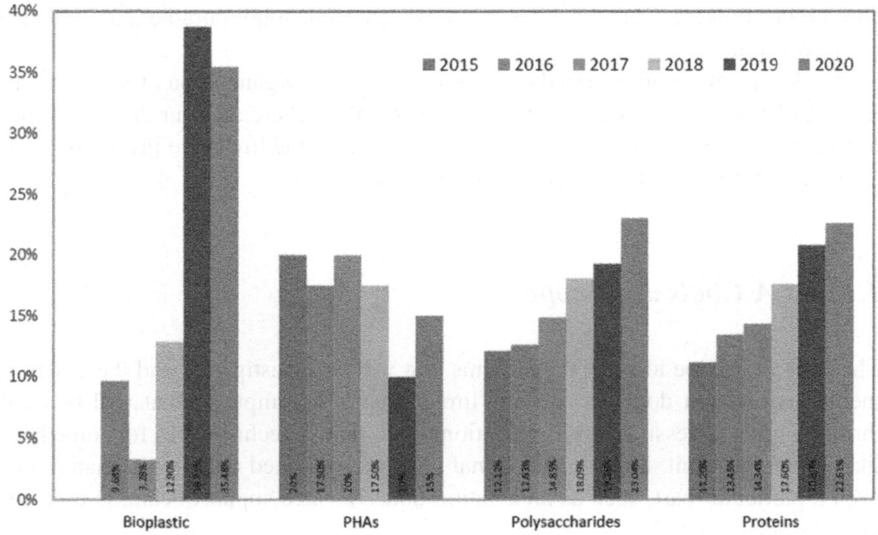

Fig. 1 Scientific publications on bioplastic using microalgae

- LCA may also assist policymakers in making better decisions. LCA was used to look at the global warming potential (GWP), energy consumption, and material input/output of seven different digestate processing procedures for biogas production. The purpose of this LCA was to help policymakers make choices in the agricultural business [23].

1.3 Circular Bioeconomy Contributing to Sustainability

The global population is increasing at a tremendous rate, but the natural resources are limited to support this larger population. The government and researchers are both working to develop a branch of the resource-intensive type of economy that is more resource-efficient and sustainable. Scholars, governments, and global cooperation often suggest this new kind of methodology towards economic movement as being known as a "bioeconomy" or "bio-based economy." The main motive of this type of bioeconomy is to reduce dependency on fossil-based products, as with sustainable bio-based products, that results in reducing environmental impacts due to fossil-based products and global warming [24]. The products from renewable biological resources and the transformation of these resources and their waste streams into useful products, such as food, feed, bio-based products, and bioenergy [25]. The bioeconomy involves various economic sectors such as agriculture, forestry, fisheries, food, bio-based chemicals and materials, and bioenergy. The World Bank clarified some socioeconomic elements, stating that "the ocean economy approach attempts to boost economic development, social inclusion, and to preserve or improve living standards while preserving the environmental health of the seas and coastal regions" [26]. The two main factors on which bioeconomy has been built are as follows: (i) reduce the amount of biowaste that is exploited for the recovery of energy and materials; and (ii) improving bioenergy capacity by the utilization of more efficient genera, biotechnology advancements, and novel extraction and processing techniques. The biggest obstacle in moving towards industrial bio-based goods is ensuring the secure and adequate availability of bioenergy to fulfill the criteria of food and feed supply while being competitively priced. According to this perception, algae cultivation doesn't need or want rich soils or arable land, and it may be cultivated in marginal areas without affecting food and feed yield [27]. The green economy and the circular economy are important for geographic, commercial, end-up-wasting, ecologic, global warming, as well as research policies all have a role in increasing global demand for products and reducing the dependency on fossil resources. The Office of Science and Technology Policy as well as the Executive President's office came up with the first report of the United States' national green economy planning with assistance from a number of different government agencies, and it mostly talks about how to make renewable energy [28].

2 The Studies for the Production of Bioplastics from Microalgae

Studies for the development of bioplastics from algae may be divided into two main categories. Bioplastics are the composites which are prepared from microalgae biomass by blending, bio- or fossil-based polymers and additives. The production of such types of composites is achieved by thermomechanical processes like compression molding. The alternative technique relies on the production of biopolymers like polyhydroxybutyrates (PHBs) and starches inside microalgae cells, and their products may be retrieved and reprocessed for bioplastic production. In this scenario, the microalgae cells aren't employed directly to make biocomposite materials. However, improvements in technology and effective usage are still needed to allow commercialization, industrialization, and scale-up [29].

Species or microalgae strain	Product	Materials ratio	Particle size	Characterization	Refs.
S. platensis and C. vulgaris	100% algae-based plastics Hybrid blends with PE and glycerol	Glycerol 0–30% (by weight)	53–75 μm	57% protein	[30]
S. platensis	Bioplastic biofilm	Compatibilizer concentartion: 0–6%		60% protein	[31]

(continued)

(continued)

Species or microalgae strain	Product	Materials ratio	Particle size	Characterization	Refs.
Spirulina	PBS/*Spirulina* composites	(Varying from 15 to 50% loading) and PBS with and without PBS-g-MAH		60% protein (on dry weight basis)	[32]
Chlorella	PP from *Chlorella* and MPP (maleic anhydride modified polypropylene)	MPP/*Chlorella*: 0.5	50 μm		[33]
C. sp	*Chlorella*/PE composites	*Chlorella*/MPE (Modified PE): 10–40% *Chlorella*/UPE (Unmodified PE): 10–40%	~1 mm		[34]
Chlorogloea fritschii	Bioplastic poly-3-hydroxybutyrate			PHB levels at 14–17% (*w/w* DW)	[35]
Phaeodactylum tricornutum	Bioplastic PHB			PHB levels of up to 10.6% of algal dry weight	[36]
Nannocloropsis gaditana	Biocomposites: biomass and PBAT	Ratios of biomass: 10, 20, 30			[37]

From 2015 to March 2021, 32 research topics concerning "microalgae" were found using the Web of Science (March 2021), as indicated in Fig. 1. The topic "food" has the highest number of papers in the literature connected to microalgae, with 8104, followed by "chemicals" with 7164 and "biofuels" with 6586. These keywords cover a wide range of themes; however, more particular markets appear to have been overlooked thus far, as they only match a few publications. In the disciplines of "proteins for food" (2123 publications), "pigments" (1400 publications), and "biogas," for example, numerous papers have been published (747 publications) so far. From 2015 to 2021, research into the application of microalgae as feedstocks in the context of the biorefinery and the circular economy has been increased. However, so much needs to be examined in terms of developing process systems capable of meeting the circular economy and bioenergy requirements as such a sustainable way [33].

At present, attention from scientists, economists, and industrialists has been shifted towards making the circular economy model (CEM) a reality in the pursuit of resolving issues that are concurrent with world consumption and demand. This model is constructed and conducted majorly upon reserve, revival and reusability approach that aims to scale down on the utilisation of feedstock and natural deposits [1]. The recent circular economy concept has been focusing on the biorefinery of sustainable resources and biomass that would aid in reducing waste disposition and emissions of greenhouse gases (GHG) [38]. According to Zetterholm et al. [4], biorefineries are pivotal in the development of a developing net-zero community.

2.1 Production of Biopolymer Using Microalgae Cells

Algae may be found in a variety of habitats, including freshwater springs, lichens, and marine environments, and they have autotrophic, mixotrophic, and heterotrophic lives. Nutrient deprivation or other cellular stress can be used to control the development of algae in the lab or on a wide scale to promote the creation of enormous amounts of fats, carbohydrates, and hydrocarbons [39]. Microalgae are cultivated in 2 steps: the first one is a continuously cultured with plenty of nitrogen and nutritional media, and the second is when a part of the culture has reached the highest cell density, which is a mode in which salt stress is generated. The cells are now in the process of producing biopolymers. Chemical, enzymatic, biochemical, and mechanical treatment techniques can be used to transform algal biomass containing intrinsic biopolymers or precursors into marketable products [40, 41].

2.1.1 Microalgal Polysaccharides

After protein and pigment extractions, microalgal biomass that has been defatted may still be used to extract carbohydrates and polyester blends. In a typical microalgal biorefinery process, glycerol, a by-product of biodiesel manufacturing, and leftover microalgal biomass can be used to make cellulose, starch, PHA, and biocomposites [42]. Algal biorefineries will be revolutionized and decarbonized as a result of this technique, allowing for a sustainable economy and a cyclic bioeconomy [43].

Many microalgal taxa include cellulose and hemicellulose as structural polysaccharides, *chlorella vulgaris* has been shown to have up to 47.5% of cellulose content. Cellulose nanocrystals obtained from *Dunaliella tertiolecta's* oceanic refined biomass were discovered to be an excellent coupling agent and biofiller. The biocomposites that result have higher modulus and tensile strength. Microalgal cellulose retrieval and usage are still in their infancy, but the microalgal biorefinery technique has great promise for reducing environmental effects and energy magnitude in cellulose manufacturing as compared with present industrialized procedures [44].

Polymeric compounds found outside of cells: Extracellular Polymeric Substances (EPS) are higher-density biopolymers produced by microalgae by the use of various types of methods, which include excretions, secretions, and cell lysis. Lipids, polysaccharides, proteins, various photosynthates, and tiny quantities of DNA make up microalgal EPS, with polysaccharides and proteins accounting for 75–89% of the total [45].

EPS is used in a biogranulation method for waste stream treatment that uses EPS as a significant element of the bio-granule matrix materials. EPS may also be used to make important metabolites and might be used as a medication, antioxidant, or growth regulator [46].

During a brief time of nitrogen deficiency, *Chlorella sorokiniana* utilised starch as a major energy storing component. After two days of inoculation, the starch content in mixotrophic cells reached a high of 27%. Under nitrogen deficient conditions, the starch content of the *Chlorella vulgaris* P12 strain reached up to 41% of the dry weight [47]. Under proof-of-concept trials, in sulphur-depleted conditions, the microalgae *Chlamydomonas reinhardtii* 11-32A strain produced a 49% w/w starch-to-biomass ratio, and the biomass showed strong plasticization ability when combined with glycerol [48].

2.1.2 Microalgal Polyesters

Since 1966, researchers have been interested in PHA, a family that includes polyester blends made by cyanobacteria and found from algae, since it possesses chemical characteristics similar to manufactured polymers. PHA is extracted from specific microalgal species and accumulates as granules in cyanobacterial cytoplasm [49]. PHA concentration level of 2.2 gL^{-1} were obtained when *Haloferax mediterranei* was cultured in *Ulva sp.* hydrosylate [50]. Microbial PHA synthesis for a variety of industrial uses, including food packaging, has already been commercialized, but there is still a demand for it [51].

3 LCA Studies on Bioplastic Production from Microalgae

There aren't many LCA (Life Cycle Assessment) studies especially for microalgae-derived bioplastics. Bussa et al. [52] examined PLA production from microalgae against plant-based sources and discovered that the microalgae approach has great environmental improvement potential in terms of land utilization and terrestrial ecotoxicity. Beckstrom et al. examined the greenhouse gas intensities of several microalgae culture methods for bioplastic manufacture and found that cyclic flow photobioreactors had lower effect values than open raceway ponds and mixed systems [53].

However, the findings of these studies do not reveal how well microalgae-based bioplastics perform in comparison to conventional alternatives. However, LCA research on microalgae cultivation in general might suggest certain tendencies. Compared to fossil fuels, microalgal production systems offer a significantly higher potential for greenhouse gas emission reductions. Data uncertainty plagues LCAs on biofuel generation from microalgae, resulting in widely disparate conclusions [54]. Draaisma et al. [55] discovered that microalgae-based food commodity production is efficient in land usage but not so well in other potential areas, like freshwater requirements.

In general, the environmental advantages of microalgae-based production remain ambiguous, while studies frequently mention microalgal production systems' improvement potential. Synergies might be produced, for example, through biorefineries that produce several goods and their improvement in the techniques of cultivation. It's also possible that by using microalgal waste for bioplastic synthesis, total

LCA scores might improve. Microalgae production technologies now excel mostly in terms of decreased land use [29].

4 Algae Products and Circularity

Products made from algae can be used for human utilization, animal feedstock, farming, power, pharmaceutical drugs, beauty products, raw substances for the chemical sector, and bioremediation, which represents not just the different types of industries that use macro and microalgae, as well as the bioremediation of water and soil, and the reuse of algae biomasses after their first usage. Algae products that are used in food and feed are the most common. Commonly, agar, carrageenan, and alginate, are three of the most common thickeners and gelling agents. They are used in soups, dairy products, fruit preserves, ice cream, and other desserts, as well as in brewing. Currently, it is utilized as a vegetarian substitute for gelatin. Various types of high-value blends are extracted from microalgae, and they have a remarkable contribution to nutraceuticals.

Production of energy is the second major and crucial utilization of algae, followed by food and feed. Processing of algal biomass is done by anaerobic digestion to produce transport fuel, which is used in modified engines with the help of gasification, liquefaction, and pyrolysis, all thermochemical processes that can be used to produce liquid fuels. Biodiesel, kerosene, fats, hydrocarbons, and carbohydrates are some of the compounds extracted [56].

4.1 Production of Materials

The first step in the life cycle of a material's is the synthesis of component fibers and the matrix are depicted. The transformation of unprocessed resources into usable styles may differ noticeably in CED because of differences in the CED of their unit operations involving the formation of fiber and matrix. The following are the findings of the literature review: CED production, GHG emissions, and the cost of basic reinforcement and matrices for FRPs are all factors to be considered. For similar items, both the review and the LCA database provide a wide range of data for environmental footprint, cost, and also CED, but the degree of the variances varies.

Its objective, on the other hand, is to emphasize the need to select the most representative value and compare it to other studies with care. Various countries depend on varied energy source proportions for electrical power, such as fossil energy, natural gas, renewable power, and nuclear energy. Associated with energy demand participation from grid electricity for example, Japan creates a significant proportion of global CF and has relatively high GHG emissions (484 gCO_2e/kWh). Sweden, on the other hand, produces fewer emissions per MJ of electricity produced due to its reliance on renewable energy sources [57].

According to the authors, this has a substantial impact on the accuracy of the composite material impact and is the most important source of value variation, thus needing careful attention early in the composite LCA process. The use of resources from energy, water, or capital are depending mainly upon by the infrastructures, technologies, and methods used; as the variety of manufacturing processes develops, so does the range of materials used. Resources degradation for the production process of raw materials may also be influenced by the scale of the economy's small and big industrial businesses [58]. Multinational companies, such as iron and steel producers, often have processes that have optimized their energy consumption, resulting in considerable saving energy, as compared with smaller rivals. It may be difficult to compare relatively new CFRP manufacturing procedures to metal production techniques that have been refined through decades of process improvement.

4.2 Manufacturing of Polymers/Composites

Carbon fiber-reinforced polymers (CFRP) may be made in a variety of ways, the application's design requirements determine which method is used. Table 1 illustrates the CED and average manufacturing capacity with the most prevalent CFRP production techniques; these figures are normally provided for the production processes alone, excluding those needed to create component material.

The bulk of energy is used to apply pressure and heat during matrix curing and fiber adhesion. Production frequency and product variety are not considered in the estimations, despite the fact that they have implications in downstream processes that might result in environmental impact has increased significantly. For example, pultrusion, is considered a low-energy approach limited to non-complex structures with basic cross-sections. LCE is possibly able to help with this issue by including the intricacy and processability parameters into the technological component of the assessment [58].

4.3 Make Use of

The period when a component is running in its intended application is referred to as the utilization phase of its life cycle. The CED, environmental effects, and expenditures related to each application's utilization phase may be divided into those spent during normal use as well as those spent through maintenance efforts. When evaluating a vehicle's CED and emissions, for example, lifespan, distance traveled, and fuel usage are significant characteristics to consider. When compared to fuel usage, any maintenance or repair contributions are negligible [14]. This phase consumes the majority of a vehicle's life cycle energy, accounting for 60–84% of the overall energy spent. Due to the effects of vehicle weight, that is fuel economy [59]. In

Table 1 Growth of microalgae in different wastewaters, and probable polymer products that could be synthesized from the resultant biomass

Type of wastewater	Composition	Species	Refs.	Probable polymer and composites
Dairy wastewater	Chloride, sulphide, nitrate, nitrite, phosphate, TDS, TSS, lactose	*C. pyrenoidosa*	[76]	Algae-polymer composites cellulose
	Phosphates, nitrate, nitrogen-ammonia, TDS, TSS	*Scenedesmus quadricauda, Tetraselmis suecica*	[77]	Defatted algae-based composites PHA PLA Starch
Slaughterhouse, wastewater	Nitrate, phosphate, COD, BOD, TDS, Iron, sulphide, hardness	*C. pyrenoidosa*	[78]	Defatted algae-based composites algae-polymer composites
Municipal wastewater	COD, TN, TP, ammonia	*Chlorella, Haematococcus, Scenedesmus, Chlamydomonas, Chloroccum*	[79]	Algae-polymer composites cellulose
	COD, TN, TP, ammonia	*Chlorella minutissima*	[80]	Defatted algae-based composites
Municipal, dairy, pulp and paper wastewater mixture	Nitrate, ammonium phosphate	*Selenastrum minutum*	[81]	PHA
Municipal wastewater centrate	Nitrogen, phosphorous, COD, metal ions	*Chlorella* sp.	[82]	PLA starch

comparison, the manufacturing process accounts for just 4–7% of the entire lifetime energy consumption of a mild steel passenger cars built with current technology [60].

Because lower powertrain demands such as rolling resistance as well as acceleration are directly proportional to vehicle weight, 75% of fuel usage is directly related to vehicle weight, and every 10% weight reduction results in a 6–8% increase in fuel efficiency. As a result, the light weighting advantages of CFRP are much more apparent in the transportation industry, where the maximum number of usage stages is if the utilized period is long enough, the PMB may be compensated by fuel usage reductions realized during the usage phase [61].

4.4 End of Life's Scenario (EOL)

The last phase of such an LCA examines how a material is treated as it approaches the end of its useful life. This requires melt reprocessing of EOL scrap for steel and aluminum, which produces between 95 and 100% [62]. The most popular end-of-life alternatives for CFRP are landfills, incineration, and, most recently, recycle operations. The most popular examples are mechanical grinding and fiber reclamation. Chemical procedures (solvolysis [63], acidic digesting [64], solvolysis of supercritical fluid extraction [65]) and thermal techniques (pyrolysis [66], microwave pyrolysis [67], fluidized bed pyrolysis [68]) are now the two main options for higher value component recovery but the amount of manufacturing varies from commercial to laboratory. Pimenta & Pinho and Oliveux et al. discussed several possible approaches for remanufacturing recycled fibers into new feedstock material [69]. The market for CF is presently modest due to the paucity of commercial-scale remanufacturing techniques. The great majority of CFRP recycling life cycle assessments (LCAs) concentrate just on fiber reclamation, with no realistic remanufacturing steps included [63].

5 Microalgal Biopolymers from Wastewater Cultivated Biomass

Microalgae productivity is affected by light, temperature, the amount and kind of energy source and essential nutrients, the existence of predators, the density of the culture, the species cultivated, as well as other culture parameters like pH or salinity. A cultivation plan is established based on specific uses, species, culture medium, and circumstances, as well as an overall cultivation strategy [70].

Biopolymer synthesis from microalgae on a large scale necessitates the development of particular techniques that take into account the aforementioned characteristics. The major input expenses for biopolymer manufacturing from microalgae are nutrient supply, extraction method, and energy source. While heat and light intensity in large outdoor growing systems are difficult to control, while the synthetic medium for bulk biomass productiont is not cost-effective, implying that wastewater streams might be used for long-term biomass accumulation [71]. Numerous studies have demonstrated the value of using agricultural wastes such as dairy-derived liquid digestate (DLD), distillation wastes, and domestic waste also including wastewaters for microalgal cultivation in conjunction with sustainable treating wastewater, biomass transformation, and biocomposites production, whether diluted or undiluted [72]. *Chlamydomonas* sp. exhibited huge biomass production of 3.1 g/l in sterile waste from a local wastewater treatment plant, as well as resistance to a range of temperatures and light conditions [73].

Higher concentrations of unsterilized DLD inhibited the development of *Chlorella vulgaris*, however 25% DLD allowed the microalgae to grow at a rate of 0.69 d^{-1} [74].

On the 15th day after incubating, the dry basis output of *Chlorella pyrenoidosa* was reported to be 6.8 g/L in 75% unprocessed dairy effluent, which was four times the growth rate in the culture media. By using dairy effluent as a growth medium, they are able to boost biomass production while eliminating nitrogen and phosphates from the effluent. The biomass generated was subsequently converted into biofuels. Sustainable algae-based biofineries have been proposed for treating wastewater, biocomposite and biofuel synthesis, CO_2 fixation, and electricity production [75].

Combining the synthesis of biopolymers and biofuel in algae biorefineries based on wastewater might lower the costs and speed up the deployment of algal biofuels. Furthermore, this improves greenhouse gas emission reductions, resulting in decarbonization and potentially leading to the carbon-negative route. As a consequence, testbed algal biorefineries using wastewater for the closed-loop production of bioplastics, biofuel, and high-value bio-based products must be established in the near future, as integrated, large-scale research is limited [41]. Table 1 summarizes the numerous wastewater utilized in microalgae production, their content, and the potential polymer products (this list is merely suggestive and not complete).

5.1 PHA Manufacturing from Several Industries Waste Streams

The selection of appropriate raw materials for biocomposite manufacturing is critical since it may have a significant influence on the environmental impact of the process. PHA synthesis using glucose from maize as a source has been reported to have a deleterious effect on the eco-balance as a result of photochemical haze, eutrophication, and acidification related to corn farming. As part of the effort for transition to a circular bioeconomic model, treating wastewater facilities are increasingly being seen as end-of-pipe activities inside bio refineries systems [83].

The essential method in this scenario is to use renewable, economical, and widely accessible carbon substrates that can support both microbial growth and PHA synthesis at a low cost. Microorganisms may recover PHA from a number of carbon sources, including low-cost, convoluted waste outputs and fatty acids [84]. Several scientific studies have shown this to be true: waste sources which are used to make PHAs include domestic wastewater, kitchen wastes, molasses, olive oil milling wastes, palm oil processing industrial effluents, lingo-cellulosic residue, cannery waste, biofuel industry waste, waste cooking oil, paper processing industry wastewater and effluent, and dairy effluent.

Reduced distance for raw material transportation is critical for reducing fuel demands and gaseous emissions, thus the location of the production site as well as the amount of readily available resources must be considered. To be adopted, these innovative waste-to-product methods must fulfil three key criteria: they must be more environmentally friendly than standard manufacturing processes, create enough revenue to be economically viable, and be socially acceptable [85]. Bio-plastics may be combined using industrial advancement in regards of output, yielding, production efficiency, downstream processing (DSP) (product separation, refining), and waste stream recycling to provide environmental security as well as economic benefits [86].

5.2 *Biofuel Industry Crude Glycerol*

In recent years, increasing biodiesel production has led to a dramatic reduction in the price of glycerol, a key byproduct of biodiesel production. Glycerol has therefore gained popularity as a viable white biotechnology substrate [87]. Because glycerol has more atoms of carbon than carbohydrates, cells that employ it maintain a more condensed physiological condition. These carbon atoms help in the synthesis of intracellular polymers by providing a route. As compared with the clean culture, the use of such MMC for PHA production is seen as a strategy to reduce the environmental impact of the process. On the other hand, studies has shown the use of MMC as a substrate for preparative applications employing biological effluent as a substrate [88].

5.3 *Wastewater & Activated Sludge from Pulp and Paper Mills*

Paper and pulp production has developed to become one of the world's most significant industrial sectors. An organic management facility for these industries generates tonnes of additional sludge per day on average. Furthermore, these mills typically generate huge amounts of effluent, especially from raw materials, which has the potential to harm the aquatic ecosystem [89]. Depending on the pulping technique, added chemicals, and quantity of water employed, the substrates may also comprise non-compostable adsorbable biological halogen (AOX), organic compounds, phenols, color, and other compounds [90]. Microbes that assist organisms identify carbon stores inside the body are subjected to selection of pressure as a result of these dynamic conditions in the manufacturing of PHA, activated sludge is expected to be less expensive as compared to the pure culture. This is the case because reactor sterilization is not required, and wastewater organic substances may be considered at a minimal cost. Activated sludge from synthetic wastewater has been used to study PHA production in a wide range of composites [91].

5.4 *Whey from Dairy Industry*

The global output of whey is about 1.4 tonnes per year [92]. It has two advantages for PHA manufacturing because it is a low-cost raw material. Lactose, the main carbon source in whey, could be used to help develop and expand products. The full whey lactose is thought to be the most cost-effective way to make PHA. By transforming polluted whey into useful items, the usage of excess whey combines cost-effective benefits with environmental sustainability [93].

5.5 Food Industry Agro-food Wastes

Massive amounts of food waste have resulted from an increase in demand for food production as a result of population growth, as well as the food processing sector associated with it. Food waste contains complex proteins, carbohydrates, lipids, and nutraceuticals [94]. Lignocellulosic biomass is made up of three main components: hemicellulose, cellulose, and lignin, and is classified as agro-food waste. It is the fermentation of a wide range of metabolites using a specific microorganism. Lignocellulosic elements, as well as waste from food and agriculture, were used as sources for the production of PHA [95].

5.6 Recycling from the Waste Stream

PHA production is also increased using the feed-forward method. During the previous reaction cycle, biomass from the next reaction step is introduced into the culture of the following reaction step. It's beneficial since it reduces carbon emissions by eliminating the demand for new biomass. Wei et al. [96] employed pyrolysis to make bio-oil and charcoal from residual bacterial biomass (RBB). Bio-oil and charcoal production were 28% and 48%, respectively. Koller [97] used halophile *Haloferax mediterranei* to produce PHA from whey from the dairy sector and then recycled the waste stream for future industrial uses.

6 The Role of Downstream Processing (DSP) in the Circular Bioeconomy

To understand how the PHA manufacturing process fits into the circular bioeconomy, it's necessary to look at the complete process. Bioplastics like this should be used to outperform traditional plastics in terms of total environmental impact. The synthesis of PHA necessitates the use of DSP, and the separation of PHA from algal biomass must be efficient and adequate. After fermentation, there have been two methods for recovering PHAs: disintegrating the organic matter with acids, alkalis, surfactants, and enzymes to separate PHA granules; or retrieving PHAs from microalgal biomass using a solvent that changes the cell membrane permeability and solubilizes the biopolymer inside the cells [98].

PHA recovery may be a major financial achievement as well as have a considerable impact on the manufacturing process total environmental footprint. Innovative extraction techniques that utilize recyclable, ecologically acceptable solvents such as lactic acid esters [99] are required.

Quantitative release of PHA granules can be achieved using techniques such as ultrasonication and enzymatic digestion, which do not require the use of too much solvents. The PHA recovery strategies are selected based on the circumstances like the strains that produce PHA, the required product purification, the accessibility of material for separation technologies, and the permissible molar size of the biopolymer [100].

6.1 Downstream Processing LCA

It aids academics, policymakers, and decision-makers by focusing on the most important process stages or variables that result in higher GHG emissions or power consumption. However, research shows that attempts to use LCA techniques to uncover and estimate the impact on the environment of PHA synthesis focus on isolated components of manufacturing, including solely biopolymer production, energy requirements, or emissions of carbon dioxide, that might or might not be in agreement with one another [99]. PHA extraction using dimethyl carbonate (DMC) was subjected to a life cycle assessment (LCA) and compared to halogenated hydrocarbons [101].

DMC-based extraction has been found to outperform halogenated hydrocarbons in terms of environmental performance. Four models were explored using the DMC procedure: extraction using microbial sludge or dehydrated biomass and recovery using evaporation of the solvent or polymeric precipitation. Extraction of dry microalgae biomass or precipitation extraction of PHB were determined to be the most promising methods. When LCA was used to start comparing treated by NaOH or H_2SO_4 (4.08 kg CO_2 and 6.27 kg CO_2 equivalents per hour) respectively, to sodium hypochlorite digestion (29.46 kg CO_2 equivalents per hour), sodium hypochlorite digestion seemed to have the maximum carbon footprint (29.46 kg CO_2 equivalents per hour) [102].

The least expensive recovery method (1.12 $/kg PHA) was sodium hypochlorite, followed by sulphuric acid (1.22 $/kg PHA). All approaches result in a decrease in the molecular weight of the polymers when compared to normal chloroform extraction. The treatment done by sulphuric acid was the most effective of all the extraction methods, with a high purity of 98%, and recovery rate of 79% and minimum GHG emissions [102].

The alkali treatment has the lowest GHG emissions and NREU of 106 MJ per kilogram of PHB production, whereas the treatments by solvent have the maximum GHG emissions and NREU of 4.3 kg CO_2 equivalent and 156 MJ per kilogram of PHB, as per their LCA. PHA synthesis using purified glycerol as a substrate and cultures that are monoseptic does not give much environmental gain over typical polymers, according to the various studies, because of the significant energy usage during the manufacturing process. As a consequence, identifying ecological hotspots early in the process is crucial. The Sustainable Process Index (SPI) is one tool for doing so, since it indicates key features connected to PHA manufacturing and its environmental effects, including process output, energy usage, and CO_2 emissions [99].

Energy consumption is a significant source of environmental pressure in a method, while process yield refers to the weight placed on the quantity of product by the process's pressure. The clean technology tool that focuses on reducing waste and emissions while increasing productivity is another model approach for environmental evaluation. Material utilization, wastewater avoidance, reducing excess heat, and emissions into the atmosphere are examples of areas where industries may enhance their material and energy flow. According to this concept, PHA production must go for zero-emission, which implies no wastewater outflow, no greenhouse gas emissions, or no solid waste [99].

However, applying the Cleaner Synthesis ideas to biotechnological applications, particularly in the domain of PHA generation, requires expertise and knowledge. It will aid in the optimization of future PHA manufacturing processes, reducing waste and conserving energy. Such studies and methodologies are very valuable in demonstrating how bioplastics production may be included into long-term sustainability patterns. Other techniques to assess sustainability used in similar studies include carbon emissions, carbon efficiency, health & security scoring, and microalgae biomass usage efficiency.

7 Biocomposite Manufacture based Techno-economic Analysis

To comprehend PHA manufacture utilizing waste materials in the perspective of the circular economy, a techno-economic evaluation must be included in the process. Any method's industrial feasibility, as well as the main process factors that determine manufacturing costs, would be shown by techno-economic studies. It helps academics build a cost-effective solution from an industry standpoint by identifying bottlenecks in a process. *Haloferax mediterranei* was employed in one of the studies to manufacture PHA from waste sludge from rice-based industries for the production of biofuel [103].

A PHA concentration of 13.12 g/L was achieved in 135 h, with a 63% (w/w) PHA content. To desalinate the waste stillage medium, a stirrer with axial and transverse vanes was utilized in a cylindrical agitator with an embedded heater. Salts were found, retrieved, and re-used for PHA synthesis during desalination (99.3%). For an annual output of 1890 tonnes, PHA was estimated to cost $2.05 per kilogramme. Desalination was the most costly part. *Cupriavidus necator* was employed as the microorganism in another investigation, and citric acid waste was used as the feedstock of carbon during the fermentation [104].

After 42 h of culture, a biomass composition of 61.6 g/L was obtained, with a PHB composition of 68.8%. According to a techno-economic analysis, increasing the PHB concentration in the fermentation process ranging from 42.5 to 96.6 g/L using various processing modification techniques reduces the system upstream operational costs (plant having capacity of 2000 metric tons) by 1.62 to 0.93 $/kg PHB whilst also lowering the unit production cost ranging from 4.28 to 3.5 $/kg PHB. According to the findings, the amount of PHB in the fermenter has a significant economic impact. PHB was recovered via a surfactant-hypochlorite chemical method from industrial wastewater containing a microbial colony [105].

In a techno-economic study, the unit cost of manufacturing was estimated to be 4.1–6.8 $/kg PHB, and this was less than the prior study's usual cost of $7.5/kg. Raw material prices were lowered from 30 to 22% of operating expenses, proving that methane is a cost-effective carbon source. It is found that the cost of drying biomass has been key cost component. Acetone loss must be maintained to a minimum during extraction and retrieval of PHB. Costs produced are at 3.2–5.4 $/kg of PHB, if PHB producing microorganisms that arethermophilic methanotrophs are discovered and exploited. Soybean oil, wasted cooking oil (WCO), processed and raw glycerol were investigated as carbohydrates for PHA synthesis in *C. necator* [106].

PHA concentrations in various substrates were 20.73, 11.05, 31.07, and 25.01 g/l after 72 h. Dichloromethane and ethanol were used to extract PHA. These substrates had production costs for one kilogram of PHB as 1.63, 1.18, 0.48, and 0.36 USD, respectively. The study also found that raw glycerol was the best carbon source for PHA synthesis, and that the cost of PHA production was strongly dependent on the amount of PHA produced during fermentation and the price of carbon substrates. PHA output, PHA composition, PHA yields (in terms of carbon substrates), and carbon substrate price are all major cost determining factors in PHA synthesis during upstream processing. The cost of carbon sources is reduced by using domestic as well as industrial waste and a mixed microbe that does not need sterilization.

The PHA content, on the other hand, is the most important factor since it influences the final PHA output as well as the downstream process efficiency. Plant capacity has an impact on the cost of generating PHA. In a study, the unit cost of production of a PHA facility was reduced from $4.29/kg to $2.71/kg when the capacity was raised from 2000 to 10,000 tonnes per year [104].

Aside from process characteristics, the manufacturer must be aware of plant capacity as well as demands. Four techniques for pretreatment of fermented organic material were investigated before being extracted by employing propylene carbonate as a solvent. Ultrasonication (10 kHz), thermal pre-treatment (95 °C for 45 min), high intensity pressure (90 MPa), and no prior-treatment were all investigated for fermented biomass. The ultimate PHB extraction efficiency for the four pre-treatment operations was 92.2, 92.1, 97.8, and 81.7%, respectively. For the above mentioned pre-treatment processes, one kilogram of PHB production cost were 4.46, 4.28, 4.28, and 4.72 USD respectively. High pressure and heat pre-treatment were the most cost-effective pre-treatment processes in PHB's DSP. Another study used a mixed microbial population to make PHB [85]. Chemical treatments of biomasses using 0.2 M NaOH and 0.2% (w/v) surfactants, surfactant and NaOCl (sodium hypochlorite), and dichloromethane were all examined for DSP (solvent) with a manufacturing cost for one kilogram was 1.54$, when the alkali treatment was found to be the most cost-effective, and 2.15$ when processing done by using a solvent-based technique and was found to be the least cost-effective.

8 Conclusion

Synthetic plastics are more feasible but are more hazardous than bioplastics. Dependence on synthetic plastics will negatively impact the environment and health. However, PHAs, PLAs and starch do not produce any health hazards while degrading, they reduce carbon dioxide emissions during their formation and degrade to organic matter after being discarded. Thereby making it as environmentally sustainable. Therefore, with the growing demand, manufacturing of biopolymers and fibers needs development and research for their feasible applications and in the evolution of new technologies.

References

1. Ghisellini, P., Cialani, C., & Ulgiati, S. (2016). A review on circular economy: The expected transition to a balanced interplay of environmental and economic systems. *Journal of Cleaner Production, 114*, 11–32.
2. Ness, D. (2008). Sustainable urban infrastructure in China: Towards a factor 10 improvement in resource productivity through integrated infrastructure systems. *The International Journal of Sustainable Development & World Ecology, 15*(4), 288–301.
3. Cherubini, F. (2010). The biorefinery concept: Using biomass instead of oil for producing energy and chemicals. *Energy Conversion and Management, 51*(7), 1412–1421.
4. Zetterholm, J., Bryngemark, E., Ahlström, J., Söderholm, P., Harvey, S., & Wetterlund, E. (2020). Economic evaluation of large-scale biorefinery deployment: A framework integrating dynamic biomass market and techno-economic models. *Sustainability, 12*(17), 7126.
5. Modolo, R., Benta, A., Ferreira, V., & Machado, L. (2010). Pulp and paper plant wastes valorisation in bituminous mixes. *Waste Management, 30*(4), 685–696.
6. Horton, A. A., Walton, A., Spurgeon, D. J., Lahive, E., & Svendsen, C. (2017). Microplastics in freshwater and terrestrial environments: Evaluating the current understanding to identify the knowledge gaps and future research priorities. *Science of the Total Environment, 586*, 127–141.
7. Madadi, R., Tabatabaei, M., Aghbashlo, M., Zahed, M. A., & Pourbabaee, A. A. (2018). Biodiesel from microalgae. In *Waste to Wealth* (pp. 277–318). Springer.
8. Mehta, P., Singh, D., Saxena, R., Rani, R., Gupta, R. P., Puri, S. K., & Mathur, A. S. (2018). High-value coproducts from algae—An innovational way to deal with advance algal industry. *Waste to Wealth* (pp. 343–363). Springer.
9. Papadaki, S. G., Kyriakopoulou, K. E., & Krokida, M. K. (2016). Life cycle analysis of microalgae extraction techniques. *Chemical Engineering Transactions, 52*, 1039–1044.

10. Clift, R., Doig, A., & Finnveden, G. (2000). The application of life cycle assessment to integrated solid waste management: Part 1—Methodology. *Process Safety and Environmental Protection, 78*(4), 279–287.
11. Zhang, C. (2015). The environmental impacts of fibre-reinforced polymer composites in construction. *Proceedings of the Institution of Civil Engineers-Construction Materials, 168*(6), 276–286.
12. Bachmann, J., Hidalgo, C., & Bricout, S. (2017). Environmental analysis of innovative sustainable composites with potential use in aviation sector—A life cycle assessment review. *Science China Technological Sciences, 60*(9), 1301–1317.
13. Merugula, L. A., Khanna, V., & Bakshi, B. R. (2010). Comparative life cycle assessment: Reinforcing wind turbine blades with carbon nanofibers. In *Proceedings of the 2010 IEEE International Symposium on Sustainable Systems and Technology*. IEEE.
14. Kim, H. C., & Wallington, T. J. (2013). Life-cycle energy and greenhouse gas emission benefits of lightweighting in automobiles: Review and harmonization. *Environmental Science & Technology, 47*(12), 6089–6097.
15. Aolaid, A., & Teder, J. (2006). The dismantling of end-of-life vehicles in Estonia before putting into effect the Directive 2000/53/EC (3 pp). *Environmental Science and Pollution Research, 13*(2), 140–142.
16. Das, S. (2011). Life cycle assessment of carbon fiber-reinforced polymer composites. *The International Journal of Life Cycle Assessment, 16*(3), 268–282.
17. Suzuki, T., & Takahashi, J. (2005). Prediction of energy intensity of carbon fiber reinforced plastics for mass-produced passenger cars. In *Proceedings of 9th Japan International SAMPE Symposium*.
18. Klöpffer, W. (2012). The critical review of life cycle assessment studies according to ISO 14040 and 14044. *The International Journal of Life Cycle Assessment, 17*(9), 1087–1093.
19. Meng, F., McKechnie, J., Turner, T., & Pickering, S. (2017). Energy and environmental assessment and reuse of fluidised bed recycled carbon fibres. *Composites Part A: Applied Science and Manufacturing, 100*, 206–214.
20. Azapagic, A. (1999). Life cycle assessment and its application to process selection, design and optimisation. *Chemical Engineering Journal, 73*(1), 1–21.
21. Bennion, E. P., Ginosar, D. M., Moses, J., Agblevor, F., & Quinn, J. C. (2015). Lifecycle assessment of microalgae to biofuel: Comparison of thermochemical processing pathways. *Applied Energy, 154*, 1062–1071.
22. Quinn, J. C., Smith, T. G., Downes, C. M., & Quinn, C. (2014). Microalgae to biofuels lifecycle assessment—Multiple pathway evaluation. *Algal Research, 4*, 116–122.
23. Rehl, T., & Müller, J. (2011). Life cycle assessment of biogas digestate processing technologies. *Resources, Conservation and Recycling, 56*(1), 92–104.
24. Özçimen, D., İnan, B., Koçer, A. T., & Vehapi, M. (2018). Bioeconomic assessment of microalgal production. *Microalgal Biotechnology, 195*.
25. Europeia, C. (2012). Communication from the Commission to the European Parliament, the Council, the European Economic and Social Committee and the Committee of the Regions. In *A Roadmap for moving to a competitive low carbon economy in* (Vol. 2050).
26. Bank, W., Economic, U. N. D. o., & Affairs, S. (2017). *The potential of the blue economy: Increasing long-term benefits of the sustainable use of marine resources for small island developing states and coastal least developed countries*. World Bank.
27. Trentacoste, E. M., Martinez, A. M., & Zenk, T. (2015). The place of algae in agriculture: Policies for algal biomass production. *Photosynthesis Research, 123*(3), 305–315.
28. Ronzon, T., & M'Barek, R. (2018). Socioeconomic indicators to monitor the EU's bioeconomy in transition. *Sustainability, 10*(6), 1745.
29. Onen Cinar, S., Chong, Z. K., Kucuker, M. A., Wieczorek, N., Cengiz, U., & Kuchta, K. (2020). Bioplastic production from microalgae: A review. *International Journal of Environmental Research and Public Health, 17*(11), 3842.
30. Zeller, M. A., Hunt, R., Jones, A., & Sharma, S. (2013). Bioplastics and their thermoplastic blends from Spirulina and Chlorella microalgae. *Journal of Applied Polymer Science, 130*(5), 3263–3275.

31. Dianursanti, Noviasari, C., Windiani, L., & Gozan, M. (2019). Effect of compatibilizer addition in Spirulina platensis based bioplastic production. In *AIP Conference Proceedings*. AIP Publishing LLC.
32. Zhu, N., Ye, M., Shi, D., & Chen, M. (2017). Reactive compatibilization of biodegradable poly (butylene succinate)/Spirulina microalgae composites. *Macromolecular Research, 25*(2), 165–171.
33. Zhang, Q., Wang, J., Lyu, H., Zhao, Q., Jiang, L., & Liu, L. (2019). Ball-milled biochar for galaxolide removal: Sorption performance and governing mechanisms. *Science of the Total Environment, 659*, 1537–1545.
34. Otsuki, T., Zhang, F., Kabeya, H., & Hirotsu, T. (2004). Synthesis and tensile properties of a novel composite of Chlorella and polyethylene. *Journal of Applied Polymer Science, 92*(2), 812–816.
35. Monshupanee, T., Nimdach, P., & Incharoensakdi, A. (2016). Two-stage (photoautotrophy and heterotrophy) cultivation enables efficient production of bioplastic poly-3-hydroxybutyrate in auto-sedimenting cyanobacterium. *Scientific Reports, 6*(1), 1–9.
36. Hempel, F., Bozarth, A. S., Lindenkamp, N., Klingl, A., Zauner, S., Linne, U., Steinbüchel, A., & Maier, U. G. (2011). Microalgae as bioreactors for bioplastic production. *Microbial Cell Factories, 10*(1), 1–6.
37. Torres, S., Navia, R., Campbell Murdy, R., Cooke, P., Misra, M., & Mohanty, A. K. (2015). Green composites from residual microalgae biomass and poly (butylene adipate-co-terephthalate): Processing and plasticization. *ACS Sustainable Chemistry & Engineering, 3*(4), 614–624.
38. Boccia, F., Di Donato, P., Covino, D., & Poli, A. (2019). Food waste and bio-economy: A scenario for the Italian tomato market. *Journal of Cleaner Production, 227*, 424–433.
39. Kakarla, R., Choi, J.-W., Yun, J.-H., Kim, B.-H., Heo, J., Lee, S., Cho, D.-H., Ramanan, R., & Kim, H.-S. (2018). Application of high-salinity stress for enhancing the lipid productivity of Chlorella sorokiniana HS1 in a two-phase process. *Journal of Microbiology, 56*(1), 56–64.
40. Kartik, A., Akhil, D., Lakshmi, D., Gopinath, K. P., Arun, J., Sivaramakrishnan, R., & Pugazhendhi, A. (2021). A critical review on production of biopolymers from algae biomass and their applications. *Bioresource Technology, 329*, 124868.
41. Mohan, S. V., Hemalatha, M., Chakraborty, D., Chatterjee, S., Ranadheer, P., & Kona, R. (2020). Algal biorefinery models with self-sustainable closed loop approach: Trends and prospective for blue-bioeconomy. *Bioresource Technology, 295*, 122128.
42. Kostag, M., & El Seoud, O. A. (2021). Sustainable biomaterials based on cellulose, chitin and chitosan composites-A review. *Carbohydrate Polymer Technologies and Applications, 2*, 100079.
43. Rai, P., Mehrotra, S., Priya, S., Gnansounou, E., & Sharma, S. K. (2021). Recent advances in the sustainable design and applications of biodegradable polymers. *Bioresource Technology, 325*, 124739.
44. Zanchetta, E., Damergi, E., Patel, B., Borgmeyer, T., Pick, H., Pulgarin, A., & Ludwig, C. (2021). Algal cellulose, production and potential use in plastics: Challenges and opportunities. *Algal Research, 56*, 102288.
45. Babiak, W., & Krzemińska, I. (2021). Extracellular polymeric substances (EPS) as microalgal bioproducts: A review of factors affecting EPS synthesis and application in flocculation processes. *Energies, 14*(13), 4007.
46. Xiao, R., & Zheng, Y. (2016). Overview of microalgal extracellular polymeric substances (EPS) and their applications. *Biotechnology Advances, 34*(7), 1225–1244.
47. da Maia, J. L., Cardoso, J. S., da Silveira Mastrantonio, D. J., Bierhals, C. K., Moreira, J. B., Costa, J. A. V., & de Morais, M. G. (2020). Microalgae starch: A promising raw material for the bioethanol production. *International Journal of Biological Macromolecules, 165*, 2739–2749.
48. Mathiot, C., Ponge, P., Gallard, B., Sassi, J.-F., Delrue, F., & Le Moigne, N. (2019). Microalgae starch-based bioplastics: Screening of ten strains and plasticization of unfractionated microalgae by extrusion. *Carbohydrate Polymers, 208*, 142–151.

49. García, G., Sosa-Hernández, J. E., Rodas-Zuluaga, L. I., Castillo-Zacarías, C., Iqbal, H., Parra-Saldívar, R. (2021). Accumulation of PHA in the microalgae Scenedesmus sp. under nutrient-deficient conditions. *Polymers, 13*(1), 131.
50. Ghosh, S., Gnaim, R., Greiserman, S., Fadeev, L., Gozin, M., & Golberg, A. (2019). Macroalgal biomass subcritical hydrolysates for the production of polyhydroxyalkanoate (PHA) by Haloferax mediterranei. *Bioresource Technology, 271*, 166–173.
51. Zhao, X., Cornish, K., & Vodovotz, Y. (2020). Narrowing the gap for bioplastic use in food packaging: An update. *Environmental Science & Technology, 54*(8), 4712–4732.
52. Bussa, M., Eisen, A., Zollfrank, C., & Röder, H. (2019). Life cycle assessment of microalgae products: State of the art and their potential for the production of polylactid acid. *Journal of Cleaner Production, 213*, 1299–1312.
53. Beckstrom, B. D., Wilson, M. H., Crocker, M., & Quinn, J. C. (2020). Bioplastic feedstock production from microalgae with fuel co-products: A techno-economic and life cycle impact assessment. *Algal Research, 46*, 101769.
54. Medeiros, D. L., Sales, E. A., & Kiperstok, A. (2015). Energy production from microalgae biomass: Carbon footprint and energy balance. *Journal of Cleaner Production, 96*, 493–500.
55. Draaisma, R. B., Wijffels, R. H., Slegers, P. E., Brentner, L. B., Roy, A., & Barbosa, M. J. (2013). Food commodities from microalgae. *Current Opinion in Biotechnology, 24*(2), 169–177.
56. Vigani, M. (2020). The bioeconomy of microalgae-based processes and products. In *Handbook of microalgae-based processes and products* (pp. 799–821). Elsevier.
57. Chua, M. H., Smyth, B. M., Murphy, A., & Butterfield, J. (2015). Understanding aerospace composite components' supply chain carbon emissions. In *Proceedings of the Irish Manufacturing Conference (IMC32)*, Belfast, UK.
58. Song, Y. S., Youn, J. R., & Gutowski, T. G. (2009). Life cycle energy analysis of fiber-reinforced composites. *Composites Part A: Applied Science and Manufacturing, 40*(8), 1257–1265.
59. Wheatley, A., Warren, D., & Das, S. (2013). Development of low-cost carbon fibre for automotive applications. In *Advanced Composite Materials for Automotive Applications: Structural Integrity and Crashworthiness* (pp. 51–73).
60. Hottle, T., Caffrey, C., McDonald, J., & Dodder, R. (2017). Critical factors affecting life cycle assessments of material choice for vehicle mass reduction. *Transportation Research. Part D, Transport and Environment, 56*, 241.
61. Vaidya, U., & Chawla, K. (2008). Processing of fibre reinforced thermoplastic composites. *International Materials Reviews, 53*(4), 185–218.
62. Bushi, L., Skszek, T., & Reaburn, T. (2019). New ultralight automotive door life cycle assessment. *The International Journal of Life Cycle Assessment, 24*(2), 310–323.
63. Oliveux, G., Dandy, L. O., & Leeke, G. A. (2015). Degradation of a model epoxy resin by solvolysis routes. *Polymer Degradation and Stability, 118*, 96–103.
64. Dang, W., Kubouchi, M., Yamamoto, S., Sembokuya, H., & Tsuda, K. (2002). An approach to chemical recycling of epoxy resin cured with amine using nitric acid. *Polymer, 43*(10), 2953–2958.
65. Piñero-Hernanz, R., García-Serna, J., Dodds, C., Hyde, J., Poliakoff, M., Cocero, M. J., Kingman, S., Pickering, S., & Lester, E. (2008). Chemical recycling of carbon fibre composites using alcohols under subcritical and supercritical conditions. *The Journal of Supercritical Fluids, 46*(1), 83–92.
66. Meyer, L. O., Schulte, K., & Grove-Nielsen, E. (2009). CFRP-recycling following a pyrolysis route: Process optimization and potentials. *Journal of Composite Materials, 43*(9), 1121–1132.
67. Lester, E., Kingman, S., Wong, K. H., Rudd, C., Pickering, S., & Hilal, N. (2004). Microwave heating as a means for carbon fibre recovery from polymer composites: A technical feasibility study. *Materials Research Bulletin, 39*(10), 1549–1556.
68. Pickering, S. J., Kelly, R. M., Kennerley, J., Rudd, C., & Fenwick, N. (2000). A fluidised-bed process for the recovery of glass fibres from scrap thermoset composites. *Composites Science and Technology, 60*(4), 509–523.

69. Pimenta, S., & Pinho, S. T. (2011). Recycling carbon fibre reinforced polymers for structural applications: Technology review and market outlook. *Waste Management, 31*(2), 378–392.
70. Karan, H., Funk, C., Grabert, M., Oey, M., & Hankamer, B. (2019). Green bioplastics as part of a circular bioeconomy. *Trends in Plant Science, 24*(3), 237–249.
71. Dobroth, Z. T., Hu, S., Coats, E. R., & McDonald, A. G. (2011). Polyhydroxybutyrate synthesis on biodiesel wastewater using mixed microbial consortia. *Bioresource Technology, 102*(3), 3352–3359.
72. Horue, M., Berti, I. R., Cacicedo, M. L., & Castro, G. R. (2021). Microbial production and recovery of hybrid biopolymers from wastes for industrial applications-a review. *Bioresource Technology, 340*, 125671.
73. Klassen, V., Blifernez-Klassen, O., Bax, J., & Kruse, O. (2020). Wastewater-borne microalga Chlamydomonas sp.: A robust chassis for efficient biomass and biomethane production applying low-N cultivation strategy. *Bioresource Technology, 315*, 123825.
74. Zhu, S., Feng, S., Xu, Z., Qin, L., Shang, C., Feng, P., Wang, Z., & Yuan, Z. (2019). Cultivation of Chlorella vulgaris on unsterilized dairy-derived liquid digestate for simultaneous biofuels feedstock production and pollutant removal. *Bioresource Technology, 285*, 121353.
75. Kothari, R., Pathak, V. V., Kumar, V., & Singh, D. (2012). Experimental study for growth potential of unicellular alga Chlorella pyrenoidosa on dairy waste water: An integrated approach for treatment and biofuel production. *Bioresource Technology, 116*, 466–470.
76. Madadi, R., Maljaee, H., Serafim, L. S., & Ventura, S. P. (2021). Microalgae as contributors to produce biopolymers. *Marine Drugs, 19*(8), 466.
77. Daneshvar, E., Zarrinmehr, M. J., Hashtjin, A. M., Farhadian, O., & Bhatnagar, A. (2018). Versatile applications of freshwater and marine water microalgae in dairy wastewater treatment, lipid extraction and tetracycline biosorption. *Bioresource Technology, 268*, 523–530.
78. Azam, R., Kothari, R., Singh, H. M., Ahmad, S., Ashokkumar, V., & Tyagi, V. (2020). Production of algal biomass for its biochemical profile using slaughterhouse wastewater for treatment under axenic conditions. *Bioresource Technology, 306*, 123116.
79. Cho, D.-H., Choi, J.-W., Kang, Z., Kim, B.-H., Oh, H.-M., Kim, H.-S., & Ramanan, R. (2017). Microalgal diversity fosters stable biomass productivity in open ponds treating wastewater. *Scientific Reports, 7*(1), 1–11.
80. Bhatnagar, A., Bhatnagar, M., Chinnasamy, S., & Das, K. (2010). Chlorella minutissima—A promising fuel alga for cultivation in municipal wastewaters. *Applied Biochemistry and Biotechnology, 161*(1), 523–536.
81. Gentili, F. G. (2014). Microalgal biomass and lipid production in mixed municipal, dairy, pulp and paper wastewater together with added flue gases. *Bioresource Technology, 169*, 27–32.
82. Wang, L., Min, M., Li, Y., Chen, P., Chen, Y., Liu, Y., Wang, Y., & Ruan, R. (2010). Cultivation of green algae Chlorella sp. in different wastewaters from municipal wastewater treatment plant. *Applied Biochemistry and Biotechnology, 162*(4), 1174–1186.
83. De Vegt, O., Werker, A., Fetter, B., Hopman, R., Krins, B., & Winters, R. (2012). PHA from waste water. Transformation of residual materials and waste water into valuable bioplastics. *Bioplastics Magazine, 7*(4), 26–28.
84. Fukui, T. (1998). Efficient production of polyhydroxyalkanoates from plant oils by Alcaligenes eutrophus and its recombinant strain. *Applied Microbiology and Biotechnology, 49*(3), 333–336.
85. Fernandez Dacosta, C. (2018). *Alternative sources of fossil carbon: Ex-ante assessment of novel technologies using waste as a source*. Utrecht University.
86. Yadav, B., Pandey, A., Kumar, L. R., & Tyagi, R. D. (2020). Bioconversion of waste (water)/residues to bioplastics-A circular bioeconomy approach. *Bioresource Technology, 298*, 122584.
87. Kawaguchi, H., Ogino, C., & Kondo, A. (2017). Microbial conversion of biomass into bio-based polymers. *Bioresource Technology, 245*, 1664–1673.
88. Coats, E. R., Loge, F. J., Wolcott, M. P., Englund, K., & McDonald, A. G. (2008). Production of natural fiber reinforced thermoplastic composites through the use of polyhydroxybutyrate-rich biomass. *Bioresource Technology, 99*(7), 2680–2686.

89. Pellegrin, V., Juretschko, S., Wagner, M., & Cottenceau, G. (1999). Morphological and biochemical properties of a Sphaerotilus sp. isolated from paper mill slimes. *Applied and Environmental Microbiology, 65*(1), 156–162.
90. Buyukkamaci, N., & Koken, E. (2010). Economic evaluation of alternative wastewater treatment plant options for pulp and paper industry. *Science of the Total Environment, 408*(24), 6070–6078.
91. Takabatake, H., Satoh, H., Mino, T., & Matsuo, T. (2000). Recovery of biodegradable plastics from activated sludge process. *Water Science and Technology, 42*(3–4), 351–356.
92. Peters, D. (2006). Carbohydrates for fermentation. *Biotechnology Journal: Healthcare Nutrition Technology, 1*(7–8), 806–814.
93. Povolo, S., Toffano, P., Basaglia, M., & Casella, S. (2010). Polyhydroxyalkanoates production by engineered Cupriavidus necator from waste material containing lactose. *Bioresource Technology, 101*(20), 7902–7907.
94. Nielsen, C., Rahman, A., Rehman, A., Walsh, M., & Miller, C. (2017). Food waste conversion to microbial polyhydroxyalkanoates. *Microbial Biotechnology, 10*(6), 1338–1352.
95. Reddy, M. V., Nikhil, G., Mohan, S. V., Swamy, Y., & Sarma, P. (2012). Pseudomonas otitidis as a potential biocatalyst for polyhydroxyalkanoates (PHA) synthesis using synthetic wastewater and acidogenic effluents. *Bioresource Technology, 123*, 471–479.
96. Wei, L., Liang, S., Guho, N. M., Hanson, A. J., Smith, M. W., Garcia-Perez, M., & McDonald, A. G. (2015). Production and characterization of bio-oil and biochar from the pyrolysis of residual bacterial biomass from a polyhydroxyalkanoate production process. *Journal of Analytical and Applied Pyrolysis, 115*, 268–278.
97. Koller, M. (2015). Recycling of waste streams of the biotechnological poly (hydroxyalkanoate) production by Haloferax mediterranei on whey. *International Journal of Polymer Science, 2015.*
98. Kosseva, M. R., & Rusbandi, E. (2018). Trends in the biomanufacture of polyhydroxyalkanoates with focus on downstream processing. *International Journal of Biological Macromolecules, 107*, 762–778.
99. Koller, M., Gasser, I., Schmid, F., & Berg, G. (2011). Linking ecology with economy: Insights into polyhydroxyalkanoate-producing microorganisms. *Engineering in Life Sciences, 11*(3), 222–237.
100. Mannina, G., Presti, D., Montiel-Jarillo, G., & Suárez-Ojeda, M. E. (2019). Bioplastic recovery from wastewater: A new protocol for polyhydroxyalkanoates (PHA) extraction from mixed microbial cultures. *Bioresource Technology, 282*, 361–369.
101. Righi, S., Baioli, F., Samorì, C., Galletti, P., Tagliavini, E., Stramigioli, C., Tugnoli, A., & Fantke, P. (2017). A life cycle assessment of poly-hydroxybutyrate extraction from microbial biomass using dimethyl carbonate. *Journal of Cleaner Production, 168*, 692–707.
102. Dietrich, K., Dumont, M.-J., Del Rio, L. F., & Orsat, V. (2017). Producing PHAs in the bioeconomy—Towards a sustainable bioplastic. *Sustainable Production and Consumption, 9*, 58–70.
103. Bhattacharyya, A., Jana, K., Haldar, S., Bhowmic, A., Mukhopadhyay, U. K., De, S., & Mukherjee, J. (2015). Integration of poly-3-(hydroxybutyrate-co-hydroxyvalerate) production by Haloferax mediterranei through utilization of stillage from rice-based ethanol manufacture in India and its techno-economic analysis. *World Journal of Microbiology and Biotechnology, 31*(5), 717–727.
104. Pavan, F. A., Junqueira, T. L., Watanabe, M. D., Bonomi, A., Quines, L. K., Schmidell, W., & de Aragao, G. M. (2019). Economic analysis of polyhydroxybutyrate production by Cupriavidus necator using different routes for product recovery. *Biochemical Engineering Journal, 146*, 97–104.
105. Dacosta, C. F., Posada, J. A., & Ramirez, A. (2015). Large scale production of polyhydroxyalkanoates (PHAs) from wastewater: A study of techno-economics, energy use and greenhouse gas emissions. *International Journal of Environmental, Chemical, Ecological, Geological and Geophysical Engineering, 9*(5), 433–438.

106. Leong, Y. K., Show, P. L., Lin, H. C., Chang, C. K., Loh, H.-S., Lan, J.C.-W., & Ling, T. C. (2016). Preliminary integrated economic and environmental analysis of polyhydroxyalkanoates (PHAs) biosynthesis. *Bioresources and Bioprocessing, 3*(1), 1–9.

Recent Developments in Water Hyacinth Fiber Composites and Their Applications

Melbi Mahardika, Hairul Abral, and Devita Amelia

1 Introduction

The increase of environmental awareness makes humans develop various green materials. One resource that is often used in making those materials is natural fibers. Natural fibers are abundantly available, inexpensive, and have good mechanical properties [1–3]. One of the readily available sources of natural fiber is water hyacinth. Water hyacinth was a weed plant that commonly grows above water's surface in lakes, rivers, and other pretty wide and deep puddles. Originally from South America, water hyacinth (*Eichhornia crassipes*) is one of the most invasive plants in the world. This plant can cause significant ecological and socio-economic effects [4]. It also can decrease the amount of light and the solubility of oxygen in the water [5]. The growth rate and resistance of water hyacinths are so high that their growth is difficult to control and harms the ecosystem. Water hyacinth can grow up to 2 times its weight in 6–28 days and replicate in 4–58 days [6].

The high growth rate of water hyacinth makes water hyacinth a renewable and sustainable resource. Furthermore, a good fiber can be extracted and used for several purposes from water hyacinth in many recent years. Water hyacinth fiber also showed good compatibility with any composite matrix.

M. Mahardika
Research Center for Biomass and Bioproducts, National Research and Innovation Agency (BRIN), Bogor 16911, Indonesia

H. Abral (✉)
Department of Mechanical Engineering, Andalas University, Padang 25163, Indonesia
e-mail: habral@yahoo.com

D. Amelia
Department of Chemical Engineering, Universitas Indonesia, Depok 16424, Indonesia

Research Collaboration Center for Nanocellulose, BRIN—Andalas University, Padang 25163, Indonesia

Sapuan S. M. and I. Ahmad (eds.), *Composites from the Aquatic Environment*, Composites Science and Technology, https://doi.org/10.1007/978-981-19-5327-9_11

Water hyacinth fiber was the highest percentage of cellulose content which is one of the main factors responsible for improving thermal stability and tensile properties. Previous research reported that applying nanocellulose from water hyacinth fiber in yam starch-based biocomposite showed the highest thermal stability, low moisture absorption, and improved mechanical properties [7, 8]. Adding 1 wt% nanocellulose from water hyacinth fiber in thermoplastic starch and with 1 h ultrasonication increased mechanical properties by 185% compared to pure thermoplastic starch film [9]. Previous studies have reported that they have succeeded in making polylactic acid (PLA) biocomposites with water hyacinth fiber and nano-silica as reinforcement using the screw extruder method and compression molding machine for electronic applications [10]. The composites produced after adding water hyacinth fiber particles with various proportions showed an increase in tensile strength and superior water absorption. Synthesis and characterization of Polyvinyl alcohol (PVA)/water hyacinth-based hydrogel successfully produced without external cross-linker but using gamma radiation used for adsorption to remove the dye and heavy metals [11]. Adding water hyacinth to the biocomposite significantly escalated the swelling capacity, improved mechanical properties, and water uptake. Tapioca starch-based biocomposite with 10% water hyacinth fiber reinforcement treated with ultrasonication during the gelatinization process can improve the mechanical properties, moisture resistance, and thermal stability of the biocomposite [12]. Making biodegradable foam from starch foam with 5 wt% water hyacinth powder as reinforcement can increase mechanical properties and reduce moisture content [13]. Water hyacinth fiber pulp added as reinforcement in tapioca starch-based biocomposite increases tensile strength and tensile modulus due to good fiber distribution in the matrix, and interface bonding between water hyacinth fiber and thermoplastic starch [14]. This chapter shows how good the water hyacinth fiber is as a composite reinforcement. The discussion will start from the availability of the fiber, properties, fiber extraction, composite fabrication, and applications of water hyacinth composite. This chapter also explores the possibilities of strengthening water hyacinth fiber as an alternative material for various applications for sustainable development.

2 Water Hyacinth Fibers

Water hyacinth fibers are generally extracted from inflated and hollow petioles. To get water hyacinth fiber, takes some special steps. Water hyacinth can grow into 30–40 cm in length and float on waterbodies. The plants have black and fibrous root, and the leaves are basal and straightforward with hollow petioles that helps the plant float on the water surface. Water hyacinth also has around 5–6 petals and purple colors. The most widely used water hyacinth is the leaf stalk, extracted from its fiber for various needs. Figure 1 shows the parts of the water hyacinth plant.

Fig. 1 Water hyacinth plant parts [15]

2.1 *Availability*

Although originally from South America, water hyacinth has spread into over 50 countries, mainly in Southeast Asia, the United States, central and western Africa, and Central America, especially on the amazon river [16]. This plant can easily be found in tropical and sub-tropical waterbodies with high water nutrient concentrations due to agricultural runoff, deforestation, and poor wastewater treatment [4]. The vast amount of water hyacinth makes this plant bring the disadvantages to the environment so need to be handled. Water hyacinth is an aquatic plant that can grow and adapt quickly in different climates, water levels, water currents, and changes in nutrient availability, pH, temperature, and removal of various toxic metals in water [16]. Even with water hyacinth's high and fast growth, it can damage the aquatic environment and is considered a weed. In Indonesia, water hyacinth grows in abundance in almost all lakes, rivers, swamps, wetlands, water reservoirs, and shallow ponds. Besides that, water hyacinth can also be used for various applications such as biogas, handcraft, medical needs, and resources to develop composite material. The use of water hyacinth will increase its economic value due to the renewable and sustainability of this aquatic plant.

2.2 *Water Hyacinth Fiber Composition*

Water hyacinth fiber is composed of cellulose, hemicellulose, and lignin. The levels of chemical composition in water hyacinth vary depending on the source of the water hyacinth fiber. This difference in chemical composition can be caused by differences in the growing location of the water hyacinth. Several previous studies that have reported the chemical composition of water hyacinth fiber are shown in Table 1. A study reported that water hyacinth has a composition of 60% cellulose, 8%

hemicellulose, and 17% lignin [17]. However, several other studies reported different results, especially in cellulose levels, 43% [18] and 25.64% [6].

The chemical composition of water hyacinth from Thailand reported in previous studies was cellulose 57%, hemicellulose 25.6%, and lignin 4.1% [19]. Water hyacinth fiber obtained from the Udon Thani area of Thailand contains 52.1% cellulose, 17.6% hemicellulose, and 8.7% lignin [20]. Istirokhatun point et al. 2015 reported the potential utilization of cellulose from water hyacinth fiber for membranes from the Rawa Pening Lake area, Indonesia [21]. Water hyacinth from Pathumthani province, Thailand contains cellulose, hemicellulose, and lignin content of 50.4%, 19.5%, and 2.25%, respectively [22]. Water hyacinth fiber from Phra Nakhon Si Ayutthaya province, Thailand, contains 57% cellulose, 25.6% hemicellulose, and 4.1% lignin [23]. Water hyacinth fiber originating from the Nile River area in Kafr El-Zayat, Al-Gharbiyah, Egypt has been successfully obtained for cellulose using

Table 1 Several previous studies of the chemical composition of water hyacinth fibers and various fibers from various countries

Fiber	Source	Cellulose (%)	Hemicellulose (%)	Lignin (%)	Refs.
Water hyacinth	Indonesia	43.0	29.0	7.0	[7, 9]
	Indonesia	85.9	4.5	5.4	[27]
	50 Kota district, Indonesia	67.0	20.6	4.1	[14]
	Nigeria	45.5	21.8	8.3	[5]
	Pathumthani province, Thailand	50.4	19.6	2.3	[22]
	Payakumbuh, Indonesia	64.1	15.1	10.5	[27]
	Chao Phraya River, Bangkok, Thailand	65.1	15.2	11.4	[28]
	Phra Nakhon Si Ayutthaya province, Thailand	57.0	25.6	4.1	[23]
	Kumphawapi, Udon Thani, Thailand	52.1	17.6	8.7	[20]
	Tietê River, in São Manuel	26.0	27.4	8.8	[29]
Pineapple leaf	Indonesia	62.5	13.9	15.9	[30]
Garlic straw	France	41.0	18.0	6.3	[31]
Oil palm trunk	Malaysia	50.7	77.4	11.7	[32]
Coconut palm petioles	China	33.3	33.6	19.9	[33]
Sugar palm	Malaysia	43.8	7.24	33.2	[34]

the free hazard chemical method and low cost [24]. Water hyacinth fibers from Lake Victoria and RSW from a slaughterhouse in Nairobi had a chemical composition of cellulose, hemicellulose, and lignin of 33%, 23%, and 10%, respectively [25]. Water hyacinth fiber from Yang-Shuwan Lake in Yichang City, China contains 24.5% cellulose, 34.1% hemicellulose, and 8.6% lignin [26]. From several previous studies, Indonesia and Thailand have the highest cellulose content of water hyacinth. However, several treatments can be carried out to obtain pure cellulose or any composition from water hyacinth. Research from Oyeoka et al. reports on water hyacinth fibers from Onopa Town Stream at Epie Atisa III, Yenagoa L.G.A, Bayelsa State, Nigeria has cellulose, lignin, hemicellulose, and extractives of 45.5%, 8.3%, 21.8%, and 5.8%, respectively [5].

3 Modification of Surface on Water Hyacinth Fibers

Water hyacinth fibers are generally extracted from inflated and hollow petioles. To modify the surface and get high cellulose on water hyacinth fiber using extraction methods, such as chemical, mechanical, and chemo-chemical treatment (Fig. 2) [35]. Before the water hyacinth fiber is extracted, the water hyacinth stems are separated from the leaves and roots, then washed to clear up dirt and dried at room temperature. After the separation process, the water hyacinth plant stems were cut into smaller sizes using the mechanical treatment (crushing machine and a mixer) [36, 37]. Chemical treatments of water hyacinth fiber used alkalization, bleaching, and acid hydrolysis. Alkalization is fundamentally a delignification process to eliminate lignin and hemicellulose from water hyacinth [38]. The bleaching process is required to optimize the whiteness level and remove the remaining lignin and hemicellulose [39]. Alkalization and bleaching as pretreatment in the extraction of water hyacinth fibers. The extraction of water hyacinth fibers used chemicals at a specific concentration, temperature, pressure, and time [30, 40]. Mechanical treatment can use pretreatment (crushing machine or mixer) and post-treatment (homogenizer, ultrasonication, ultrafine grinding). The disadvantage of using chemical and chemo-chemical treatment is that it damages the environment and certain chemicals are relatively expensive. However, using chemical treatment is that it contains pure cellulose for natural fibers sourced from plants. Water hyacinth has a lower lignin content than other cellulose sources, around 4% [19]. It allows the use of fewer chemicals and less energy to modify the fiber surface to obtain pure cellulose.

3.1 Chemical Treatment

The extraction of water hyacinth fibers used chemicals treatment divided into two stages: pretreatment using alkalization and bleaching; and extraction using acid hydrolysis. Alkalization as pretreatment can remove hemicellulose and lignin. The

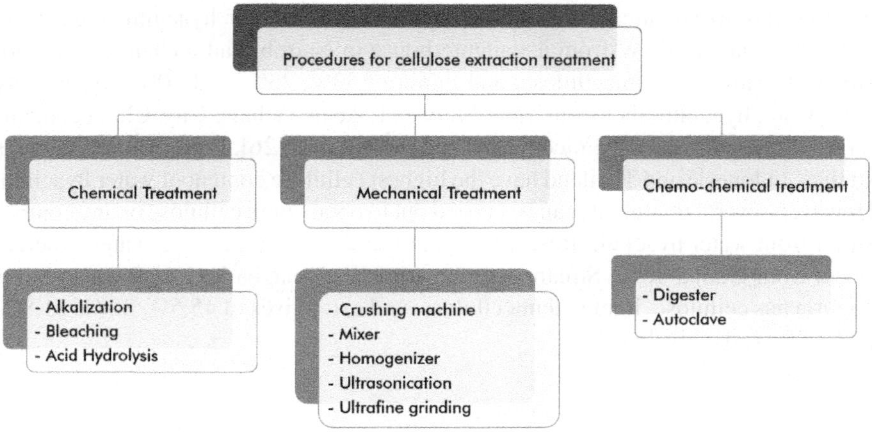

Fig. 2 Procedures for cellulose extraction treatment

reaction mechanisms of lignin removal are cleavage of α- and β-ether bonds in phenolic units and of β-ether linkages in nonphenolic units [38]. Alkali treatment increases the hydroxyl (OH) in the fiber to form chemical and physical bonds with polymer chains. In physics, the hydroxyl groups of cellulose fibers form hydrogen bonds with the hydroxyl groups of polymer chains. In chemistry, cellulose–OH fibers react with alkali (NaOH) to form–O–Na+ fibers, which bind to polymer chains [41]. The advantages of alkali treatment are relatively moderate temperatures and pressures compared to other pretreatment technologies and can increase the cellulose crystallinity index (CI) value. However, it is not environmentally friendly due to the high generation and disposal of chemical waste [38, 39]. In this process, the type of alkaline solution and its concentration affect the degree of swelling of the cellulose fiber [42]. Type of alkaline solution was used, such as NaOH and KOH. Temperature, time, and particle size can also affect alkalization. Table 2 shows alkalization results on water hyacinth fiber in previous studies.

Table 2 Alkalization results on water hyacinth fiber in previous studies

Source	Alkalization	Alkalization			Chemical composition (%)			Refs.
		Concentration	T (°C)	t (h)	Cellulose	Hemicellulose	Lignin	
Thailand	KOH	4% (wt)	90	2	89.5	5.3	0.4	[19]
Payakumbuh, Indonesia	NaOH	18%	170	2	68.96	14.27	8.38	[27]
50 Kota district, Indonesia	NaOH	25%	130	6	67.0	3.5	3.9	[14]
Kumphawapi, Udon Thani, Thailand	NaOH	5% (wt)	75	4	76.57	14.91	1.57	[20]

After alkalization, the proportion of cellulose increases, and the proportions of lignin and hemicellulose decrease. According to research Abral et al., cellulose composition increased 53.0–56.2% after alkalization using 5% NaOH during 1 h [43]. Tapinchai et al. stated that alkalization using 4% (wt) NaOH resulted in the composition of cellulose 91.1% [19]. This process can increase tensile strength because the bond between matrix and fiber is getting better due to the loss of the barrier wall in lignin and hemicellulose [44]. Darmanto resulted in tensile strength with 12% NaOH for 12 h increase 54.5–60.2 MPa [45]. The treatment of WH fibers with NaOH solution decreased the tensile and flexural strength when there was an increase in the concentration and alkali treatment duration. It was because increasing the NaOH concentration and duration of the treatment impacted the chemical structure of cellulose, causing the cellulose molecular chains to lose their crystalline structure and damaging the fibers' inter-laminar bonding, resulting in decreased fiber strength [43].

Furthermore, the water hyacinth fiber was treated with bleaching to remove the remaining lignin and hemicellulose after alkaline treatment. Bleaching also purposes to increase the degree of whiteness in water hyacinth fibers. Bleaching chemicals generally used sodium chlorite ($NaClO_2$), sodium hypochlorite (NaClO), hydrogen peroxide (H_2O_2), and acetic acid [38, 46, 47]. NaClO is reactive and hazardous to transport. The use of hydrogen peroxide is more environmentally friendly and cost-effective than other methods, with its effectiveness depending on the low lignin and extractive content in the plant cell. Table 3 shows the results of previous studies for the bleaching process on water hyacinth fibers [38, 46]. There are several alternatives for the bleaching process, such as (1) bleaching [29], (2) alkalization-bleaching, and (3) bleaching-alkalization-bleaching [47]. In the study of Bronzato et al. [29], water hyacinth fiber was treated with bleaching using acetic acid and hydrogen peroxide without alkalization by performing in a vertical autoclave, under the pressure of 2.5 kgf/cm^2 for 1 h. After the bleaching process, the results showed a decrease in cellulose content using acetic acid and hydrogen peroxide from 27% to 14.92% and 12.18%, respectively [29]. The cellulose composition decreases due to the depolymerization of polysaccharides, resulting from breaking glycosidic bonds [48]. NaClO is more effective at removing lignin than hydrogen peroxide, resulting in lower residual lignin in cellulose samples [19]. According to Sun et al. [47], the cellulose extraction process with bleaching-alkalization-bleaching. First bleaching, NaClO ultrapure water was mixed with a weight ratio of 1:3 and adjusted to pH 4 using glacial acetic acid. Bleaching was carried out at room temperature and stirred overnight, followed by alkalization using NaOH (weight 1%) at room temperature (22 C) for 2 h with continuous stirring. After alkalizing, proceed to the second bleaching using NaClO for 2 h and stirring continuously. The results obtained from the FTIR analysis were that the peak intensity of cellulose increased after NaOH and the second bleaching treatment because the hemicellulose and lignin materials were removed [47]. Visualization of water hyacinth fiber after pretreatment (alkalization and bleaching) can be seen in Fig. 3.

After alkalization and bleaching, water hyacinth fibers were treated with acid hydrolysis, which aims to hydrolyze the amorphous regions of the cellulose.

Table 3 The results of previous studies for the bleaching process on water hyacinth fibers

Source	Pretreatment	Bleaching	Bleaching			Chemical composition (%)			Refs.
			Concentration	T (°C)	t (h)	Cellulose	Hemicellulose	Lignin	
Thailand	Alkalization	$NaClO_2$	10 g	80	2	85.00	11.70	0.20	[19]
Payakumbuh, Indonesia	Alkalization	$NaClO_2$: Acetic acid	5%: 1	70	2	83.06	5.09	6.74	[27]
Phra Nakhon Si Ayutthaya province, Thailand	Alkalization	NaClO	10% (v/v)	80 75	2 3	82.5	4.10	1.80	[23]
Kumphawapi, Udon Thani, Thailand	Alkalization	$NaClO_2$	2% (wt)	80	5	81.11	11.17	0.84	[20]
Tietê River, in São Manuel	NA	H_2O_2	1.7 mol/L	NA	1	12.20	22.30	7.40	[29]

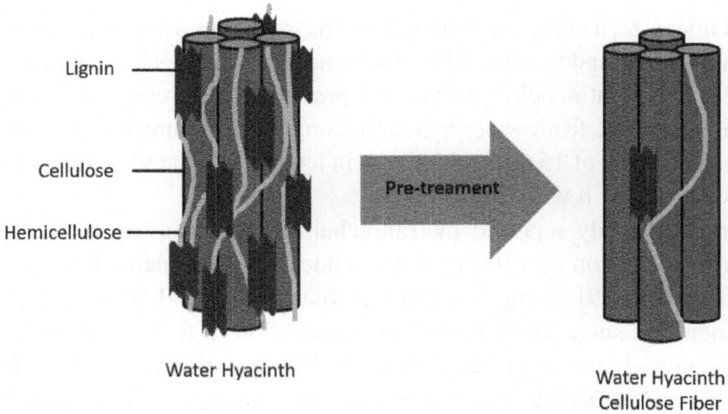

Fig. 3 Water hyacinth fiber after pretreatment

Hydrochloric acid (HCl) and sulfuric acid (H_2SO_4) are the most widely used acid for acid hydrolysis. Sulfuric acid can esterify hydroxyl groups by sulfate ions creating crystalline regions of cellulose fibers to form stable colloidal dispersions of nanocellulose materials in the remaining reaction mixture [49]. The main controlling factors that affect the properties of the obtained nanocellulose are reaction time, temperature, and acid concentration. Acid hydrolysis is strongly influenced by reaction time, temperature, and acid concentration [50]. Based on the research of Syafri et al. [27], acid hydrolysis for water hyacinth fiber using 30% H_2SO_4 for 30 min at room temperature. After acid hydrolysis, the chemical composition obtained is more significant than after pretreatment with alkalization and bleaching. The chemical composition after acid hydrolysis is 85.85% cellulose, 4.49% hemicellulose, and 5.55% lignin [27]. After acid hydrolysis, the crystallinity index of water hyacinth fibers increased by 16% compared to fibers that had undergone bleaching treatment because the non-cellulosic content was removed from the bleaching process [18]. According to Pakciam et al. [46], a high yield of cellulose was obtained in the acid hydrolysis process with 5 M HCl at 60 °C for half an hour and 3.5 M HCl at 60 °C for half an hour. In SEM analysis, water hyacinth fiber morphological changes occurred from untreated and treated (NaOH and acid hydrolyzed). The untreated samples had a smooth surface, and after treatment, the fibers became loose due to the breakdown of hemicellulose and lignin [46].

3.2 *Mechanical Treatment*

The disadvantage of mechanical treatment was the high lignin and other non-cellulose components. Mechanical treatment was used a crushing machine, mixer, homogenizer, and ultrasound [18, 36, 37]. The dried water hyacinth fiber was crushed into a crushing machine to obtain fiber in powder. Mechanical sieves of various sizes obtain a uniform fiber size, usually microparticles. The treatment was continued with a high-shear homogenizer and ultrasound to obtain nanocellulose. The size range after this mechanical treatment is below 100 nm. A previous study reported that the production of nanocellulose from water hyacinth using an ultrasonic crusher succeeded in obtaining a diameter of 15 nm and 147 nm in length [7]. The ultrasonication process was carried out for 1 h with 600 W.

The previous study reported by Tanpichai et al. used a Super Mass Colloider grinder with a rotation of 1500 rpm to produce cellulose nanofibers with a diameter of 10–30 nm [19]. Ultrasonic crusher treatment at 600 W for 1 h after chemical treatment has succeeded in obtaining nanocellulose from water hyacinth fibers with an average diameter of 10–20 nm [9]. Nanocrystalline cellulose from *Eichhornia crassipes (Mart.) Solms* (water hyacinths) have been successfully extracted through alkalization, bleaching, acid hydrolysis, and sonication [46]. After mechanical treatment, ultrasonication of the suspension used ultrasonic waves at 130 W at 20 kHz for 60 min using Bandelin electronics UW 2070–Heinrichstraße 3–4 D-12207 Berlin. The resulting nanocrystalline cellulose suspension has an average diameter of 93 nm by particle size analyzer (PSA) testing [46]. Nanofibrilled cellulose from water hyacinth fiber was successfully produced using a high-speed homogenizer treatment for 40 min after chemical treatments such as alkaline-treated water hyacinth fibers and bleaching process [23]. The average width of nanofibrillated cellulose after high-speed homogenizer treatment was 16.8 nm. From the measurement of the tensile strength of nanofibrillated cellulose paper, it increased 18 times compared to water hyacinth fiber without high-speed homogenizer treatment [23]. The tool used for mechanical treatment of the homogenizer is Homogenizer T25 ULTRA-TURRAX, IKA Works, Inc., Germany, with a speed of 20,000 rpm for 40 min [23]. So mechanical treatments such as Super Mass Colloider Grinding, high-speed homogenizer, and Ultrasonication succeeded in producing nanocellulose from water hyacinth with an average diameter below 100 nm.

3.3 Chemo-chemical Treatment

This process used a combination of chemicals and mechanical treatment to obtain pure cellulose in nanoscale sizes called nanocellulose. Chemical treatment was applied to obtain pure cellulose and chemical treatment to reduce the size of the fiber. Water hyacinth treated with enzymatic treatment, bleached, sulfuric acid, and ultrasonication for 10 min can produce an average diameter of cellulose nanocrystals of 20–50 nm and 13.8 MPa tensile strength of the suspension made of films [5]. Digested is used to get water hyacinth powder using 5% NaOH with rotation and heating for 3 h [5]. Treatment of 25% NaOH solution on water hyacinth fiber with chemo-chemical treatment using a digester at a temperature of 130 °C pressure of 2 bar for 6 h succeeded in increasing the cellulose content from 42.8 to 67%, reducing the hemicellulose content significantly from 20.6 to 3.5%, and reduced lignin to 3.9% [14].

3.4 Water Hyacinth Fiber Composite

Natural fiber-based composites such as water hyacinth fiber are commonly used using solution casting, screw extruder, compression molding, and injection molding methods [5, 10, 13]. Table 4 shows composites sourced from various biopolymers and synthetics as a matrix with water hyacinth fiber as reinforcement.

From previous research, water hyacinth fiber extracted to obtain pure cellulose can be used for various applications as an auspicious candidate material to replace synthetic fiber.

Table 4 Water hyacinth fiber-reinforced composites with the various matrix

Water hyacinth fiber	Matrix	Preparation method of composites	Application	Refs.
Cellulose nanocrystals	Polyvinyl alcohol/gelatin	Solution casting	Packaging	[5]
Nanocellulose	Bengkuang starch	Solution casting with ultrasonic bath treatment	Bioplastic-based starch	[7, 9]
Powder of water hyacinth	Polylactic acid (PLA)	Hydraulic injection molding machine	Electronic applications	[10]
Powder of water hyacinth	Polyvinyl alcohol	Gamma radiation method	Hydrogel for adsorption to remove the dye and heavy metals	[11]
Powder of water hyacinth	Tapioca starch	Solution casting	Bioplastic-based starch	[12, 14]
Cellulose microfibers	Sago starch	Solution casting	Bioplastic-based starch	[27]
Powder of water hyacinth	Native cassava starch	Compression molding	Food packaging applications	[13]
Dried of water hyacinth fiber	Polyester	Casting technique	Composites	[51]
Fibers	Bio-based epoxy	Casting technique	Develop fully biobased, sustainable, ecofriendly Composites	[28]
Fibers	Poly (lactic acid)	Internal melt Mixer and compression molding machine	Biodegradable composites	[52]

4 Conclusion

This chapter provides information regarding the utilization of water hyacinth fiber into cellulose which has the potential to be used for various applications. The treatments given to produce cellulose include chemical treatments such as alkalization, bleaching, and acid hydrolysis; however, there are still many technical and economic problems that must be considered before cellulose from water hyacinth fiber can be successfully commercialized. From its studied properties, cellulose from water hyacinth can be used as a substitute for synthetic fibers such as glass and carbon fibers. The primary studies in this chapter are as follows:

(i) The chemical composition of water hyacinth fiber is cellulose, hemicellulose, lignin, pectin, and wax as the main components
(ii) Chemical treatments such as alkalization, bleaching, acid hydrolysis are used to obtain pure cellulose from water hyacinth fiber.
(iii) The treatments used to obtain nanocellulose from water hyacinth fibers used mechanical treatments such as ultrasonication, high-speed homogenizer, and super mass collider grinder.
(iv) The cellulose fiber composite of water hyacinth with polymer matrix exhibits the promised properties such as mechanical properties, thermal resistance, and water and gas vapor resistance.
(v) Therefore, cellulose from water hyacinth fiber can be considered an alternative even superior to cellulose from other plant sources and more environmentally friendly than synthetic fibers.

References

1. Sanjay, M. R., Madhu, P., Jawaid, M., Senthamaraikannan, P., Senthil, S., & Pradeep, S. (2018). Characterization and properties of natural fiber polymer composites: A comprehensive review. *Journal of Cleaner Production, 172*, 566–581.
2. Naveen, J., Jawaid, M., Amuthakkannan, P., & Chandrasekar, M. (2019). Mechanical and physical properties of sisal and hybrid sisal fiber-reinforced polymer composites. In *Mechanical and physical testing of biocomposites, fibre-reinforced composites and hybrid composites* (pp. 427–440). Elsevier.
3. Rahman, M., Das, S., & Hasan, M. (2018). Mechanical properties of chemically treated banana and pineapple leaf fiber reinforced hybrid polypropylene composites. *Advances in Materials and Processing Technologies, 4*(4), 527–537.
4. Villamagna, A. M., & Murphy, B. R. (2010). Ecological and socio-economic impacts of invasive water hyacinth (Eichhornia crassipes): A review. *Freshwater Biology, 55*(2), 282–298. https://doi.org/10.1111/j.1365-2427.2009.02294.x
5. Oyeoka, H. C., Ewulonu, C. M., Nwuzor, I. C., Obele, C. M., & Nwabanne, J. T. (2021). Packaging and degradability properties of polyvinyl alcohol/gelatin nanocomposite films filled water hyacinth cellulose nanocrystals. *Journal of Bioresources and Bioproducts, 6*(2), 168–185.
6. Guna, V., Ilangovan, M., Anantha Prasad, M. G., & Reddy, N. (2017). Water hyacinth: A unique source for sustainable materials and products. *ACS Sustainable Chemistry & Engineering, 5*(6), 4478–4490.

7. Syafri, E., et al. (2019). Effect of sonication time on the thermal stability, moisture absorption, and biodegradation of water hyacinth (Eichhornia crassipes) nanocellulose-filled bengkuang (Pachyrhizus erosus) starch biocomposites. *Journal of Materials Research and Technology, 8*(6). https://doi.org/10.1016/j.jmrt.2019.10.016
8. Asrofi, M., Abral, H., Kasim, A., Pratoto, A., Mahardika, M., & Hafizulhaq, F. (2018). Characterization of the sonicated yam bean starch bionanocomposites reinforced by nanocellulose water hyacinth fiber (whf): The effect of various fiber loading. *Journal of Engineering Science and Technology, 13*(9), 2700–2715.
9. Asrofi, M., Abral, H., Kasim, A., Pratoto, A., Mahardika, M., & Hafizulhaq, F. (2018). Mechanical properties of a water hyacinth nanofiber cellulose reinforced thermoplastic starch bionanocomposite: Effect of ultrasonic vibration during processing. *Fibers, 6*(2), 40.
10. Panda, P. K., et al. (2021). Exploration on mechanical behaviours of hyacinth fibre particles reinforced polymer matrix-based hybrid composites for electronic applications. *Advances in Materials Science and Engineering, 2021.*
11. Hossain, M., Afroz, S., Islam, M. U., Alam, A. K. M., Khan, R. A., & Alam, A. (2021). Synthesis and characterization of polyvinyl alcohol/water-hyacinth (Eichhornia crassipes) based hydrogel by applying gamma radiation. *Journal of Polymer Research, 28*(5), 1–12.
12. Asrofi, M., Abral, H., Putra, Y. K., Sapuan, S. M., & Kim, H.-J. (2018). Effect of duration of sonication during gelatinization on properties of tapioca starch water hyacinth fiber biocomposite. *International Journal of Biological Macromolecules, 108*, 167–176.
13. Chaireh, S., Ngasatool, P., & Kaewtatip, K. (2020). Novel composite foam made from starch and water hyacinth with beeswax coating for food packaging applications. *International Journal of Biological Macromolecules, 165*, 1382–1391.
14. Abral, H., et al. (2018). Characterization of tapioca starch biopolymer composites reinforced with micro scale water hyacinth fibers. *Starch/Staerke, 70*(7–8), 1–8. https://doi.org/10.1002/star.201700287
15. Gaurav, G. K., Mehmood, T., Cheng, L., Klemeš, J. J., & Shrivastava, D. K. (2020). Water hyacinth as a biomass: A review. *Journal of Cleaner Production, 277*, 122214.
16. Jones, J. L., Jenkins, R. O., & Haris, P. I. (2018). Extending the geographic reach of the water hyacinth plant in removal of heavy metals from a temperate Northern Hemisphere river. *Science and Reports, 8*(1), 1–15.
17. Abdel-Fattah, A. F., & Abdel-Naby, M. A. (2012). Pretreatment and enzymic saccharification of water hyacinth cellulose. *Carbohydrate Polymers, 87*(3), 2109–2113. https://doi.org/10.1016/j.carbpol.2011.10.033
18. Asrofi, M., et al. (2018). Isolation of nanocellulose from water hyacinth fiber (WHF) produced via digester-sonication and its characterization. *Fibers and Polymers, 19*(8), 1618–1625.
19. Tanpichai, S., Biswas, S. K., Witayakran, S., & Yano, H. (2019). Water hyacinth: A sustainable lignin-poor cellulose source for the production of cellulose nanofibers. *ACS Sustainable Chemistry & Engineering, 7*(23), 18884–18893. https://doi.org/10.1021/acssuschemeng.9b04095
20. Pakutsah, K., & Aht-Ong, D. (2020). Facile isolation of cellulose nanofibers from water hyacinth using water-based mechanical defibrillation: Insights into morphological, physical, and rheological properties. *International Journal of Biological Macromolecules, 145*, 64–76. https://doi.org/10.1016/j.ijbiomac.2019.12.172
21. Istirokhatun, T., Rokhati, N., Rachmawaty, R., Meriyani, M., Priyanto, S., & Susanto, H. (2015). Cellulose isolation from tropical water hyacinth for membrane preparation. *Procedia Environmental Sciences, 23*, 274–281. https://doi.org/10.1016/j.proenv.2015.01.041
22. Chonsakorn, S., Srivorradatpaisan, S., & Mongkholrattanasit, R. (2018). Effects of different extraction methods on some properties of water hyacinth fiber. *Journal of Natural Fibers.*
23. Tanpichai, S., et al. (2021). Extraction of nanofibrillated cellulose from water hyacinth using a high speed homogenizer. *Journal of Natural Fibers*, 1–21.
24. Salahuddin, N., Abdelwahab, M. A., Akelah, A., & Elnagar, M. (2021). Adsorption of Congo red and crystal violet dyes onto cellulose extracted from Egyptian water hyacinth. *Natural Hazards, 105*(2), 1375–1394. https://doi.org/10.1007/s11069-020-04358-1

25. Omondi, E. A., Ndiba, P. K., & Njuru, P. G. (2019). Characterization of water hyacinth (E. crassipes) from Lake Victoria and ruminal slaughterhouse waste as co-substrates in biogas production. *SN Applied Sciences, 1*(8), 1–10.
26. Ruan, T., Zeng, R., Yin, X.-Y., Zhang, S.-X., & Yang, Z.-H. (2016). Water hyacinth (Eichhornia crassipes) biomass as a biofuel feedstock by enzymatic hydrolysis. *BioResources, 11*(1), 2372–2380.
27. Syafri, E., Wahono, S., Irwan, A., Asrofi, M., Sari, N. H., & Fudholi, A. (2019). Characterization and properties of cellulose microfibers from water hyacinth filled sago starch biocomposites. *International Journal of Biological Macromolecules.*
28. Sumrith, N., Techawinyutham, L., Sanjay, M. R., Dangtungee, R., & Siengchin, S. (2020). Characterization of alkaline and silane treated fibers of 'water hyacinth plants' and reinforcement of 'water hyacinth fibers' with bioepoxy to develop fully biobased sustainable ecofriendly composites. *Journal of Polymers and the Environment, 28*(10), 2749–2760.
29. Bronzato, G. R. F., Ziegler, S. M., Silva, R. C., Cesarino, I., & Leão, A. L. (2017). Characterization of the pre-treated biomass of Eichhornia crassipes (water hyacinth) for the second generation ethanol production. *Molecular Crystals and Liquid Crystals, 655*(1), 224–235.
30. Mahardika, M., Abral, H., Kasim, A., Arief, S., & Asrofi, M. (2018). Production of nanocellulose from pineapple leaf fibers via high-shear homogenization and ultrasonication. *Fibers, 6*(2), 28. https://doi.org/10.3390/fib6020028
31. Kallel, F., Bettaieb, F., Khiari, R., García, A., Bras, J., & Chaabouni, S. E. (2016). Isolation and structural characterization of cellulose nanocrystals extracted from garlic straw residues. *Industrial Crops and Products, 87*, 287–296. https://doi.org/10.1016/j.indcrop.2016.04.060
32. Lamaming, J., Hashim, R., Sulaiman, O., Leh, C. P., Sugimoto, T., & Nordin, N. A. (2015). Cellulose nanocrystals isolated from oil palm trunk. *Carbohydrate Polymers, 127*, 202–208. https://doi.org/10.1016/j.carbpol.2015.03.043
33. Xu, C., Zhu, S., Xing, C., Li, D., Zhu, N., & Zhou, H. (2015). Isolation and properties of cellulose nanofibrils from coconut palm petioles by different mechanical process. *PLoS ONE, 10*(4), e0122123.
34. Ilyas, R. A., et al. (2019). Sugar palm (Arenga pinnata (Wurmb.) Merr) cellulosic fibre hierarchy: A comprehensive approach from macro to nano scale. *Journal of Materials Research and Technology, 8*(3). https://doi.org/10.1016/j.jmrt.2019.04.011
35. Abdullah, N. A., et al. (2021). Nanocellulose from agricultural waste as an emerging material for nanotechnology applications–an overview. *Polimery, 66*(3), 157–168.
36. Ajithram, A., Jappes, J. T. W., & Brintha, N. C. (2021). Investigation on utilization of water hyacinth aquatic plants towards various bio products—Survey. *Materials Today: Proceedings, 45*, 2040–2045.
37. Potluri, R., & Rao, M. S. (2019). Water absorption and density tests on the water hyacinth-based partial green composite. In *Advances in Manufacturing Technology* (pp. 77–88). Springer.
38. Pinto, E., et al. (2021). Cellulose processing from biomass and its derivatization into carboxymethylcellulose: A review. *Scientific African*, e01078.
39. Ng, L. Y., Wong, T. J., Ng, C. Y., & Amelia, C. K. M. (2021). A review on cellulose nanocrystals production and characterization methods from Elaeis guineensis empty fruit bunches. *Arabian Journal of Chemistry, 14*(9), 103339.
40. Asrofi, M., Abral, H., Kasim, A., & Pratoto, A. (2017). XRD and FTIR studies of nanocrystalline cellulose from water hyacinth (Eichornia crassipes) fiber. *Journal of Metastable and Nanocrystalline Materials, 29*, 9–16.
41. Verma, A., Parashar, A., Jain, N., Singh, V. K., Rangappa, S. M., & Siengchin, S. (2020). Surface modification techniques for the preparation of different novel biofibers for composites. In *Biofibers and biopolymers for biocomposites* (pp. 1–34). Springer.
42. Heinze, T., El Seoud, O. A., & Koschella, A. (2018). *Cellulose derivatives: Synthesis, structure, and properties*. Springer.
43. Abral, H., Putra, H., Sapuan, S. M., & Ishak, M. R. (2013). Effect of alkalization on mechanical properties of water hyacinth fibers-unsaturated polyester composites. *Polymer—Plastics Technology and Engineering, 52*(5), 446–451.

44. Lendvai, L., Karger-Kocsis, J., Kmetty, Á., & Drakopoulos, S. X. (2016). Production and characterization of microfibrillated cellulose-reinforced thermoplastic starch composites. *Journal of Applied Polymer Science , 133*(2).
45. Darmanto, J. (2020). The effect of fiber pretreatment on physical and mechanical properties in water hyacinth fiber composite. *International Journal of Emerging Trends in Engineering Research, 8*, 4799–4805. https://doi.org/10.30534/ijeter/2020/117882020
46. Packiam, K. K., Murugesan, B., Sundaramoorthy, P. M. K., Srinivasan, H., & Dhanasekaran, K. (2021). Extraction, purification and characterization of nanocrystalline cellulose from Eichhornia crassipes (Mart.) Solms: A common aquatic weed water Hyacinth. *Journal of Natural Fibers*, 1–12.
47. Sun, D., et al. (2020). A process for deriving high quality cellulose nanofibrils from water hyacinth invasive species. *Cellulose, 27*(7), 3727–3740.
48. Yue, Y., Han, J., Han, G., Zhang, Q., French, A. D., & Wu, Q. (2015). Characterization of cellulose I/II hybrid fibers isolated from energycane bagasse during the delignification process: Morphology, crystallinity and percentage estimation. *Carbohydrate Polymers, 133*, 438–447.
49. Sharma, A., Thakur, M., Bhattacharya, M., Mandal, T., & Goswami, S. (2019). Commercial application of cellulose nano-composites—A review. *Biotechnology Reports, 21*, e00316.
50. Phanthong, P., Reubroycharoen, P., Hao, X., Xu, G., Abudula, A., & Guan, G. (2018). Nanocellulose: Extraction and application. *Carbon Resources Conversion, 1*(1), 32–43.
51. Abral, H., et al. (2014). Mechanical properties of water hyacinth fibers–polyester composites before and after immersion in water. *Materials and Design, 58*, 125–129.
52. Buasri, A., Chaiyut, N., Petsungwan, T., Boonyuen, Y., & Moonmanee, S. (2012). Effect of surface treatment on interfacial and properties of water hyacinth fiber reinforced poly (lactic acid) composites. *Advanced Materials Research, 463*, 449–452.

Collagen Based Composites Derived from Marine Organisms: As a Solution for the Underutilization of Fish Biomass, Jellyfish and Sponges

M. M. Harussani, S. M. Sapuan, M. Iyad, H. K. Andy Wong, Z. I. Farouk, and A. Nazrin

1 Introduction

Collagen is by far the most common animal protein, making up over 30% of total protein in an animal's body [1]. It is the principal structural substance of all connective tissues, including bones, skin, ligaments, cartilage, and tendons, as well as interstitial tissues of all parenchymal organs. This fibrous protein is essential for the biological and structural integrity of the extracellular matrix, and also supplying mechanical strength to the tissues [2]. Collagens are composed of three long helicoidally oriented amino acid chains, for approximately 1050 chains in each helix. The triplet is the basic structure of the chains, in which glycine links with two additional amino acids and also with the repetitive pattern of $(Gly - X - Y)_n$, where the proline or hydroxyproline is frequently found at position Y [3]. Collagen has a high water absorption capacity, making it an excellent thickening, texturising, and gel-forming ingredient. Stabilisation, emulsion, foam generation, protective colloid function, adhesion and cohesion, and film-forming ability are all properties connected to its surface behaviour. Collagen is also a powerful surface-active agent that has been found to penetrate a lipid-free interface [4]. Due to its unique qualities, collagen peptides and gelatin, also known as denatured collagen, have been widely used in a variety of disciplines, including food, medicine, cosmetics, pharmaceutical industries, diagnostic imaging, leather and film industries, and therapeutic delivery [5].

M. M. Harussani · S. M. Sapuan (✉) · M. Iyad · H. K. A. Wong · Z. I. Farouk · A. Nazrin
Advanced Engineering Materials and Composites Research Centre (AEMC), Department of Mechanical and Manufacturing Engineering, Universiti Putra Malaysia, 43400 UPM Serdang, Selangor, Malaysia
e-mail: sapuan@upm.edu.my

A. Nazrin
Institute of Tropical Forestry and Forest Products (INTROP), Universiti Putra Malaysia, UPM, Serdang 43400, Selangor, Malaysia

Sapuan S. M. and I. Ahmad (eds.), *Composites from the Aquatic Environment*, Composites Science and Technology, https://doi.org/10.1007/978-981-19-5327-9_12

Because of its biocompatibility, biodegradability, accessibility, and high throughput, collagen is widely employed as a biomaterial in a variety of applications. However, it has been discovered in some circumstances that collagen obtained from terrestrial animals, such as cows and pigs, is susceptible to diseases such as bovine spongiform encephalopathy (BSE) and foot-and-mouth disease (FMD). As a result, marine organisms were extracted into collagen, which is commonly referred to as marine-derived collagen. This eliminates animal illness difficulties while also promoting the use of underutilised marine biomass for a variety of engineering and biological applications. Figure 1 depicted a high-level overview of collagen extraction from marine creatures. Marine collagen, which may be recovered from a variety of marine species such as fish, jellyfish, sponges, starfish, and other fishing wastes, has been shown to have superior chemical and physical durability and is abundant (easy to achieve). Collagen is a basic structural component of membranes in many of the organism's parts, including skin, scales, bones, swim bladders, and cartilages, thus it's easy to find in those species.

In recent years, there has been an intensive search for the use of marine collagen, especially, collagen extracted from marine sponges (spongin) and as a result, an interest in the biochemical, biophysical and molecular aspects. Marine sponges (*phylum Porifera*), for example, account for a considerable portion of benthic biomass in temperate, tropical, and polar ecosystems, and provide a wide range of ecosystem services including as shelter, food, and substrate settlement regulation [6]. Despite their numerical abundance, biomass dominance, and long lifespan in environments, their functional importance appears to be undervalued. Thus, despite plenty of marine collagen benefits towards humanity, there are still many of those that humans need

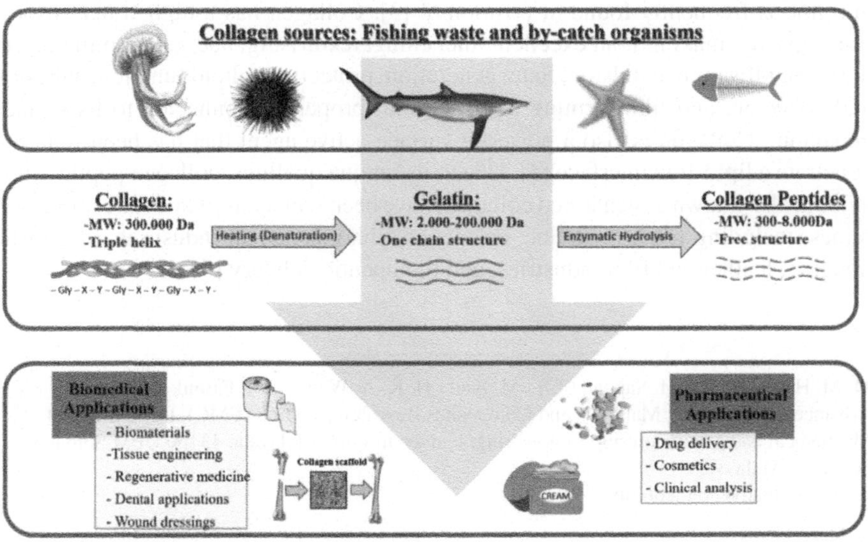

Fig. 1 The classical structure and applications of marine collagen, gelatin, and collagen peptides extracts from sustainable marine sources [3]

to discover. It is still quite difficult to find papers and journals that focus on marine collagen and its benefits. The utilization of renewable sources like fishery waste and underutilized marine organism wastes is paramount in order to promote reduced dependence on petroleum-based materials, decreased environmental pollution and improved green products [7, 8]. This shows that this source is quite rare and new to most people and that people should take the opportunity to discover and develop their knowledge on the source.

2 Renewable Sources of Marine Collagen and Its Extraction

Oceans cover greater than 70% of the planet's surface and thus are habitat to a wide variety of living things. Collagen and bioactive chemicals, which are used in a range of sectors such as pharmaceuticals, cosmetics, and food, are found in the marine environment [9]. Sea anemones, prawns, starfish, jellyfish, sponges, sea urchins, and squid are invertebrate renewable resources of marine collagen, whereas fish as well as sea mammals are vertebrate resources [10–15]. Table 1 below summarizes the information on collagen yield from specific organs of the marine species. Here, two major category of marine organisms were highlighted.

2.1 Marine Vertebrae–Fish

2.1.1 Fishery Waste–Skin

Due to its unique features, marine collagen has lately been investigated as a promising biomaterial with tremendous potential in drug delivery applications, scaffold materials in tissue engineering, composite applications, and other applications. Marine collagen has fewer dangers of disease transmission and fewer religious constraints than mammalian collagen, making it an appealing option [51]. Moreover, collagen can be found in large quantities in solid marine trash. Global fish consumption per capita is quickly increasing, rising from 9.0 kg in 1961 to 20.2 kg in 2015. This is due to increased fishery productivity fueled by fishing and aquaculture technologies [52]. As a result, a considerable amount of marine garbage has been generated. With 70–85% catch being garbage or low-value by-products [53], there is a strong motivation to extract important bioactive compounds like collagen from marine debris in terms of improving the fishing industry's environmental, economic, and social sustainability.

Furthermore, sharks are caught as bycatch when other species are fished. On the market, various sections of the body have a variety of applications. Meats are used as food, cartilage is used to make chondroitin sulphate (a food supplement used to treat

Table 1 List of renewable sources of marine collagen from various marine sources, including skins, scales, bones etc

Marine species	Organs/tissues	Collagen yield (%)	References
Vertebrate organisms			
Fishes			
Nile Tilapia (O. niloticus)	Skin	99.14	[16–18]
	Scale	27	[19–22]
Atlantic codfish (G. morhua)	Skin and bone	–	[23]
	Swim bladder	11	[24, 25]
	Skin	90	[26]
Bamboo shark (C. punctatum)	Cartilage	9	[27]
Silvertip shark (C. albimarginatus)	Skin	105	[28]
	Cartilage	–	[29]
Hammerhead shark (Sphyrna lewini)	Cartilage	5.64	[30]
Bester sturgeon (Hybrid of H. huso and A. ruthenus)	Scale, skin and swim bladder	3, 34 and 38	[31, 32]
Japanese sea bass (L. japonicus)	Skin	25	[33]
Seabass	Scale	–	[34]
Snakehead fish (Channa striata)	Scale	1.30	[35, 36]
Salmon (O. keta)	Skin	–	[37]
	Bone	–	[38]
	Scale	–	[38]
Eel fish (Evenchelys macrura)	Skin	80	[39]
Invertebrate organisms			
Sponges			
Sponge (C. reniformis)	Whole plant	10	[40–42]
Sponge (A. cannabina)	Whole plant	3	[43]
Cnidarians			
Jellyfish	Umbrella and arm	–	[44–46]
Jellyfish (A. hardenbergi)	Umbrella and arm	40	[47]
Ribbon jellyfish (Chrysaora. sp.)	Umbrella	19	[10]
Echinoderms			
Sea cucumber (H. parva)	Skin	7	[48]
Sea urchin (A. crassispina)	Skin	35	[49]
Starfish (Asterias amurensis)	Whole	5	[50]

osteoporosis and cancer), squalene is used for skin care, and skin is used to make shoes and handbags. Collagen from shark tissues was used to form a gel matrix for in vivo fibroblast culturing as a food supplement to generate functional foods, as well as to improve cryo-protection of foods [54]. Collagen was taken from the skin of the common smooth-hound, *Mustelus mustelus*, and mixed with chitosan to create a composite film that was tested as a green bioactive film to preserve nutraceutical items in another study [55]. In addition, this collagen-chitosan-based biofilm also demonstrated antioxidant activity and possible UV barrier characteristics.

Marine collagen from fish, particularly marine collagen-based scaffolds, has proven to be a viable biomaterial for a variety of biomedical applications [51]. Collagen from *Oncorhynchus keta* skin, for example, raised serum osteocalcin, femur size, mineral density, and dry weight in developing male rats [56]. In both in vivo and in vitro experiments, *Tilapia* collagen/bioactive glass (Col/BG) nanofibers were employed as wound dressings to defend from infections and enhance wound healing as well as skin regrowth [57]. In addition, collagen generated from marine species has lately been used in a variety of biomedical applications. Based on the biomimetic mineralization concept, collagen from salmon skin was combined with hydroxyapatite to create scaffolds for bone repair [58]. Human mesenchymal stem cells were able to adhere and multiply on the scaffold with interconnected holes, indicating that it was a favourable substrate for osteogenic differentiation. Tuna skin collagen peptides hydrolyzed in subcritical water showed adipogenic regulatory activity [59, 60]. Collagen hydrolysates were also produced using collagenase from unicorn leatherjacket skin (*Aluterus monoceros*) [61]. The collagen extracted was found to have anticancer, anti-diabetic, and wound-healing properties, implying that collagen extraction and hydrolysis parameters can affect bioactivity.

2.1.2 Fishery Wastes–Fish Scales

Bioactive peptides generated from collagenous resources, as well as collagen generated from fish waste, have been demonstrated to offer a variety of bioactivities with potential cosmeceutical and pharmacological applications, as well as biomaterial for tissue engineering [51, 62]. Collagen derived from fish scales was studied in order to see if it might be used in cosmetic compositions [63]. Type I collagen from the scales of the Nile tilapia, *Oreochromis niloticus*, encouraged rat odontoblast-like cells and accelerated matrix mineralization [21], suggesting that it could be used instead of type I collagen from mammals for tissue regeneration.

2.2 *Marine Invertebrate*

2.2.1 Jellyfish

The jellyfish *R. esculentum* has recently gained some interest since the type I collagen recovered from this species is very comparable to human type I collagen, making it appropriate for a variety of biological applications [64]. Furthermore, wound-healing processes were aided by peptides generated from *R. esculentum* collagen [45]. Collagen's ability to induce cell migration in injured tissue suggests that it could be used in the manufacture of synthetic cartilage tissue engineering matrix materials, primarily fibrous, hydrogel, or hybrid materials [65, 66]. Numerous efforts were made to colonise collagen-based material with connective tissue cells including such fibroblasts or endothelial cells, as well as other growth supports, in order to generate a similar biocompatibility and immunological response with commonly existing collagen-based materials. *Catastylus mosaicus*, a blubber jellyfish, was recently found to have type I collagen that efficiently supported preosteoblast development [67].

Porous collagen scaffolds were made using collagen from the huge Nomura's jellyfish, *Nemopilema nomurai* mesoglea. This collagen was found to be biocompatible with primary human fibroblasts and endothelial cells since it exhibited minimal cytotoxicity [68]. Furthermore, *N. nomurai* collagen was employed to create a 3D extremely porous hybrid collagen/hyaluronic acid scaffold that permitted fibroblast growth on its wide surface without compromising cell viability. [69]. *R. pulmo* type II collagen was recently employed to create a collagen-based biomaterial that was used to create a new adaptive device for articular cartilage repair [44, 70]. In this case, *R. pulmo* collagen was employed to make an apta-sensor for medical thrombin sensing in the bloodstream. Collagen from jellyfish contains a lot of antioxidants [71, 72]. Both jellyfish collagen and its hydrolysate were discovered to be UV radiation defenders, suggesting that they could be used in the skin care industry.

2.2.2 Echinoderms–Starfish

Collagens derived from echinoderms, such as those from sea urchin and starfish bycatch, have been described. These collagens are well-known for their unusual connective tissues, known as changeable collagenous tissues, which have lately been offered as a source of inspiration for "intelligent dynamic biomaterials" for tissue engineering [73, 74]. It also could be utilized to create collagen barrier membranes for tissue regeneration [75].

Echinoderm-derived collagen membranes are structurally and mechanically similar to commercial collagen membranes, but are significantly thinner and mechanically more robust, implying that they could be an alternative collagen source for the manufacture of effective tissue regeneration membranes. The peristomial membrane of the sea urchin, in particular, is a food industry waste that can be converted into a highly valuable by-product. It has been proposed as a source of fibrillary collagen that is both sustainable and environmentally benign for the production of membranes for regenerative medicine applications [76].

2.2.3 Sponges

Marine sponges have long been used for a variety of purposes, such as medical treatment of bacterial infections and inflammatory diseases [77]. Some sponge species were harvested and prepared as "bath sponges" for the absorptive characteristics of their fibrous skeleton throughout the Roman Empire in Mediterranean countries [78]. Moore [79] published an early study in Florida in 1910 to discuss the process and possibilities of sponge aquaculture starting from eggs or cuttings and to illustrate the existence of natural sponge beds. Figure 2 depicts the morphological organisation of a marine sponge in simplified form.

Although the knowledge is applicable to various geographies, Diaz and Rützler [80] offered prior analyses of the functional roles that sponges perform on Caribbean coastal ecosystems. Wulff [81] highlighted the following important functional roles which are strengthening coral preservation by legally enforceable live corals to the reef frame and avoiding excavating organisms from accessing their skeletons, facilitating rejuvenation of obviously deteriorated reefs by temporary fixation of carbonate rubble; rearchitecting of solid carbonate via bioerosion, nutrient recycling, and primary production through microbial symbionts.

Extraction of spongin

The spongin composite can be extracted from Demospongiae, where there are many species that give spongin composites. Araújo et al. [42] claims that the spongin-like composites collagen can be extracted from Marine Sponges (Chondrilla Caribensis and Aplysina fulva). According to their research, the samples were washed with

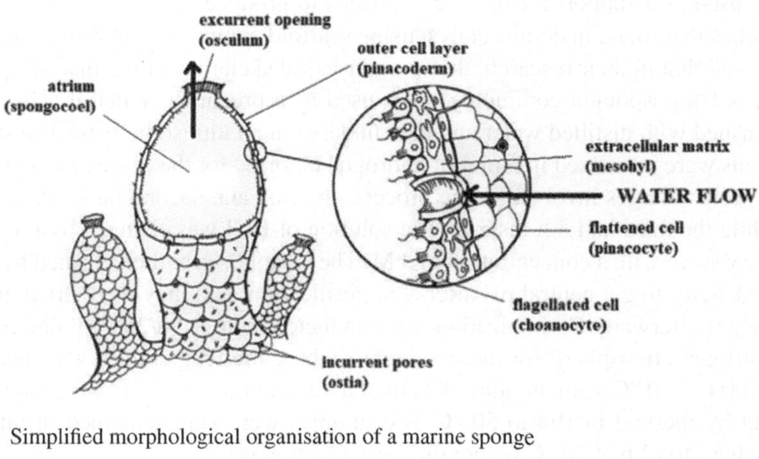

Fig. 2 Simplified morphological organisation of a marine sponge

distilled water for 3 times to remove the cell debris and were kept at -20°C. All samples were needed to undergo a pre-treatment in order to remove the excess unneeded stuff. They were needed to be cut into small pieces and kept in a stirred beaker filled with distilled water for 2 h.

Or first extraction method, Swatschek et al. [82] states that the extraction can be done by having marine sponge frozen samples to be placed in a beaker that contains 100 mM Tris-HCI buffer (10 mM EDTA, 8 M urea, 100 mM 2-mercaptoethanol, pH 9.5) and get back to pH value of 9 by using sodium hydroxide solution. The obtained solution was then stirred in a beaker for 24 h at room temperature. The solution was centrifuged at 5000 rpm for 5 min at 2°C. The substance that floated on the solution was taken out for the sake of the analysis. Acetic acid was used in the solution to get the pH value of the solution to 4 and proceeded to another centrifugation step. At this stage, the formation of the precipitate can be observed, where it was then put back in the distilled water and to be centrifuged again, then, the solution was freeze-dried for the preservation of the spongin composites.

Other than that, Berne et al. [83] claims that the pre-treated samples can be grounded into a fine powder by using cryo-milling. The free-dried marine sponge samples were placed in a grinding jar filled with nitrogen liquid and proceeded to a pre-cooling process at 30 Hz for 5 min. The cryogenic grinding was performed at 30 Hz for 2 min by repeating for 3 times and followed by cooling again with liquid nitrogen. The samples were then kept at room temperature after the grinding process. The obtained powder can be dissolved with the respective solvents, where they are0.1% typsin/ 100 mM ammonium bicarbonate with pH value of 8.5 for first method,0.1 M tris-HCI buffer with pH value of 7.5 for the second method, and deionized water at pH value of 6.8 for the third method, the process was begun with a vortex-mixer in Falcon tubes at 2500 rpm for 15 min at room temperature. The mixture was then centrifuged and the floating stuff were collected and kept frozen at -20 °C, freeze- dried and store [42].

In the study done by Ehrlich et al. [78], they claim that the materials properties of chitin found in Demospongiae, Veronica sponges, which the chitin-based scaffolds can be used as a support for metal composites to produce catalysts. There are also researches that found in doing catalyst using scaffold composites, in Żółtowska et al. [84] stated that in their research, the spongin-based skeletons of the marine sponge, which is Hippospongia communis, were used as a precursor material. They were first washed with distilled water and then underwent an ultrasonic bath. The sponge skeletons were immersed in 3 mol of hydrogen chloride for the purification process. There were 3 stages involved in the process, the first and second had a duration of 6 h while the third had 3 h instead. The solution of HCI was changed to a new one for every stage with a concentration of 3 M. The samples were then cleaned by using distilled water to get neutral pH after acid purification, and they were dried and cut into pieces afterward. Carbonization was conducted in an R 50/250/13 tube furnace in a nitrogen atmosphere for these samples, where the temperature was increased from 400 to 600 °C with an hour plateau at a heating rate of 10 °C per minute, and cooling by thermal inertia to 50 °C The samples were kept remained in nitrogen atmosphere for 2 h at 20 °C before the carbonization process.

For the modification procedure, the process underwent a few stages until the final product was obtained. The treatment of carbon materials was done by using a solution of nickel nitrate with concentration of 5 mg per litre. All samples were kept in a three-neck-round-bottom filled with 50 mL of nickel nitrate salt solution. The first stage was known as sorption, which was carried out for an hour by stirring it nonstop at speed of 800 rpm. Reduction was then carried by dropping 0.5 mol/L sodium borohydride at a rate of 5 mL/min into the solution. The reduction was resumed after finishing dropping for another 30 min at a speed of 800 rpm. The same steps of sorption and reduction were repeated 3 cycles. Lastly, the metallized materials were dried at 60 °C.

3 Properties of Marine Collagen

Collagen is typically derived from fish meat, skin, fins, scales, and waste. Purified fish collagen may be utilised in a variety of applications, including cosmetics, nutrition, medical and sport. It can be further processed to form functional biomaterials including scaffolds, sponges, gels, composites, and membranes for tissue engineering. Whales, seals, sea otters, polar bears, and other marine animals can be used to obtain marine collagen. In terms of amino acid composition and biocompatibility, collagen derived from marine animals has been shown to be comparable to collagen derived from bovine and porcine sources. Glycine, for example, is the most prevalent amino acid in collagen, making well over 30% of all amino acids. Table 2 shows the collagen derived from various marine sources.

The hydroxyproline levels have been approximated to be 35–48%, which are close to the ones in mammalian collagen [85]. These are the amino acids that are required for the formation and maintenance of the collagen-specific helical helix. However, some low glycine levels have been identified in certain marine invertebrates, such as 18.9% glycine in collagen from a marine sponge (*Chondrosia reniformis*) and 40% hydroxyproline in a research by Swatschek et al. [82]. This discrepancy can be explained by the fact that, in addition to collagen, marine tissues include other proteins such as glycoproteins, which are known to be linked to collagen and might appear as impurities in the extracted collagen, lowering its purity. This discrepancy in amino acid concentration could be due to structural and chemical variances between sources, as well as different extraction processes and biochemical studies for the same species [82].

Prior to collagen extraction, sample preparation and preservation must be done under rigorous but favorable conditions in order to produce consistent findings. Thermostability is one of the most significant parameters to consider when it comes to collagen as a biomaterial. The denaturation temperature of certain collagen from marine sources is known to be lower than the usual human physiological temperature (37 °C). Take an example of chum salmon (*Oncorhynchus keta*), a collagen gel that denatures at 19 °C, while that of shark collagen denatures approximately

Table 2 Collagen derived from marine sources

Source	Enzyme	Activity	References
Chitin	Lysozymes, cellulases, hemicellulases, proteases, and lipases	Bone repair	[87]
Jellyfish (Rhopilema esculentum)	Pepsin	Hemostatic properties collagen sponges suitable for wound dressing, cartilage repair	[12]
Dried squid	Pepsin	For biomedical applications	[15]
Crown-of-thorns starfish	Pepsin	For biomedical applications	[2]
Sponges (Chondrosia reniformis)	–	Biomaterial for tissue engineering, for dry skin cosmetic formulations, collagen nanoparticles for drug delivery systems	[40, 82, 88]
Salmon skins (Salmo salar)	–	Cosmetic applications	[63]
Hippospongia communis	–	Catalyst for reduction and oxidation	[84]
Discarded eel fish	–	3D printing of biomaterials for tissue engineering	[89]
Irish cod (Gadus morhua)	Dizym and Protamex (endoproteases)	Collagen films and coating for food packaging	[90]
Blue shark (Prionace glauca) skin	Pepsin	Chitosan- collagen composite coating to preserve red porgy (*Pagrus major*) meat	[91]
Demosponge (Aplysina aerophoba)	–	3D chitinous biocomposite scaffolds	[92]

at 30 °C. This temperature unpredictability makes some collagen-derived biomaterials hard to put into practice especially when they are meant for human medical applications [86].

4 Fabrication of Marine Collagen Based Composites

Composite materials benefit from the combination of different properties that a monolithic material cannot attain. Due to their unique properties that eliminate traditional limitations imposed by monolithic materials' physical and mechanical performance, composite materials are widely used as advanced multifunctional materials in a variety of fields, including electronics, aeronautics, medicine, automobiles, and machining tools.

Natural polymers such as collagen, chitosan, keratin, elastin, silk, elastin fibroin, and many more can be derived from nature and food waste. This is a list of natural polymers that can be altered. For potential biomedical uses, the modified compounds material can be used to make thin films, sponges, and scaffolds. The qualities of collagen-based composites are influenced by the supply of collagen as well as the technique of manufacture. Purification, fibril production, or casting and subsequent crosslinking are some of the ways for creating collagen-based material.

4.1 Extraction of Marine Collagen Prior to Composite Fabrication

Natural raw materials, as well as waste products such as demospongiae and skeleton, can be utilised to make composite materials. Berillis [93] states that collagen extraction may be done in three ways which are neutral salt solubilized collagen, acid solubilized collagen, and pepsin solubilized collagen. Jafari et al. [94] explained that depending on the marine sources, several extraction processes can be used but in general, collagen isolation is a process that involves preparation, extraction, and recovery. Washing, cleaning, and separation of animal parts are all part of the preparation process, as is size reduction by cutting or mincing the samples to make further processing easier. A chemical pre-treatment is used after the preparation to improve the effectiveness of the product.

Non-collagenous substances should be extracted and removed. Dilute acidic solvents, such as citrate buffer, 0.5 M acetic acid, or hydrochloric acid (pH 2–3) can be used as it is more efficient than neutral salt solution [93]. Different pre-treatments can be conducted based on the source materials and extraction process (alkaline or acid treatment). To do this, diluted acids and bases are used, and the collagen is partially hydrolyzed, keeping the collagen chains intact but cutting the cross-links. More delicate raw materials with less intertwined collagen fibres, such as porcine and fish skins, are more suited to the acidic procedure while the alkaline process method is utilised for thicker materials that require a stronger treatment agent penetration, such as bovine ossein or shavings [95].

Marine collagen composites can be derived from fish bones and other marine skeletons. Jafari et al. [94] stated that one of the most common methods for extracting collagen from fishbone is to use pulsed electric fields (PEF) with high intensity. A combined extraction strategy comprising semi-bionic extraction (SBE) and pulsed electric fields (PEF) treatments was used to remove collagen, chondroitin and calcium from discarded fish bones. The semi bionic extraction (SBE) method simulates the human gastrointestinal tract's digestion and absorption process by using a series of acid and alkaline extractions and varied pH ranges. The researchers extracted collagen for about 3.87 mg/mL using 22.79 kV/cm of PEF and the combination approach of PEF and SBE was shown to be effective for the extraction of calcium, collagen and chondroitin from fish skeleton. The maximum collagen yield achieved by using 1% of pepsin with 20 kV/cm of PEF strength is around 16 mg/mL. The bone collagen contains high hydroxyapatite and calcium content. Therefore, the desalting process by using EDTA or HCl was necessary to remove the hydroxyapatite and calcium during the process of pre-treatment. On the other hand, the use of HCl may damage the collagen.

4.2 Marine Collagen Composites Fabrication

The new improvement in fabrication technologies have allowed the development of marine collagen composites that resemble the complex native tissue structural hierarchy and mechanical integrity. Composites have been found to be the most promising and discerning material available everywhere in the world. As a "blue resource," marine collagen has aroused scientific and industrial interest as a "blue resource" with potential use in food, medicine, pharmaceutics, and cosmetics. The widespread availability of damaging by-products from the fish processing industry has fueled research into converting these low-cost by-products such as fish skin and scales into collagen-based goods with high added value and minimal environmental impact.

4.2.1 Collagen-Based Bi-Layered Composite Wound Dressings

Noncrosslinked, N-Ethyl-N'-(3-dimethylaminopropyl) carbodiimide/N-Hydroxy succinimide (EDC/NHS) cross-linked, or commercial Beiling collagen sponge were used as the inner layer of the bi-layered composite wound dressing. Sun et al. [96] states that they made non-crosslinked and crosslinked collagen sponges from Nile tilapia skin using 100/40 mM N-Ethyl-N'-(3-dimethylaminopropyl) carbodiimide/N-Hydroxysuccinimide (EDC/NHS) in sixhole tissue culture plates in their laboratory. For this experiment, a commercial Beiling collagen sponge was cut to the same size. The outer layer was made up of 5 cm sizes of medical spun-laced nonwoven with good air permeability, very viscous acrylic resin adhesive, and release paper. Medical-grade chitosan was dissolved in a 0.5 M acetic acid solution and combined with 30% (solute mass) glycerin to form a 3 percent solution. The above chitosan solution was applied to the outer layer of the bi-layered composite wound dressing with a triangular coating rod, and then dried at 50 °C for 5 h in an electric-heat constant-temperature drying oven.

4.2.2 Marine Algae-PLA (MAP) Composites

Galaxaura oblongata, *Corallina elongata*, *Cystoseria compressa*, *Stypopodium schimperi* and *Sargassum vulgare* are among the five varieties of algae. Red algae make up the first three, while brown algae make up the rest. Near Antalya, Turkey, *Corallina elongata* (*Rhodophyta*) and *Galaxaura oblongata* (*Rhodophyta*) were gathered while *Cystoseria compressa* (*Phaeophyta*), *Sargassum vulgare* (*Phaeophyta*), and *Stypopodium schimperi* (*Phaeophyta*) were gathered near Iskenderun, Turkey. For sampling, SCUBA and free dives between 0 and 40 m were employed, as well as vertical and horizontal scans underwater. During sampling, the "Olympus OM-DE-M5" camera was used to capture underwater photos of macroalgae. Binocular light microscopes and SZX16 stereo zoom were used to identify the materials, which were then soaked in 1000 mL of distilled water for about 72 h to eliminate any water-soluble substances. After filtration, marine algae were then ground four times at 300 rpm for ten minutes in a high-speed rotary grinder before being vacuum-dried at 50 °C for 24 h. Dried marine algae were ground again, sieved, and vacuum-dried for 12 h at 100 °C. Finally, we were able to create micro-scale MAPs containing absolute moisture levels of less than 3%.

The above-mentioned MAP samples were washed in acetone and dried in an oven at 80 °C for 24 h. After that, PLA (Goodfellow, 459–898-81) and MAP were mixed in a mechanical mixer at 150 °C for 15 min at 50 rpm. The composites were mixed, then pressed into thin plates with a bespoke hot press at 180 °C and 8 MPa for 20 min before cooling in an oven. Differential scanning calorimetry (DSC) study has determined that the glass transition temperature of algae, particularly Sargassum vulgare, is around 175 °C [97].

4.2.3 Fish Collagen/PCL Composite Scaffolds

Electrospinning is a manufacturing technology in which polymer fibres are deposited onto a collector utilising a high voltage differential between the needle and the collector [98]. In order to produce smooth and homogeneous fibres, it is necessary to use a carrier solvent with the proper viscosity, volatility, and conductivity [99]. Fluoro-alcohols, such as hexafluoro-isopropanol (HFIP), are frequently employed for electrospinning of marine collagen because of their volatile nature and ability to dissolve collagen to generate a viscous and conductive solution ideal for electrospinning. Because HFIP dissolves a wide spectrum of polymers, mechanically stronger polymers like polycaprolactone (PCL) can be added to the collagen electrospinning solution to create composite fibres that are stronger than those electrospun with simply marine collagen. Those composites, in the form of fish collagen/PCL composite scaffolds, were manufactured by Chai et al. [100] via electrospinning. Another technique to stabilise the structure of electrospun marine collagen scaffolds, increase mechanical properties, and control the degradation rate is to crosslink the scaffold after it has been spun. Zhou et al. [101] say that the scaffolds were shown to accelerate wound healing whilst the scaffolds from Hassanbhai et al. [102] were shown to have no toxicity and no long-term in vivo inflammatory responses.

4.2.4 Chitosan Based Collagen/ Gelatin Composite Scaffolds from Big Eye Snapper Priacanthus Hamrur Skin

Extraction of collagen

Acid-soluble collagen (ASC) was extracted from the prepared and stored skin [103]. For around 36 h, 50 gms of Priacanthus humrur skin was soaked in 5% Lactic acid at 4 °C to 7 °C. At room temperature, the soaked sample was homogenised with a blender. The homogenised sample was combined with 0.1 M NaOH solution and maintained at 5 °C to eliminate non-collagenous proteins. The treated skin was then rinsed in cold water many times to remove the NaOH until the wash water reached a neutral pH. The sample was centrifuged at 7800 × g for 15 min at 5 °C. After that, the precipitate was collected and freeze-dried. Collagen was extracted and stored in an airtight container.

Extraction of gelatin

According to Kołodziejska et al. [104], Priacanthus humrur fragmented skin was chopped, and 0.1 M NaCl solution was added to it three times at 4 °C for three minutes each time. Cold water and NaOH were used to rinse the skin. Gelatin was removed from the skin by swirling it in water for 60 min at 45 degrees Celsius. The insoluble material was removed by centrifugation at 10,000 × g for 30 min at 15 °C. The supernatant was taken and frozen at 20 °C. The produced samples were freeze dried and stored at 4 °C.

Development of scaffold

Scaffold materials were produced using a blending approach with diverse biopolymer compositions made from collagen, gelatin from Priacanthus humrur, and chitosan dispersion. Gelatin was dispersed in demineralised water with a mechanical stirrer and kept at 4 °C. Chitosan and collagen were dissolved in 0.5 M acetic acid solution and gelatin was dispersed in demineralised water with a mechanical stirrer and kept at 4 °C. To begin, each polymer's individual dispersion is blended together to achieve a % concentration of each polymer in the combination [103]. To remove air that had accumulated in the solution, the mixes were centrifuged at 2400 × g for 30 min. The resulting liquid was poured into cylindrical moulds and maintained at 10 °C for 20 h to allow collagen to gel before being freeze dried at -20 °C. The scaffold structures were kept at 4 °C in hermetically sealed containers.

5 Application of Marine Collagen Composites

Marine collagen composite applications in various industries are very much in the early development phases and have yet to be explored. Marine collagen has been explored for its possible uses in drug delivery systems, tissue engineering, cosmetics, and nutricosmetics, to name a few [105]. Collagenous extracts from fishes and marine sponges have attracted a lot of biotechnological interest recently, as seen by a broad range of applications in cosmetics and biomedicine. The following are some of the potential applications.

5.1 Chitosan-Marine Collagen Composite Scaffolds for Bone Regeneration

For bone regeneration, Chitin-Hydroxyapatite-collagen scaffolds (CHCS) with stable physicochemical features have been developed by Liao and Huang [106]. For tissue engineering, chitin has previously been recognised as a suitable raw material due to its rigidity and naturalness. Compressive strength is a critical criterion for bone-repair materials [107]. Mechanical strength of CHCS was modified by the chitin/Hydroxyapatite (HAP) weight ratio and collagen concentration [108]. Pure chitin's natural stiffness may not be sufficient for bone tissue creation; thus HAP was used to improve its mechanical characteristics. These findings supported previous research [109], which showed that HAP might boost the CHCS's compressive strength.

5.2 Marine Derived Nanohydroxyapatite and Their Composites for Dental Application

Hydroxyapatite nanocomposites are widely utilized in implant and biomedical applications. It is now currently under development and is under clinical trials for utilizing them as dental implants. In recent research, the heat treatment method was used to successfully recycle crab shells as a seafood waste into a highly useful biomaterial [110]. The resulting powder is a combination of needle-like hydroxyapatite (HAp) nanorods and nanospheres of carbonated hydroxyapatite (CHAp) nanoparticles. The powder has the identical properties to apatite powder that was produced from other sources such as bovine, human and sheep. This method of obtaining apatite powder from crab shells has proven to be an environmentally benign and cost-effective technology with prospective uses in the dentistry industry.

Balu et al. [111] states that, several preparation procedures, such as mechanochemical, hydrothermal, sol–gel, hydrolysis, emulsion, precipitation, pyrolysi, combustion, and sonochemical are used to make HAp nanoparticles, depending on the required uses. Senthil et al. [112] concluded that the discarded marine bones from Bluefin trevally fishes were ideal for the creation of nanosized HAp, which might be used in dental implant and bone tissue engineering applications. According to Balu et al. [111], zirconia may be used as a waste-derived hydroxyapatite nanoparticles or dopant with marine bone to build a ceramic that is especially designed for dental implant materials.

5.3 Marine Collagen for 3D Bioprinting of Scaffold Composites in Tissue Engineering

3D printing for tissue engineering is one that repairs or replaces the damaged tissue or organ when it comes to an organ failure or loss. Marine biotechnology, which is commonly known as blue biotechnology that uses the biological materials that are originally from the marine environment. Yoo et al. [113] found that the coral are a non-renewable resource and marine derived coral is able to be used as bone implant biomaterial. Therefore, having a renewable or great amount of resources is better than a non-renewable that decreases eventually. Govindharaj et al. [89] investigated the usability of the collagen extracted from discarded Eel fish skin on 3D blue biomaterial printing for tissue engineering.

For material preparation, the remaining moisture from the eel skins was retained after smashing it into small pieces. After that, 0.5 M acetic acid solution was added into the Eel skin pieces and continued stirring for around 2 days at a certain temperature. It was then proceeded to filtration to remove all the insoluble components and the precipitate was taken by adding salt. The precipitation separated acid soluble collagen (ASC) during spinning at speed of 6000 rpm for 1 h. The ASC was used for the preparation of pepsin-soluble collagen (PSC) by distribute 0.5 M of acetic acid with 1% of pepsin. Next, the mixture was then incubated for a day at a specific temperature which is 4 °C. The mixture was then isolated by spinning at 8000 rpm for 1 h at 4 °C after incubation. The supernatants were gathered while the NaCL was put in to salted out PSC. The isolated PSC was mixed in the 0.5 M of acetic acid solution and dialyzed against 0.1 M acetic acid with water at afterwards. Lastly, the collagen was lyophilized and freeze-dried to produce pepsin soluble collagen powder. To prepare for hydrogel, the collagen from the discarded Eel skin and sodium alginate were mixed and dissolved in sterile Milli-Q water and was stirred until homogenous hydrogel was formed.

The hydrogels were prepared to proceed 3D printing and Table 3 shows 4 different wt.% of Eel skin collagen in them. After proceeding to 3D printing of all samples, the 3D printed cuboidal structures are shown in Fig. 3.

For group B and C, the scaffold composites showed comparable inner structure and viscosity characteristics, furthermore, group A was similar, which showed the potential for being used as 3D printing for bio-materials for patient's tissue regeneration of specified cases. However, as shown in Fig. 1, group D showed an unstable

Table 3 The formation of the samples. Adapted from [89]

Blue biomaterials			
Sample groups	Alginate (%)	Eel skin collagen (mg/ml)	Cross linking agents ($CaCl_2$) (mM)
Group A	5	–	100
Group B	5	10	
Group C	5	20	
Group D	5	30	

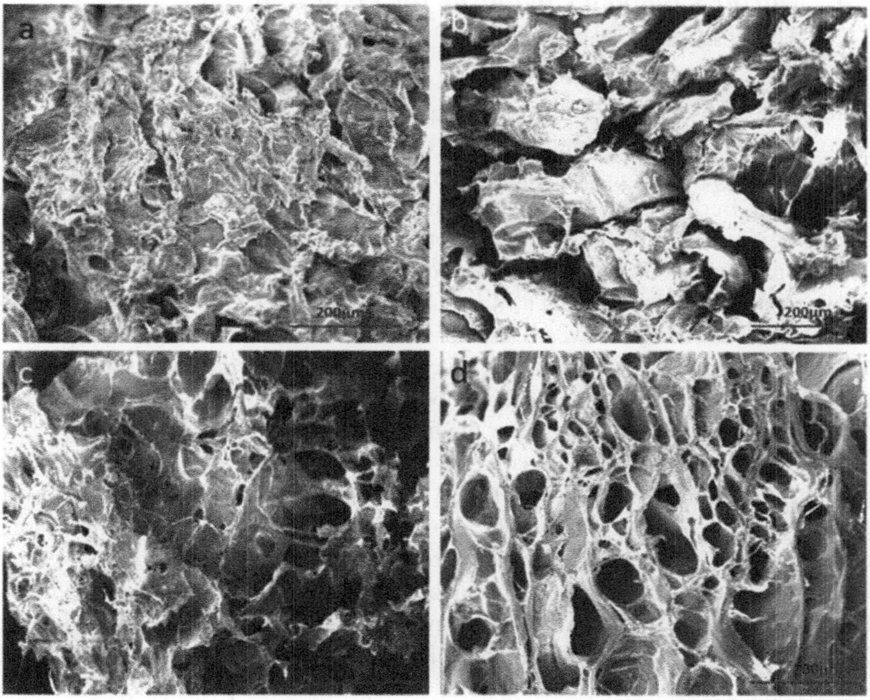

Fig. 3 The 3D printed cuboidal structures of all samples. Adapted from [89]

structure, where Chai et al. [100] explained that the inside of the gel chemical potential and the outside of the gel chemical potential should be balanced out for an equilibrium condition. With that said, it has been explained that the case of group D might be due to the increase of volume fraction of the collagen polymer that causes the decrease in viscosity caused by the change in the thermodynamic force in mixing and the stored elastic forces in the stretched polymer chains.

The use of marine resources for biomaterials is still in its early stages of development. Using the discarded eel fish, the collagen from eel skin has a simple isolation procedure, takes less time to create, and is cost effective. However, this type of collagen does not have high efficiency, which is ~4.2%. In order to increase efficiency, there is still a lot of work to proceed to find a way to achieve the highest efficiency. This valorization of marine biowastes will be a good alternative way to solve the waste disposal, Khan and Tanaka [114] stated that although there are many commercially and synthetic materials give the needed the and physiochemical and mechanical properties for tissue engineering application, meanwhile, they have inadequate biomimetic property that helps in the tissue regeneration.

5.4 *Marine Collagen Composite for Food Industry Application*

Collagen also has a wide application in the food processing industry. Although so far there is no literature available for the use of marine collagen in this application, the potential application of marine collagen, mostly in food processing, has sparked a lot of attention. Berillis [93] states that the coatings and edible films are a distinct type of packaging material that differs from other bio-based or conventional packaging They're made of edible substances like collagen. The collagen can be extracted by the fish skin to produce collagen films. O'Sullivan et al. [90] claims that collagen protein was effectively recovered from Irish cod (Gadus morhua) skins using acetic acid extraction and further processing procedures, and this collagen was then utilised to make collagen films. Collagen could be effectively recovered from a number of marine species with the potential for the film-forming ingredient.

The use of chitosan-fish skin collagen composite coating was also proven in the food preserving application. Liu et al. [91] states that the biodegradability, biocompatibility, and cost effectiveness of chitosan-fish skin collagen composite coating make it an environmentally acceptable preservation method with a lot of potential. In their study, the fish skin collagen from blue shark skin and Chitosan was called 10 B with the addition of Pepsin was employed to make a chitosan-fish skin collagen composite. Pepsin was used to produce chitosan-fish skin collagen composite. The solution was then used as a coating solution to preserve red porgy fillets which is a type of coral reef fish. The result of the study shows that the inclusion of collagen to the coating solution (at the maximum concentration tested; 0.8 percent) increased the preservation effect of the red porgy fillets.

Yanwong and Threepopnatkul [115] states that gelatin is a type of collagen-derived protein that has been hydrolyzed. Gelatin comes from the bones or skin of animals such as beef, poultry, swine, and fish. Coppola et al. [3] explained that for food preservation, gelatin films and lysozymes which are bioactive peptides should be included together. Incorporating chitosan into gelatin film forming solutions resulted in active films that inhibited the growth of relevant food poisoning microorganisms. Wu et al. [116] studied that the incorporation of fish gelatin films with cinnamon essential oil nanoliposomes resulted in a longer-lasting release effect, improved antibacterial stability, and a lower release rate. Cinnamon essential oil containing nanoliposomes is an antimicrobial carrier for generating biodegradable gelatin-based films, according to the research, as it can extend the food shelf-lifes. Yanwong and Threepopnatkul [115] studies have shown that either peppermint or citronella oil incorporated into the fish skin gelatin films could improve the antibacterial activity. When peppermint or citronella oils were applied to the fish skin gelatin film, it demonstrated a considerable increase in antimicrobial activities against E. coli and S. aureus. There are a lot of methods that can incorporate the gelatin based films from marine collagen for food preservation application. New processes and formulations for producing marine gelatin-based films with improved ultimate qualities and prospective uses, on the other hand, need to be researched further.

5.5 *Spongin-Based Scaffolds with Nickel Composites for Functional Catalyst*

Żółtowska et al. [84] claims that the modification of structured bio-carbon can be obtained from spongin-based scaffolds with nickel compounds for a functional catalyst for the use of reduction and oxidation reactions. It is said that the contribution of that research is to be able to be used in environmental protection. Throughout the research, the low temperature carbonization of spongin-based scaffolds was used to produce hierarchical 3D carbonaceous structures that maintained the original morphology of the spongin-based skeleton. The scaffolds were carried out with the modification with nickel compounds through a reduction method, which was used to obtain novel catalysts.

The prepared materials were tested in the reduction of 4-nitrophenol to analyze their catalytic properties. Emam et al. [117] claims that the reduction of 4-nitrophenol is commonly used as a determinant of the catalytic activity of mixed materials. The materials were first tested to measure the materials to show that they exhibit catalytic ability in the reduction of 4-nitrophenol, and yet the reaction time of the samples varies from 4 and 6 min. However, the reaction does not work without the presence of a catalyst. 3 prepared materials, which are known as $NiO/Ni(OH)_2/Ni$ with 3 different Ni percentages in each material where they are 18.68%, 15.19% and 26.01%. In short, they are classified as $NiO/Ni(OH)_2/Ni)_{400}$, $NiO/Ni(OH)_2/Ni_{500}$ and $NiO/Ni(OH)_2/Ni_{600}$ respectively. As a result, in the reduction of 4-nitrophenol, $NiO/Ni(OH)_2/Ni_{500}$ showed a slow reduction of the peak intensity assigned, whereas $NiO/Ni(OH)_2/Ni_{600}$ showed rapid decrease of the peak intensity after 60 s of the reaction. However, $NiO/Ni(OH)_2/Ni_{400}$ showed the highest reaction rate constant, furthermore, the time of reaction for each catalyst used has shown similar timing as shown in Table 4. The result of catalytic reduction reaction by repeating 5 times.

Table 4 Kinetic parameters of reduction of 4-nitrophenol using the prepared catalysts

Catalyst	No. of cycle	K (min^{-1})	R^2	Time of reaction (min)
$NiO/Ni(OH)_2/Ni_{400}$	1	1.022	0.949	5
	2	0.936	0.959	5
	3	0.732	0.975	6
	4	0.635	0.989	7
	5	0.513	0.994	9
$NiO/Ni(OH)_2/Ni_{500}$		0.648	0.952	6

A spongin-based fibrous scaffold, which is extracted from the marine known as demosponge *Hippospongia communis* with the modification of the structured bio-carbon obtained from it and treatment with the nickel nitrate to form a carbonized spongin-based scaffold composites have proven its ability by utilizing it as a catalyst for oxidation or reduction reactions of various phenolic compounds. With that said, this research has provided evidence that spongin-based scaffolds can be used to produce a structured carbonaceous material. Which is able to function with modification of nickel and nickel oxide.

5.6 Drug Delivery Carrier

A drug delivery system is a device that regulates the distribution of a pharmacological agent to a patient in order to achieve a therapeutic effect. For the targeted distribution of Gentamicin, Macha and Ben-Nissan [118] produced thin film composites by combining polylactic acid and coral-derived hydroxyapatite (HAp) utilising hydrothermal mediated hydroxyapatite. These thin films are applied using a dip or spray coating method to miniature neural devices, dental implants, and total or cochlear hip replacement implants, and then they can inhibit microorganisms from multiplying and generating biofilms, that can ultimately lead to post-operative, dental, neural, and orthopaedic implant diseases. Hydrothermal transformation was proposed by Lagopati and Agathopoulos [119] to manufacture hydroxyapatite scaffolds from cuttlefish bone for drug administration applications, especially in chronic osteomyelitis whereby antibiotics should be supplied for long periods of time. The study proposed drug incorporation into nano-HAp and polymer coating, drug conjugation with hydrogels and infiltration into nano-HAp pores from cuttlefish, and drug absorption on nano-HAp surfaces with a precoated polymer for prolonged drug delivery.

There are studies showing a number of ways of drug delivery in humans and animals such as delivering drugs at controlled rate, dose titration, therapeutic drug monitoring, slow delivery, and targeted delivery. Researchers have used collagen from medusa Catostylus tagi to develop a microparticulate protein delivery system. The jellyfish collagen has been chosen as a polymeric matrix to create collagen microparticles (CMPs). This can be achieved by using emulsification-gelation-solvent extraction method. The collagen matrix's strong cross linking collects therapeutic proteins and controls their release from the system while preserving biological activity.

Collagen from a marine sponge (*Porifera, dictyoceratida*) has been analyzed and turned into L-cysteine hydrochloride-containing polymeric films that can repair wounds [14]. In vitro investigations show that cysteine in the biopolymer is delivered more slowly than the pure medication at the wound site. This improves the system's suitability for bio-based medication delivery based on marine collagen. It combines cysteine's known healing qualities with the collagen network's potentials as a biocompatible carrier that may absorb excess wound exudate while releasing the medication.

5.7 Biocomposite Scaffolds Based on Chitosan for Tissue Engineering

Mutsenko et al. [92] investigated that 3D chitinous biocomposite scaffolds can be derived from the cultivated marine demosponge *Aplysina aerophoba* for tissue engineering. For tissue engineering purposes, it does not fixate on any particular species in the demosponges' family. In the study of Ehrlich et al. [78], they found out that the Verongida sponges, Porifera, which it has potential in biomedical applications especially in cartilage tissue engineering. However, Aerophoba has a higher survival rate of 80% in cultivation as compared to other species [14].

The macroporous 3D biocomposite scaffolds were produced with the treatment of alkali-acidic solutions on A. aerophoba fragments. The internal composition of the biocomposite scaffolds was as shown in Fig. 3.

Martino et al. [120] claims that the biomaterials for stem cell tissue engineering in an ideal condition, would actually support the stem cell attachment, proliferation and differentiation through their physicochemical. Besides, Engler et al. [121] and Do et al. [122] also claim that mechanical properties and the 3D spatial geometry would resemble the extracellular matrix. The viability of human Mesenchymal stem cell (hMSCs) cultured on the biocomposite scaffolds was observed for 21 days and it showed a toxic less effect that would damage the cell from the scaffolds for the time being as shown in Fig. 4. As it can be seen from the Fig. 5, the scaffold was covered by viable cells, which are green colored. Furthermore, it also indicates the absence of cytotoxic effect of the scaffold after decellularization and demineralization procedures.

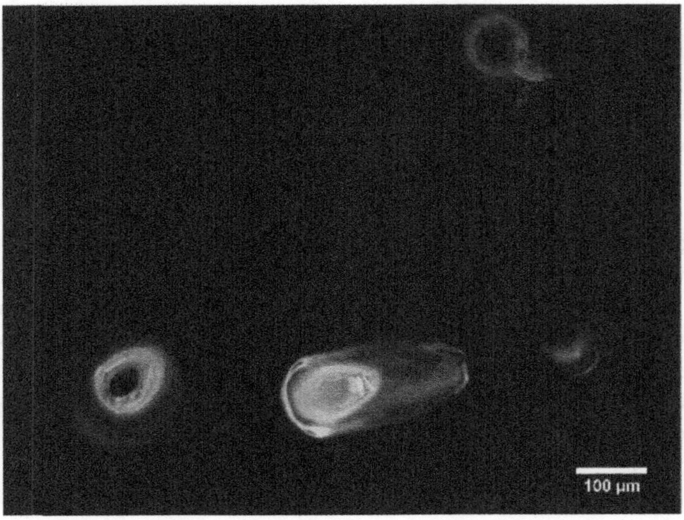

Fig. 4 The cross-section view of *A. aerophoba* chitin fibers. Adapted from [92]

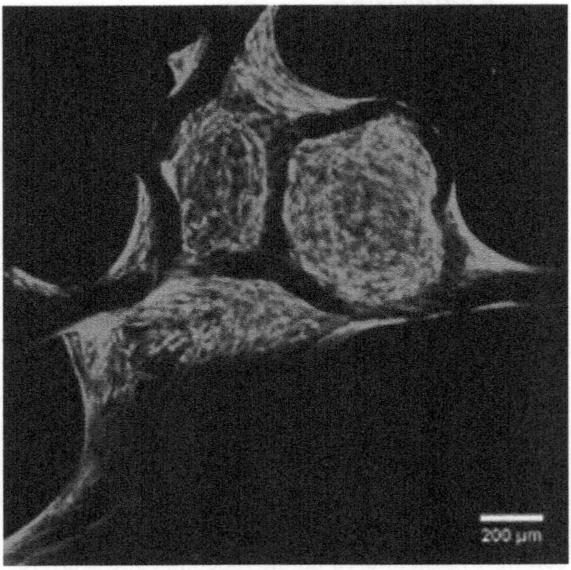

Fig. 5 Result of viability of hMSCs cultured on biocomposite scaffolds on day 21. Adapted from [92]

Alizarin red staining, which demonstrates mineralization of osteogenically induced cells, were also shown growing on the scaffolds as shown in Fig. 6. The cells filled up the space with mineralized matrix between the chitin fibers. Therefore, it has proven that the potential of the biocomposite scaffolds extracted from the Demospongiae to be used in even more applications of biomedical purpose.

Fig. 6 Alizarin red staining shown on the biocomposite scaffold. Adapted from [92]

6 Conclusion

There are many applications that have been developed by using marine collagen such as collagen extracted from spongin as there are tissue engineering, bioprinting, drug delivery carrier, cosmetic and more. Although these studies are still in the embryonic stage, where they are yet to be used or being introduced to the market, they do have potential in replacing some of the current use materials for the sake of environment, cost effective and also simpler production process in the future with future development and study.

References

1. Kiew, P. L., & Don, M. M. (2013). Modified Lowry's method for acid and pepsin soluble collagen measurement from Clarias species muscles. *Science and Reports, 2*, 1–5.
2. Tan, C. C., Karim, A. A., Latiff, A. A., Gan, C. Y., & Ghazali, F. C. (2013). Extraction and characterization of pepsin-solubilized collagen from the body wall of crown-of-thorns Starfish (Acanthaster planci). *International Food Research Journal, 20*, 3013.
3. Coppola, D., Oliviero, M., Vitale, G. A., Lauritano, C., D'Ambra, I., Iannace, S., & de Pascale, D. (2020). Marine collagen from alternative and sustainable sources: Extraction, processing and applications. *Marine Drugs, 18*, 214.
4. Hashim, P., Ridzwan, M. M. S., Bakar, J., & Hashim, M. D. (2015). Collagen in food and beverage industries. *International Food Research Journal, 22*, 1.
5. Bama, P., Vijayalakshimi, M., Jayasimman, R., Kalaichelvan, P. T., Deccaraman, M., & Sankaranarayanan, S. (2010). Extraction of collagen from cat fish (Tachysurus maculatus) by pepsin digestion and preparation and characterization of collagen chitosan sheet. *International Journal of Pharmacy and Pharmaceutical Sciences, 2*, 133–137.
6. Pawlik, J. R., & McMurray, S. E. (2020). The emerging ecological and biogeochemical importance of sponges on coral reefs. *Annual Review of Marine Science, 12*, 315–337.

7. Harussani, M. M., Sapuan, S. M., Rashid, U., Khalina, A., & Ilyas, R. A. (2022). Pyrolysis of polypropylene plastic waste into carbonaceous char: Priority of plastic waste management amidst COVID-19 pandemic. *Science of the Total Environment, 803*, 149911. https://doi.org/10.1016/J.SCITOTENV.2021.149911
8. Harussani, M. M., Rashid, U., Sapuan, S. M., & Abdan, K. (2021). Low-temperature thermal degradation of disinfected covid-19 Non-woven Polypropylene—based isolation gown wastes into carbonaceous char. *Polymers (Basel)., 13*, 3980.
9. Venkatesan, J., Anil, S., Kim, S.-K., & Shim, M. S. (2017). Marine fish proteins and peptides for cosmeceuticals: A review. *Marine Drugs, 15*, 143.
10. Barzideh, Z., Latiff, A. A., Gan, C.; Benjakul, S., Karim, A.A. (2014). Isolation and characterisation of collagen from the ribbon jellyfish (C hrysaora sp.). *international journal of food science & technology, 49*, 1490–1499.
11. Zhuang, Y., Zhao, X., & Li, B. (2009). Optimization of antioxidant activity by response surface methodology in hydrolysates of jellyfish (Rhopilema esculentum) umbrella collagen. *Journal of Zhejiang University. Science. B, 10*, 572–579.
12. Cheng, X., Shao, Z., Li, C., Yu, L., Raja, M. A., & Liu, C. (2017). Isolation, characterization and evaluation of collagen from jellyfish Rhopilema esculentum Kishinouye for use in hemostatic applications. *PLoS ONE, 12*, e0169731.
13. Khong, N. M. H., Yusoff, F. M., Jamilah, B., Basri, M., Maznah, I., Chan, K. W., & Nishikawa, J. (2016). Nutritional composition and total collagen content of three commercially important edible jellyfish. *Food Chemistry, 196*, 953–960.
14. Langasco, R., Cadeddu, B., Formato, M., Lepedda, A. J., Cossu, M., Giunchedi, P., Pronzato, R., Rassu, G., Manconi, R., & Gavini, E. (2017). Natural collagenic skeleton of marine sponges in pharmaceutics: Innovative biomaterial for topical drug delivery. *Materials Science and Engineering C, 70*, 710–720.
15. Jankangram, W., Chooluck, S., & Pomthong, B. (2016). Comparison of the properties of collagen extracted from dried jellyfish and dried squid. *African Journal of Biotechnology, 15*, 642–648.
16. Hu, Z., Yang, P., Zhou, C., Li, S., & Hong, P. (2017). Marine collagen peptides from the skin of Nile Tilapia (Oreochromis niloticus): Characterization and wound healing evaluation. *Marine Drugs, 15*, 102.
17. Song, W.-K., Liu, D., Sun, L.-L., Li, B.-F., & Hou, H. (2019). Physicochemical and biocompatibility properties of type I collagen from the skin of Nile tilapia (Oreochromis niloticus) for biomedical applications. *Marine Drugs, 17*, 137.
18. Elbialy, Z.I., Atiba, A., Abdelnaby, A., Al-Hawary, I. I., Elsheshtawy, A.; El-Serehy, H. A., Abdel-Daim, M. M., Fadl, S.E., Assar, D.H. (2020). Collagen extract obtained from Nile tilapia (Oreochromis niloticus L.) skin accelerates wound healing in rat model via up regulating VEGF, bFGF, and α-SMA genes expression. *BMC Vet Research 16*, 1–11.
19. Terada, M., Izumi, K., Ohnuki, H., Saito, T., Kato, H., Yamamoto, M., Kawano, Y., Nozawa-Inoue, K., Kashiwazaki, H., Ikoma, T. (2012) Construction and characterization of a tissue-engineered oral mucosa equivalent based on a chitosan-fish scale collagen composite. *Journal of Biomedical Materials Research—Part B Applied Biomaterials, 100*, 1792–1802.
20. Chen, J., Li, L., Yi, R., Xu, N., Gao, R., & Hong, B. (2016). Extraction and characterization of acid-soluble collagen from scales and skin of tilapia (Oreochromis niloticus). *LWT-Food Sci. Technol., 66*, 453–459.
21. Tang, J., Saito, T. (2015). Biocompatibility of novel type I collagen purified from tilapia fish scale: an in vitro comparative study. *Biomed Research International, 2015.*
22. El-Rashidy, A. A., Gad, A., Abu-Hussein, A.E.-H.G., Habib, S. I., Badr, N. A., & Hashem, A. A. (2015). Chemical and biological evaluation of Egyptian Nile Tilapia (Oreochromis niloticas) fish scale collagen. *International Journal of Biological Macromolecules, 79*, 618–626.
23. Yamada, S., Yoshizawa, Y., Kawakubo, A., Ikeda, T., Yanagiguchi, K., & Hayashi, Y. (2013). Early gene and protein expression associated with osteoblast differentiation in response to fish collagen peptides powder. *Dental Materials Journal, 32*, 233–240.

24. Sousa, R. O., Alves, A. L., Carvalho, D. N., Martins, E., Oliveira, C., Silva, T. H., & Reis, R. L. (2020). Acid and enzymatic extraction of collagen from Atlantic cod (Gadus Morhua) swim bladders envisaging health-related applications. *Journal of Biomaterials Science, Polymer Edition, 31*, 20–37.
25. Sousa, R. O., Martins, E., Carvalho, D. N., Alves, A. L., Oliveira, C., Duarte, A. R. C., Silva, T. H., & Reis, R. L. (2020). Collagen from Atlantic cod (Gadus morhua) skins extracted using CO2 acidified water with potential application in healthcare. *Journal of Polymer Research, 27*, 1–9.
26. Bai, C., Wei, Q., & Ren, X. (2017). Selective extraction of collagen peptides with high purity from cod skins by deep eutectic solvents. *ACS Sustainable Chemistry & Engineering, 5*, 7220–7227.
27. Kittiphattanabawon, P., Benjakul, S., Visessanguan, W., & Shahidi, F. (2010). Isolation and characterization of collagen from the cartilages of brownbanded bamboo shark (Chiloscyllium punctatum) and blacktip shark (Carcharhinus limbatus). *LWT-Food Science and Technology, 43*, 792–800.
28. Elango, J., Lee, J. W., Wang, S., Henrotin, Y., De Val, J. E. M. S. M Regenstein, J., Lim, S. Y., Bao, B., Wu, W. (2018) Evaluation of differentiated bone cells proliferation by blue shark skin collagen via biochemical for bone tissue engineering. *Mar. Drugs, 16*, 350.
29. Jeevithan, E., Bao, B., Bu, Y., Zhou, Y., Zhao, Q., & Wu, W. (2014). Type II collagen and gelatin from silvertip shark (Carcharhinus albimarginatus) cartilage: Isolation, purification, physicochemical and antioxidant properties. *Marine Drugs, 12*, 3852–3873.
30. Chi, C.-F., Wang, B., Li, Z.-R., Luo, H.-Y., & Ding, G.-F. (2013). Characterization of acid-soluble collagens from the cartilages of scalloped hammerhead (Sphyrna lewini), red stingray (Dasyatis akajei), and skate (Raja porosa). *Food Sci. Biotechnol., 22*, 909–916.
31. Mredha, M. T. I., Zhang, X., Nonoyama, T., Nakajima, T., Kurokawa, T., Takagi, Y., & Gong, J. P. (2015). Swim bladder collagen forms hydrogel with macroscopic superstructure by diffusion induced fast gelation. *Journal of Materials Chemistry B, 3*, 7658–7666.
32. Zhang, X., Ookawa, M., Tan, Y., Ura, K., Adachi, S., & Takagi, Y. (2014). Biochemical characterisation and assessment of fibril-forming ability of collagens extracted from Bester sturgeon Huso huso× Acipenser ruthenus. *Food Chemistry, 160*, 305–312.
33. Bao, Z., Sun, Y., Rai, K., Peng, X., Wang, S., Nian, R., & Xian, M. (2018). The promising indicators of the thermal and mechanical properties of collagen from bass and tilapia: Synergistic effects of hydroxyproline and cysteine. *Biomater. Sci., 6*, 3042–3052.
34. Krishnan, S., Sekar, S., Katheem, M. F., Krishnakumar, S., & Sastry, T. P. (2012). Fish scale collagen—a novel material for corneal tissue engineering. *Artificial Organs, 36*, 829–835.
35. Wang, J. K., Yeo, K. P., Chun, Y. Y., Tan, T. T. Y., Tan, N. S., Angeli, V., & Choong, C. (2017). Fish scale-derived collagen patch promotes growth of blood and lymphatic vessels in vivo. *Acta Biomaterialia, 63*, 246–260.
36. Hasri, N. M., & Zebua, N. (2020). Test of burn wounds healing effects of collagen from snakehead fish (Channa striata) bone in the preparation of cream on male white rats (Rattus norvegicus). *Indonesian Journal Pharmaceutical Clinical Research, 3*, 62–75.
37. Zhang, Z., Wang, J., Ding, Y., Dai, X., & Li, Y. (2011). Oral administration of marine collagen peptides from Chum Salmon skin enhances cutaneous wound healing and angiogenesis in rats. *Journal of the Science of Food and Agriculture, 91*, 2173–2179.
38. Wang, W., Itoh, S., Aizawa, T., Okawa, A., Sakai, K., Ohkuma, T., & Demura, M. (2010). Development of an injectable chitosan/marine collagen composite gel. *Biomedical Materials, 5*, 65009.
39. Veeruraj, A., Arumugam, M., Ajithkumar, T., & Balasubramanian, T. (2012). Isolation and characterization of drug delivering potential of type-I collagen from eel fish Evenchelys macrura. *Journal of Materials Science. Materials in Medicine, 23*, 1729–1738.
40. Nicklas, M., Schatton, W., Heinemann, S., Hanke, T., & Kreuter, J. (2009). Preparation and characterization of marine sponge collagen nanoparticles and employment for the transdermal delivery of 17β-estradiol-hemihydrate. *Drug Development and Industrial Pharmacy, 35*, 1035–1042.

41. Parisi, J. R., Fernandes, K. R., Avanzi, I. R., Dorileo, B. P., Santana, A. F., Andrade, A. L., Gabbai-Armelin, P. R., Fortulan, C. A., Trichês, E. S., & Granito, R. N. (2019). Incorporation of collagen from marine sponges (spongin) into hydroxyapatite samples: Characterization and in vitro biological evaluation. *Marine Biotechnology, 21*, 30–37.
42. Araújo, T. A. T., de Souza, A., Santana, A. F., Braga, A. R. C., Custódio, M. R., Simões, F. R., Araújo, G. M., Miranda, A., Alves, F., & Granito, R. N. (2021). Comparison of different Methods for Spongin-like collagen extraction from marine sponges (Chondrilla caribensis and Aplysina fulva): Physicochemical properties and in vitro biological analysis. *Membranes (Basel)., 11*, 522.
43. Tziveleka, L.-A., Ioannou, E., Tsiourvas, D., Berillis, P., Foufa, E., & Roussis, V. (2017). Collagen from the marine sponges Axinella cannabina and Suberites carnosus: Isolation and morphological, biochemical, and biophysical characterization. *Marine Drugs, 15*, 152.
44. Keller, L., Pugliano, M. (2017). Combined jellyfish collagen type ii, human stem cells and tgf-β3 as a therapeutic implant for cartilage repair. Journal of Stem Cell Research & Therapeutics, 7.
45. Felician, F. F., Yu, R.-H., Li, M.-Z., Li, C.-J., Chen, H.-Q., Jiang, Y., Tang, T., Qi, W.-Y., & Xu, H.-M. (2019). The wound healing potential of collagen peptides derived from the jellyfish Rhopilema esculentum. *Chinese Journal of Traumatology, 22*, 12–20.
46. Calejo, M. T., Almeida, A. J., & Fernandes, A. I. (2012). Exploring a new jellyfish collagen in the production of microparticles for protein delivery. *Journal of Microencapsulation, 29*, 520–531.
47. Khong, N. M. H., Yusoff, F. M., Jamilah, B., Basri, M., Maznah, I., Chan, K. W., Armania, N., & Nishikawa, J. (2018). Improved collagen extraction from jellyfish (Acromitus hardenbergi) with increased physical-induced solubilization processes. *Food Chemistry, 251*, 41–50.
48. Adibzadeh, N., Aminzadeh, S., Jamili, S., Karkhane, A. A., & Farrokhi, N. (2014). Purification and characterization of pepsin-solubilized collagen from skin of sea cucumber Holothuria parva. *Applied Biochemistry and Biotechnology, 173*, 143–154.
49. Nagai, T., & Suzuki, N. (2000). Partial characterization of collagen from purple sea urchin (Anthocidaris crassispina) test. *International Journal of Food Science & Technology, 35*, 497–501.
50. Lee, K., Park, H. Y., Kim, Y. K., Park, J. Il, Yoon, H. D. (2009) Biochemical characterization of collagen from the starfish Asterias amurensis. *Journal of the Korean Society for Applied Biological Chemistry, 52*, 221–226.
51. Lim, Y.-S., Ok, Y.-J., Hwang, S.-Y., Kwak, J.-Y., & Yoon, S. (2019). Marine collagen as a promising biomaterial for biomedical applications. *Marine Drugs, 17*, 467.
52. Coppola, D., Lauritano, C., Palma Esposito, F., Riccio, G., Rizzo, C., & de Pascale, D. (2021). Fish waste: From problem to valuable resource. *Marine Drugs, 19*, 116.
53. Pal, G. K., & Suresh, P. (2016). V Sustainable valorisation of seafood by-products: Recovery of collagen and development of collagen-based novel functional food ingredients. *Innovative Food Science and Emerging Technologies, 37*, 201–215.
54. Limpisophon, K., Iguchi, H., Tanaka, M., Suzuki, T., Okazaki, E., Saito, T., Takahashi, K., & Osako, K. (2015). Cryoprotective effect of gelatin hydrolysate from shark skin on denaturation of frozen surimi compared with that from bovine skin. *Fisheries Science, 81*, 383–392.
55. Ben Slimane, E., & Sadok, S. (2018). Collagen from cartilaginous fish by-products for a potential application in bioactive film composite. *Marine Drugs, 16*, 211.
56. Xu, Y. J., Han, X. L., & Li, Y. (2010). Effect of marine collagen peptides on long bone development in growing rats. *Journal of the Science of Food and Agriculture, 90*, 1485–1491.
57. Zhou, T., Sui, B., Mo, X., & Sun, J. (2017). Multifunctional and biomimetic fish collagen/bioactive glass nanofibers: Fabrication, antibacterial activity and inducing skin regeneration in vitro and in vivo. *International Journal of Nanomedicine, 12*, 3495.
58. Hoyer, B., Bernhardt, A., Heinemann, S., Stachel, I., Meyer, M., & Gelinsky, M. (2012). Biomimetically mineralized salmon collagen scaffolds for application in bone tissue engineering. *Biomacromolecules, 13*, 1059–1066.

59. Abuine, R., Rathnayake, A. U., & Byun, H.-G. (2019). Biological activity of peptides purified from fish skin hydrolysates. *Fisheries and Aquatic Sciences, 22*, 1–14.
60. Lee, E. J., Hur, J., Ham, S. A., Jo, Y., Lee, S., Choi, M.-J., & Seo, H. G. (2017). Fish collagen peptide inhibits the adipogenic differentiation of preadipocytes and ameliorates obesity in high fat diet-fed mice. *International Journal of Biological Macromolecules, 104*, 281–286.
61. Kumar, L. V., Shakila, R. J., & Jeyasekaran, G. (2019). In vitro anti-cancer, anti-diabetic, anti-inflammation and wound healing properties of collagen peptides derived from unicorn leatherjacket (Aluterus monoceros) at different hydrolysis. *Turkish Journal of Fisheries and Aquatic Sciences, 19*, 551–560.
62. Nasri, M. (2019) Bioactive peptides from fish collagen byproducts: A review. *Byproducts from Agriculture and Fisheries. Adding Value Food, Feed. Pharma, Fuels*, 309–333.
63. Alves, A. L., Marques, A. L. P., Martins, E., Silva, T. H., & Reis, R. L. (2017). Cosmetic potential of marine fish skin collagen. *Cosmetics, 4*, 39.
64. Felician, F. F., Xia, C., Qi, W., & Xu, H. (2018). Collagen from marine biological sources and medical applications. *Chemistry & Biodiversity, 15*, e1700557.
65. Hoyer, B., Bernhardt, A., Lode, A., Heinemann, S., Sewing, J., Klinger, M., Notbohm, H., & Gelinsky, M. (2014). Jellyfish collagen scaffolds for cartilage tissue engineering. *Acta Biomaterialia, 10*, 883–892.
66. Pustlauk, W., Paul, B., Gelinsky, M., & Bernhardt, A. (2016). Jellyfish collagen and alginate: Combined marine materials for superior chondrogenesis of hMSC. *Materials Science and Engineering C, 64*, 190–198.
67. Rastian, Z., Pütz, S., Wang, Y., Kumar, S., Fleissner, F., Weidner, T., & Parekh, S. H. (2018). Type I collagen from jellyfish catostylus mosaicus for biomaterial applications. *ACS Biomaterials Science & Engineering, 4*, 2115–2125.
68. Song, E., Kim, S. Y., Chun, T., Byun, H.-J., & Lee, Y. M. (2006). Collagen scaffolds derived from a marine source and their biocompatibility. *Biomaterials, 27*, 2951–2961.
69. Jeong, S. I., Kim, S. Y., Cho, S. K., Chong, M. S., Kim, K. S., Kim, H., Lee, S. B., & Lee, Y. M. (2007). Tissue-engineered vascular grafts composed of marine collagen and PLGA fibers using pulsatile perfusion bioreactors. *Biomaterials, 28*, 1115–1122.
70. Sugahara, T., Ueno, M., Goto, Y., Shiraishi, R., Doi, M., Akiyama, K., & Yamauchi, S. (2006). Immunostimulation effect of jellyfish collagen. *Bioscience, Biotechnology, and Biochemistry, 70*, 2131–2137.
71. Ding, J.-F., Li, Y.-Y., Xu, J.-J., Su, X.-R., Gao, X., & Yue, F.-P. (2011). Study on effect of jellyfish collagen hydrolysate on anti-fatigue and anti-oxidation. *Food Hydrocoll., 25*, 1350–1353.
72. Yu, H., Liu, X., Xing, R., Liu, S., Li, C., & Li, P. (2005). Radical scavenging activity of protein from tentacles of jellyfish Rhopilema esculentum. *Bioorganic & Medicinal Chemistry Letters, 15*, 2659–2664.
73. Barbaglio, A., Tricarico, S., Ribeiro, A., Ribeiro, C., Sugni, M., Di Benedetto, C., Wilkie, I., Barbosa, M., Bonasoro, F., & Carnevali, M. D. C. (2012). The mechanically adaptive connective tissue of echinoderms: Its potential for bio-innovation in applied technology and ecology. *Marine Environment Research, 76*, 108–113.
74. Barbaglio, A., Tricarico, S., Di Benedetto, C., Fassini, D., Lima, A. P., Ribeiro, A. R., Ribeiro, C. C., Sugni, M., Bonasoro, F., Wilkie, I. (2013) The smart connective tissue of echinoderms: a materializing promise for biotech applications. *Cah Biol Mar, 54*, 713–720.
75. Ferrario, C., Leggio, L., Leone, R., Di Benedetto, C., Guidetti, L., Coccè, V., Ascagni, M., Bonasoro, F., La Porta, C. A. M., & Carnevali, M. D. C. (2017). Marine-derived collagen biomaterials from echinoderm connective tissues. *Marine Environment Research, 128*, 46–57.
76. Benedetto, C., & Di; Barbaglio, A., Martinello, T., Alongi, V., Fassini, D., Cullorà, E., Patruno, M., Bonasoro, F., Barbosa, M.A., Carnevali, M.D.C. (2014). Production, characterization and biocompatibility of marine collagen matrices from an alternative and sustainable source: The sea urchin Paracentrotus lividus. *Marine Drugs, 12*, 4912–4933.
77. Holm, P. (2012) Oceans past: management insights from the history of marine animal populations; Routledge; ISBN 113656036X.

78. Ehrlich, H., Deutzmann, R., Brunner, E., Cappellini, E., Koon, H., Solazzo, C., Yang, Y., Ashford, D., Thomas-Oates, J., & Lubeck, M. (2010). Mineralization of the metre-long biosilica structures of glass sponges is templated on hydroxylated collagen. *Nature Chemistry, 2*, 1084–1088.
79. Maslin, M., Gaertner-Mazouni, N., Debitus, C., Joy, N., & Ho, R. (2021). Marine sponge aquaculture towards drug development: An ongoing history of technical, ecological, chemical considerations and challenges. *Aquac. Reports, 21*, 100813.
80. Diaz, M. C., & Rützler, K. (2001). Sponges: An essential component of Caribbean coral reefs. *Bulletin of Marine Science, 69*, 535–546.
81. Wulff, J. (2001). Assessing and monitoring coral reef sponges: Why and how? *Bulletin of Marine Science, 69*, 831–846.
82. Swatschek, D., Schatton, W., Kellermann, J., Müller, W. E. G., & Kreuter, J. (2002). Marine sponge collagen: Isolation, characterization and effects on the skin parameters surface-pH, moisture and sebum. *European Journal of Pharmaceutics and Biopharmaceutics, 53*, 107–113.
83. Berne, S., Kalauz, M., Lapat, M., Savin, L., Janussen, D., Kersken, D., Ambrožič Avguštin, J., Zemljič Jokhadar, Š, Jaklič, D., & Gunde-Cimerman, N. (2016). Screening of the Antarctic marine sponges (Porifera) as a source of bioactive compounds. *Polar Biology, 39*, 947–959.
84. Żółtowska, S., Bielan, Z., Zembrzuska, J., Siwińska-Ciesielczyk, K., Piasecki, A., Zielińska-Jurek, A., & Jesionowski, T. (2021). Modification of structured bio-carbon derived from spongin-based scaffolds with nickel compounds to produce a functional catalyst for reduction and oxidation reactions: Potential for use in environmental protection. *Science of the Total Environment, 794*, 148692.
85. Yamada, S., Yamamoto, K., Ikeda, T., Yanagiguchi, K., Hayashi, Y. (2014). Potency of fish collagen as a scaffold for regenerative medicine. *Biomed Research International, 2014*.
86. Ehrlich, H., Ilan, M., Maldonado, M., Muricy, G., Bavestrello, G., Kljajic, Z., Carballo, J. L., Schiaparelli, S., Ereskovsky, A., Schupp, P. (2010). Three-dimensional chitin-based scaffolds from Verongida sponges (Demospongiae: Porifera). Part I. Isolation and identification of chitin. *International Journal of Biological Macromolecules, 47*, 132–140.
87. Xing, F., Chi, Z., Yang, R., Xu, D., Cui, J., Huang, Y., Zhou, C., & Liu, C. (2021). Chitin-hydroxyapatite-collagen composite scaffolds for bone regeneration. *International Journal of Biological Macromolecules, 184*, 170–180.
88. Pozzolini, M., Bruzzone, F., Berilli, V., Mussino, F., Cerrano, C., Benatti, U., & Giovine, M. (2012). Molecular characterization of a nonfibrillar collagen from the marine sponge Chondrosia reniformis Nardo 1847 and positive effects of soluble silicates on its expression. *Marine Biotechnology, 14*, 281–293.
89. Govindharaj, M., Roopavath, U. K., & Rath, S. N. (2019). Valorization of discarded Marine Eel fish skin for collagen extraction as a 3D printable blue biomaterial for tissue engineering. *Journal of Cleaner Production, 230*, 412–419.
90. O'Sullivan, A., Shaw, N. B., Murphy, S. C., Van de Vis, J. W., van Pelt-Heerschap, H., & Kerry, J. P. (2006). Extraction of collagen from fish skins and its use in the manufacture of biopolymer films. *Journal of Aquatic Food Product Technology, 15*, 21–32.
91. Liu, J., Shibata, M., Ma, Q., Liu, F., Lu, Q., Shan, Q., Hagiwara, T., & Bao, J. (2020). Characterization of fish collagen from blue shark skin and its application for chitosan-collagen composite coating to preserve red porgy (Pagrus major) meat. *Journal of Food Biochemistry, 44*, e13265.
92. Mutsenko, V. V., Bazhenov, V. V., Rogulska, O., Tarusin, D. N., Schütz, K., Brüggemeier, S., Gossla, E., Akkineni, A. R., Meißner, H., & Lode, A. (2017). 3D chitinous scaffolds derived from cultivated marine demosponge Aplysina aerophoba for tissue engineering approaches based on human mesenchymal stromal cells. *International Journal of Biological Macromolecules, 104*, 1966–1974.
93. Berillis, P. (2015). Marine collagen: Extraction and applications. *Reports of Trends Biochemistry and Molecular Biology Microbiology*, 1–13.

94. Jafari, H., Lista, A., Siekapen, M. M., Ghaffari-Bohlouli, P., Nie, L., Alimoradi, H., & Shavandi, A. (2020). Fish collagen: Extraction, characterization, and applications for biomaterials engineering. *Polymers (Basel)., 12*, 2230.
95. Schmidt, M.M., Dornelles, R. C. P., Mello, R. O., Kubota, E. H., Mazutti, M. A., Kempka, A. P., Demiate, I. M. (2016). Collagen extraction process. *International Food Research Journal, 23*.
96. Sun, L., Li, B., Jiang, D., & Hou, H. (2017). Nile tilapia skin collagen sponge modified with chemical cross-linkers as a biomedical hemostatic material. *Colloids Surfaces B Biointerfaces, 159*, 89–96.
97. Kolsi, R. B. A., Salah, H. Ben, Jardak, N., Chaaben, R., Jribi, I., El Feki, A., Rebai, T., Jamoussi, K., Allouche, N., Blecker, C. (2017). Sulphated polysaccharide isolated from Sargassum vulgare: Characterization and hypolipidemic effects. *Carbohydrate Polymers. 170*, 148–159.
98. Ghorani, B., & Tucker, N. (2015). Fundamentals of electrospinning as a novel delivery vehicle for bioactive compounds in food nanotechnology. *Food Hydrocolloids, 51*, 227–240.
99. Bazrafshan, Z., & Stylios, G. K. (2019). Spinnability of collagen as a biomimetic material: A review. *International Journal of Biological Macromolecules, 129*, 693–705.
100. Chai, Q., Jiao, Y., & Yu, X. (2017). Hydrogels for biomedical applications: Their characteristics and the mechanisms behind them. *Gels, 3*, 6.
101. Zhou, C., Shi, Q., Guo, W., Terrell, L., Qureshi, A. T., Hayes, D. J., & Wu, Q. (2013). Electrospun bio-nanocomposite scaffolds for bone tissue engineering by cellulose nanocrystals reinforcing maleic anhydride grafted PLA. *ACS Applied Materials & Interfaces, 5*, 3847–3854.
102. Hassanbhai, A. M., Lau, C. S., Wen, F., Jayaraman, P., Goh, B. T., Yu, N., & Teoh, S.-H. (2017). In vivo immune responses of cross-linked electrospun tilapia collagen membrane. *Tissue Engineering Part A, 23*, 1110–1119.
103. Tylingo, R., Mania, S., Panek, A., Piątek, R., & Pawłowicz, R. (2016). Isolation and characterization of acid soluble collagen from the skin of african catfish (Clarias gariepinus), salmon (Salmo salar) and baltic cod (Gadus morhua). *Journal of Biotechnology and Biomaterials, 6*, 2.
104. Kołodziejska, I., Skierka, E., Sadowska, M., Kołodziejski, W., & Niecikowska, C. (2008). Effect of extracting time and temperature on yield of gelatin from different fish offal. *Food Chemistry, 107*, 700–706.
105. Subhan, F., Ikram, M., Shehzad, A., & Ghafoor, A. (2015). Marine collagen: An emerging player in biomedical applications. *Journal of Food Science and Technology, 52*, 4703–4707.
106. Liao, J., & Huang, H. (2019). Green magnetic hydrogels synthesis, characterization and flavourzyme immobilization based on chitin from Hericium erinaceus residue and polyvinyl alcohol. *International Journal of Biological Macromolecules, 138*, 462–472.
107. Zhao, D., Zhu, T., Li, J., Cui, L., Zhang, Z., Zhuang, X., & Ding, J. (2021). Poly (lactic-co-glycolic acid)-based composite bone-substitute materials. *Bioactive Materials, 6*, 346–360.
108. Weinand, C., Neville, C. M., Weinberg, E., Tabata, Y., & Vacanti, J. P. (2016). Optimizing biomaterials for tissue engineering human bone using mesenchymal stem cells. *Plastic and Reconstructive Surgery, 137*, 854–863.
109. Liu, M., Zheng, H., Chen, J., Li, S., Huang, J., & Zhou, C. (2016). Chitosan-chitin nanocrystal composite scaffolds for tissue engineering. *Carbohydrate Polymers, 152*, 832–840.
110. Bhattacharjee, B. N., Mishra, V. K., Rai, S. B., Parkash, O., & Kumar, D. (2019). Structure of apatite nanoparticles derived from marine animal (crab) shells: An environment-friendly and cost-effective novel approach to recycle seafood waste. *ACS Omega, 4*, 12753–12758.
111. Kumar Balu, S., Andra, S., Jeevanandam, J., & Sampath, V. (2021). Emerging marine derived nanohydroxyapatite and their composites for implant and biomedical applications. *Journal of the Mechanical Behavior of Biomedical Materials, 119*, 104523.
112. Senthil, R., Vedakumari, S. W., & Sastry, T. P. (2018). Hydroxyapatite and demineralized bone matrix from marine food waste–a possible bone implant. *American Journal of Materials Synthesis and Processing, 3*, 1–6.

113. Yoo, Y.-W., Park, G. J., & Lee, W. K. (2016). Surface modification of coralline scaffold for the improvement of biocompatibility and bioactivity of osteoblast. *Journal of Industrial and Engineering Chemistry, 33*, 33–41.
114. Khan, F., & Tanaka, M. (2018). Designing smart biomaterials for tissue engineering. *International Journal of Molecular Sciences, 19*, 17.
115. Yanwong, S., Threepopnatkul, P. (2015). Effect of peppermint and citronella essential oils on properties of fish skin gelatin edible films. In: *Proceedings of the IOP Conference Series: Materials Science and Engineering; IOP Publishing*; Vol. 87, p. 12064.
116. Wu, J., Liu, H., Ge, S., Wang, S., Qin, Z., Chen, L., Zheng, Q., Liu, Q., & Zhang, Q. (2015). The preparation, characterization, antimicrobial stability and in vitro release evaluation of fish gelatin films incorporated with cinnamon essential oil nanoliposomes. *Food Hydrocoll., 43*, 427–435.
117. Emam, H. E., El-Zawahry, M. M., & Ahmed, H. B. (2017). One-pot fabrication of AgNPs, AuNPs and Ag-Au nano-alloy using cellulosic solid support for catalytic reduction application. *Carbohydrate Polymers, 166*, 1–13.
118. Macha, I. J., & Ben-Nissan, B. (2018). Marine skeletons: Towards hard tissue repair and regeneration. *Marine Drugs, 16*, 225.
119. Lagopati, N., & Agathopoulos, S. (2019). *Hydroxyapatite scaffolds produced from cuttlefish bone via hydrothermal transformation for application in tissue engineering and drug delivery systems* (pp. 179–205). Springer.
120. Martino, M. M., Mochizuki, M., Rothenfluh, D. A., Rempel, S. A., Hubbell, J. A., & Barker, T. H. (2009). Controlling integrin specificity and stem cell differentiation in 2D and 3D environments through regulation of fibronectin domain stability. *Biomaterials, 30*, 1089–1097.
121. Engler, A. J., Sen, S., Sweeney, H. L., & Discher, D. E. (2006). Matrix elasticity directs stem cell lineage specification. *Cell, 126*, 677–689.
122. Do, A., Khorsand, B., Geary, S. M., & Salem, A. K. (2015). 3D printing of scaffolds for tissue regeneration applications. *Advanced Healthcare Materials, 4*, 1742–1762.

Recent Advances in Composites from Seaweeds

Shristy Gautam and Aishwarya Mogal

1 Introduction

Seaweeds are macroscopic algae living in deep seas, in back water, and estuaries found attached to dead corals, pebbles, rocks, shells, and so on [1]. They are classified on the basis of their colour, red seaweed (Rhodophyta), green seaweed (chlorophyta), and brown seaweed (phaeophyta) [2]. They are widely available, adaptable to a variety of locations, inexpensive, and simple to produce in a natural environment. They can also be harvested throughout the year [3]. All the three seaweed contains approximately 1–5% of lipids, 50% carbohydrates, and 7–73% minerals with high protein content in green and red seaweed that is 8–47% and low protein content in brown seaweed that is 4–24% [4]. Vitamins and minerals are abundant in seaweeds, making them an obvious health benefit. As a result, seaweeds (especially red and brown seaweeds) are commonly utilised as herbal medicine and nutrition for humans. It is possible to use them into fresh salads and soups as well as in cookies, dinners, and condiments because they are both edible and rich in essential elements [3]. Carbohydrates present in seaweeds are used to produce plastics with high quality, and which are environmentally friendly, non-toxic, inexpensive [5]. Polysaccharides in green seaweed consist of cellulose, rhamnan, ulvan, and galactan [6], while red seaweeds contain carrageenan, agar, xylan, cellulose, and porphyrin [7], and brown seaweeds contain fucoidan, laminarian, and alginate [8] which are used to make bio-plastics. Bioplastics made from seaweeds generally degrade in 4–6 weeks in soil [9]. Some of the examples used for the production of bioplastics are Ulva, Codium (green seaweed), *Gelidium*, *Gracilaria*, and *Porphyra* (red seaweed), and *Laminaria*, *Lessonia*, and *Macrocystis* (brown seaweed) [10, 11]. There are various methods for the production of bioplastics from seaweeds which are dependent on the species used,

S. Gautam (✉) · A. Mogal
Department of Molecular Biology and Genetic Engineering, School of Bioengineering and Biosciences, Lovely Professional University, Phagwara, Punjab, India
e-mail: shristy.g@gmail.com

Sapuan S. M. and I. Ahmad (eds.), *Composites from the Aquatic Environment*,
Composites Science and Technology, https://doi.org/10.1007/978-981-19-5327-9_13

type of solvents used, environmental impact, cost, amount, and time required. The methods like super-critical fluid extraction, microwave assisted extraction, enzyme assisted extraction, photo-bleaching extraction, ultrasound assisted extraction and so on can be used to obtain bioplastic film [12]. There are various applications of green, red and brown seaweeds in different fields like plastic, paper, wastewater treatment, pharmaceuticals, manure, fertilizers, biofuels, food, medicines, cosmetics, bioremediation, agricultural, and so on [2].

Petroleum based plastics are carbon based polymers which are non-biodegradable and pollute the environment [1]. Plastics if degrade releases harmful toxins like dioxins which increase the global warming [11]. Hence, to decrease the release of harmful toxins, to save the environment, an alternative to traditional plastics that is green material should be adopted which are renewable and biodegradable [13]. Green materials include bioplastics, and bio-composites which are bio-based materials like plant proteins, feed stocks and agricultural waste. Mechanical and barrier qualities, controlled drug release, and adsorption efficiency are advantages of composites over pure polymer-based materials [3]. Increasing population results in more production of this green materials, this would be burden on the food, arable land, water supply, and would lead to the competition. To solve this problem, algae would play an excellent alternative to produce green materials. Alga, also known as water plant or seaweeds are present in waste as well as marine, and fresh water environment [14] are autotrophic organisms which transform CO_2 into biomass including carbohydrates, proteins, lipids, and fats. They are microscopic organisms and can be classified as microalgae and macro-algae. Both this alga is excellent source of vitamins, proteins, fats, carbohydrates, fibers, lipids, and secondary compounds. Macro-algae are multicellular organisms having chlorophycease, phaeophyceae, rhodophyceae, while microalgae are microscopic having diatoms, blue green algae, and dinoflagellates [15]. This chapter reviews important green materials bio-composites derived from macro-algae, seaweeds.

2 Seaweeds Based Composites

Seaweeds have been used by humans since the dawn of recorded history, and some of the earliest examples may be found in ancient texts. It was around 1658 when Agar was first discovered in Japan, and it was not until 1859 that its first chemical study was performed. Until the fifteenth century, carrageenan was commonly employed as an ingredient in food products. In contrast, seaweed cultivation has just recently become an industrial crop with its rapid growth and technological advancements over the past half century. It has been proven that growing seaweed as a crop can help address the long-term issue of environmental sustainability in an effective and fast manner [16]. In contrast to petroleum-based polymers, edible and degradable polymers are created from renewable and edible materials such polysaccharides, proteins, and lipids, which decompose more quickly. Using this polymer in food and biomaterial products or as food coatings or packaging films because of its degradable qualities and preservation capabilities is a huge benefit to the environment and health [17]. In terms of edible polymers, polysaccharide holds a lot of promise due to its low cost, wide availability, biocompatibility, and minimal impact on the environment [18, 19]. As an edible film, it's been widely utilised in agricultural products to extend the shelf life of fruit and vegetables, reduce oil or fat absorption in fried food, avoid flavour loss, and extend the shelf life of frozen meals [19, 20].

The use of seaweed in energy, food, tissue engineering, biosensors, and drug delivery has received a lot of attention recently [21]. Seaweed is a versatile material that has many potential uses. There are a number of different green and economical polysaccharide materials to choose from; seaweed is just one more choice that comes from the ocean rather than land [22]. Products produced from seaweed, such as alginate, carrageenan, and agar, have unique film-forming capabilities [1, 23, 24]. Even so, as compared to non-renewable polymers like conventional ones, seaweed films have comparatively low water vapour barrier and mechanical strength qualities. As a result, seaweed is frequently combined with other ingredients in order to enhance the qualities of seaweed films.

Petroleum-based polymers have replaced traditional packaging materials because of their low cost, excellent barrier qualities, and high mechanical properties. In contrast, present landfill procedures are not able to breakdown these packing materials, resulting in environmental issues [25]. In contrast, biopolymer films often decay faster in natural environments than petroleum-based films. A variety of materials,

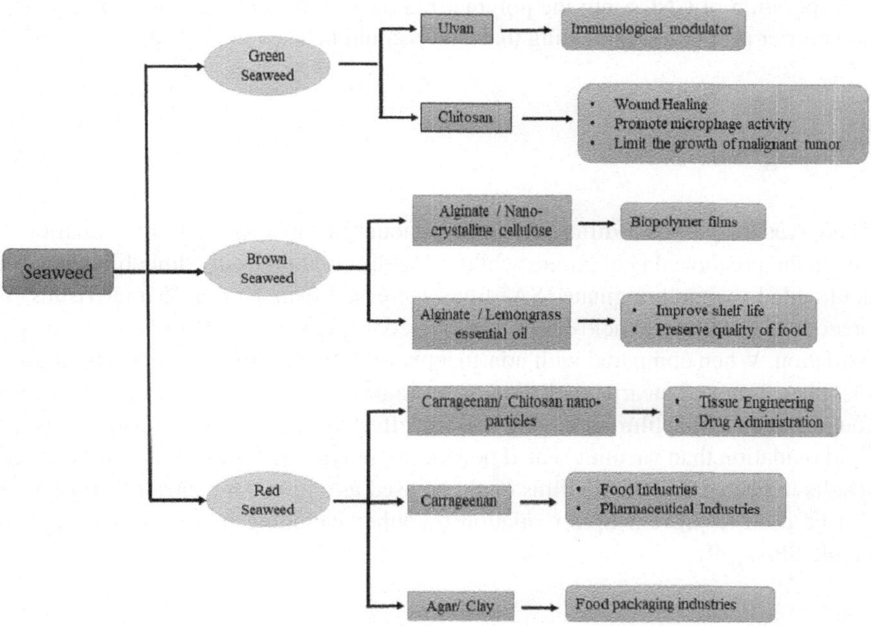

Fig. 1 Application of different seaweeds in composites

including starch, alginate, and chitosan, have been used in the study of biopolymer films by researchers [26–28] (Fig. 1).

2.1 Alginate/Nanocrystalline Cellulose

Cell walls of brown seaweeds are rich in carbohydrates, which could be employed as the biopolymer film's backbone [27, 29–31]. Using biopolymer packaging films instead of petroleum-based ones has long been regarded an interesting alternative [28, 32, 33]. All industries use packaging films, which not only protect and preserve the product during distribution, but they also provide convenience and communication to the end consumer. Brown seaweed's cell wall contains alginate, which has a linear molecular structure capable of forming a strong polymer matrix and structures, for example. There are two types of brown seaweed that can be found all over the world: Kombu (*Laminaria japonica*) and Sargassum (*Sargassum natans*) [34, 35]. Brown seaweeds were used to make seaweed biopolymer films, and the leftovers from that process were used to extract cellulose nanocrystals (CNC). Biopolymer films have been hampered in their ability to scale because of their poor physical, structural, and barrier properties compared to petroleum-based packaging material [29, 30]. Biopolymers, on the other hand, have some inherent difficulties that can be overcome by employing CNCs to strengthen biopolymer sheets [24, 36, 37]. Incorporation of CNCs into the polymer matrix significantly improved mechanical and barrier properties, indicating that CNCs could be used in packaging [37, 38].

2.2 Alginate/Starch

There was a significant difference in the amount of moisture loss and oxidation of fats in the precooked beef patties while using different types of edible films. Stearic acid added to starch-alginate (SA) films increased their barrier characteristics. In precooked beef, stearic-acid-based coatings-controlled moisture loss better than lipid oxidation. When compared with non-tocopherol films for lipid oxidation inhibition, tocopherol-treated stearic acid films were found to be more effective [39]. It was found that all edible film packaging was less effective at preventing moisture loss and lipid oxidation than vacuum-sealed pouches of polyester. Tocopherol film treatment results indicated that edible films could be used as antioxidant carriers. Tocopherol can be used to prevent lipid oxidation by either including it into or coating it on edible films [39].

2.3 Carrageenan/Locust Bean Gum/Organically Modified Nanoclay

A clay is a naturally occurring material that is mostly composed of finely ground minerals. With a high surface area-to-aspect ratio, montmorillonite (MMT) is one of the most commonly used layered silicates [40]. There are positive silicate layers in the inter-lamellar space of MMT, which function as counterions to the negative ions in the inter-lamellar space [41]. Alkylammonium cations, which can be added to silicate layers to increase their hydrophobicity, are a common approach for clay modification. An organoclay is a type of modified organic clay [41].

Cloisite 30B (C30B), a organically modified clay was found to have enhanced physical and antibacterial properties when dispersed in a biopolymer matrix containing biodegradable films made of combinations of κ-carrageenan, LBG, and Cloisite 30B. Films containing κ-car/LBG–C30B have an inhibiting impact exclusively on *Listeria monocytogenes* bacteria. Composite films made of κ-car/LBG–C30B polymers can be used to increase the shelf life and safety of the food. The incorporation of montmorillonite (MMT) into the polymer matrix has improved the physical properties of the films [40, 42]. Adding clays to biodegradable films may open up new avenues for improving biopolymer characteristics for food packaging. Biopolymers' mechanical, thermal, and barrier properties have been shown to benefit by the addition of clays, even at extremely low concentrations [43]. To improve the usage of biopolymers for food packaging, clays play a significant role.

2.4 Carrageenan/Grapefruit Seed Extract

Sulfated polysaccharide produced from red seaweed (Rhodophycae) called carrageenan has been studied extensively in the food and pharmaceutical industries for its gelling, stabilising, and emulsifying properties. Carrageenan is water-soluble. Superior film formation, water barrier, and mechanical properties were all demonstrated for the κ-carrageenan [44]. Carrageenan's random coil shape takes on a double helical structure as the film solution cools, resulting in a more compact form of the material [45]. Incorporating antimicrobial or antioxidant substances into polymeric films may help to extend the shelf-life of perishable items by protecting against the growth of microorganisms and the loss of vitamins and enzyme browning [46].

Plant extracts, which can be used to manufacture antimicrobial agents for food packaging because of their potential antifungal, antibacterial, and antioxidant properties, are particularly interesting in the food packaging sector. Extract from grapefruit seeds, pulp, or peels (*Citrus paradise*) known as grapefruit seed extract (GSE) has been demonstrated to be anti-inflammatory. Antifungal, antibacterial, antiviral and cancer-fighting capabilities have been shown to be present in it. Antifeedant characteristics have also been found to be present [47]. Polyphenolic components discovered in GSE include flavonoids, citric acid, ascorbic acid, tocopherol, limonoids, and other trace components [48]. When GSE was added to whey protein isolate, the shelf life of fish products was increased [49]. An antibacterial agent, GSE, was added to carrageenan to create active composite films, which were tested for physical and mechanical properties. When GSE was added to a composite film made of carrageenan, it drastically changed its characteristics. Carrageenan/GSE composite films offer a lot of potential for usage as UV screening films for UV-sensitive foods because of their high transparency and good UV barrier qualities. Food packaging applications, on the other hand, demand more research into enhancing the mechanical properties of the packaging material.

2.5 Agar/Nanoclay

Rhodophyceae algae, popularly known as "red seaweeds," produce Agar, a fibrous carbohydrate that can be found in many different types of seaweed. At concentrations as low as 0.04%, it produces visible gels from -d-galactopyranosyl linked (1 $\rightarrow$ 4) to a 3,6-anhydro—L-galactopyranosyl unit that is partially sulphated [50]. Biopolymers such as milk protein and starch have also been combined with agar to create environmentally friendly packaging materials such foams, films, and coatings [19, 51, 52] and starch [53]. The brittleness of pure biopolymer, as well as other qualities including low heat stability, medium gas barrier capabilities, and moderate water resistance, are typically inadequate for food packaging applications [40, 54]. Bionanocomposites, a novel class of materials that has been offered as a potential solution for improving the mechanical and barrier properties of biopolymer-based packaging materials, may be an option [43, 55–57]. A nanocomposite is a hybrid material consisting of a polymer matrix and nanoscale fillers with at least one dimension in the nanometer range. 2:1 layered silicate (or 2:1 phyllosilicates) clays like montmorillonite, saponite, or hectorite are extensively used as nanofiller in the packaging industry since they are environmentally friendly, non-toxic, and abundant in nature. A well-developed nanocomposite in which nano clays are equally spread in the polymer matrix in the state of intercalated or exfoliated demonstrates considerable advantages in mechanical, gas barrier, solvent resistance, and optical properties at low filler content (less than 5% by weight) [55–58]. Using the solvent intercalation process, agar and varied amounts of natural montmorillonite clay (Cloisite Na^+) were employed to create well-developed bio nanocomposite films. The agar/clay nanocomposite films might be employed as ecologically friendly food packaging materials with better water vapour barrier and mechanical qualities, as well as regulated water resistance properties, or as hydrogels with high water holding capacity and improved gel strength [50].

2.6 Alginate/Lemongrass Essential Oil

Essential oils have sparked a lot of interest as a natural alternative to artificial preservatives [59]. Lemongrass (*C. citratus*) is a long perennial grass that grows in tropical and subtropical climates [60]. Lemongrass essential oil has antibacterial properties against a variety of pathogens, including molds, yeasts, and gram positive and negative bacteria [60]. There has been few research on the effects of lemongrass essential oil combined into edible coatings for freshly cut fruit, such as melon [61] and apple [62]. During low temperature storage, an alginate-based edible coating containing 0.3% (w/v) lemongrass significantly ($p < 0.05$) reduced respiration rate, loss of weight, total plate counts (TPC), yeast and mold counts while maintaining firmness, colour, sensory characteristics, and morphological properties of fresh-cut pineapple. Thus, an alginate-based edible coating formulation containing 0.3% (w/v) lemongrass has the potential to improve the shelf life and retain the quality of freshly cut pineapple [63].

2.7 Agar/Nanocrystalline Cellulose/Savory Essential Oil

A study found that combining savory essential oil (SEO), a natural antibacterial, with agar-based nanocomposite film produced an active film. The nanocomposite films' microstructure, physical, mechanical, colour, and antibacterial characteristics were all impacted by this inclusion. According to the findings, agar-based nanocomposite films containing SEO can be utilized as active packaging to improve food safety and shelf life [64].

2.8 Carrageenan/Chitosan Nano-particles

Tissue engineering and drug administration are two areas where biomedical researchers are increasingly looking to natural polymers. It is possible to make both linear and branched polymers out of polysaccharides. These polysaccharides are essential in the creation of the cell membrane and in intracellular communication [65]. An exoskeleton-building material known as chitosan is a natural polymer that is formed of repeating units of chitin, the primary component of crustaceans' exoskeleton, as well as D-glucosamine [66, 67]. Chitosan is a polysaccharide that shares structural characteristics with glycosaminoglycans (GAGs), which are important components of connective tissues, and because of that, it has been studied for a variety of biomedical applications, including dentistry, wound healing, tissue engineering, and orthopaedics [67]. As a polymer derived from red seaweed, carrageenan is made of galactose and anhydro galactose units, which are joined together by glycosidic bonds. These nanoparticles, which are made of chitosan and carrageenan, can be utilised to control and sustain the release of drugs. These hydrophilic nanoparticles are formed through ionic interactions between positively charged chitosan amino groups and oppositely charged carrageenan sulphates. There are no organic solvents or other severe conditions used in this technique, which could harm the integrity of the medicine to be released. Ovalbumin, a macromolecule employed as a model protein in one of the studies, which was successfully linked to the proposed drug delivery systems, as proven by the physicochemical characterisation of the system. In an in vitro release experiment, chitosan/carrageenan nanoparticles showed a steady and controlled release of the protein over three weeks. A promising sign of their biocompatibility and safety came from tests on fibroblast-like cells using the newly created nanocarriers, which showed low toxicity. Because of their low toxicity, nanoparticles can be used in a wide range of medicinal applications, including as transporters for drugs and in tissue engineering [68].

2.9 Alginate/O-Carboxymethyl Chitosan/ **Cissus Quadrangularis** *(CQ) Extract*

Alginate (Alg) and o-carboxymethyl chitosan (O-CMC) were mixed with Cissus quadrangularis (CQ) extract to generate a "herbal scaffold" (Alg/O-CMC/CQ-E) via lyophilization. In one study, researchers revealed that the 'herbal scaffold' they generated possesses the properties needed for a scaffold for bone tissue engineering applications. Phytosteroids identified in CQ extract boosted human mesenchymal stem cell (hMSC) proliferation, osteogenic differentiation, and biomineralization. As a result of the aforesaid findings, it can be inferred that the Alg/O-CMC/CQ-E scaffold has an excellent osteo-inducing property that would make it an ideal choice for bone tissue engineering. Using a flow per fusion bioreactor to deliver a variety of physiological stimulations necessary for bone regeneration, future research with the newly created Alg/O-CMC/CQ-E scaffold is planned. CQ extract was successfully combined with alginate and O-CMC for bone tissue engineering applications for the first time. An immunologically inert, negatively charged hydrogel known as alginate could be used to regenerate a wide range of tissues [69, 70]. Alginate's carboxylic acid makes it a good candidate for alteration because of its versatility. In the presence of divalent cations, it forms an ionotropic gel. When alginate is broken down, simple sugars are released, which are totally absorbed [71]. In one of the study evaluating the efficiency of several matrices, sodium alginate was found to be an excellent scaffold for the creation of a pre-vascularized bone transplant [72]. O-carboxy methyl chitosan (O-CMC), a carboxy methyl derivative of chitosan, has been used to improve the biological performance of alginate with other materials such as chitosan. Biocompatibility, biodegradability, and hydrophilicity are among the unique characteristics of O-CMC [73, 74]. The three-dimensional hydrophilic "herbal scaffold" created by CaCl2 chemical crosslinking was used to retain the microporous structure. Mesenchymal stem cells from umbilical cord blood were a cell model employed for cell adhesion, proliferation and mineralization on Alg/O-CMC as well as Alg/O-CMC/CQ-E for this research [75].

2.10 Carrageenan/Graphene Oxide

In order to repair or replace bone, bone tissue engineering relies on a combination of scaffolds, cells, and physiologically active substances. Structure–function correlations in bone tissue engineering have resulted in the development of bioactive substitutes and synthetic materials. Hydroxyapatite ($Ca_{10}(PO_4)_6(OH)_2$), the primary constituent of natural bone, has long been employed in bone replacement systems due to its high biocompatibility and bioactivity. Because of its low tensile strength and fracture toughness, Hydroxyapatite (HA) is not suitable for many practical applications. Other materials, such as polymer, alumina (Al_2O_3), zirconia, silicon carbide, titanium (Ti), or titanium alloys [76], have been combined with HA to improve mechanical properties. We all know that the extracellular matrix (ECM) [77] is made of HA, collagen fibrils and proteins. In spite of this, the procedure remains a bit unclear [78]. The presence of charged proteins and mechanical signals produced by self-assembled collagen [79] may be the mechanism for the mineralization of HA. This could be the case. The creation of bio-interfaces, which allow natural components to be easily integrated into a variety of synthetic biomaterials, remains a major difficulty in tissue engineering [77, 80]. Biomaterials that can induce and assemble bone-like apatite that is similar to that found in natural bone are essential for bone tissue engineering. Carrageenan was used to functionalize graphene oxide (GO). Carrageenan (GO-Car) was added to the composite, which was then used as a substrate for hydroxyapatite (HA) mineralization in a biomimetic and cell-mediated manner. The presence of carrageenan on the surface of the GO allowed for the nucleation of HA. Cell adhesion, growth, and morphology were examined in relation to the GO-effect Car's on the MC3T3-E1 cells. GO-Car enhances cell differentiation and mineralization of HA in vitro. In the study, the GO-Car hybrid was found to be a great material for bone regeneration and implantation [80].

2.11 Polyelectrolyte Nanofibrous Membranes Made of Ulvan and Chitosan

Using natural polysaccharides, ulvans from the green seaweed *Ulva rigida*, and chitosan, a new class of biomaterials has been discovered by researchers. Ulvan, chitosan alone, as well as ulvan/chitosan polyelectrolytes membranes, have been synthesised and described in the study. Green seaweed cell walls were used to extract Ulvan, which is an anionic sulfated polysaccharide with a complicated structure. ULVAN's primary ingredients include the acids sulfated, rhamnose, xylose, glucuronic, and iduronic. Ulvan has been shown to be an anticoagulant, antioxidant, anticancer, and immunological modulator, amongst other properties. Besides that, it has the ability to lower low-density lipoprotein cholesterol (LDL-cholesterol), which in turn has the effect of lowering the atherogenic index [81–83]. Chitosan is a deacetylated form of chitin, which is a high molecular weight polysaccharide comprised of -(1,4)-2-acetamido-2-deoxy-d-glucose and -(1,4)-2-amino-2-deoxy-d-glucose units. Biologically renewable, biodegradable, biocompatible, non-antigenic, and non-toxic, this natural cationic polymer has a wide range of qualities. It is also bio-functional. There is evidence that chitosan can speed wound healing, promote macrophage activity, and limit the growth of malignant cells [84, 85]. Creating supramolecular structures and membrane stabilisers is made possible through electrostatic interactions between two polymers with oppositely charged backbones, like anionic ulvan and cationic chitosan. Changing the weight ratio of the two polysaccharides can change the porosity. They may have been able to bind to the 7F2 osteoblasts because the nanofibrous structure resembling the extracellular matrix mimics the fibrous part of that structure [86].

2.12 *Agar/Gelatin*

A lot of attention has recently been paid to the co-hydrogels' behaviour [85]. Crosslinked polymeric networks made up of two or more polymers are known as co-hydrogels. It is possible to modify the composition of the co-hydrogels to modulate their characteristics. Since gelatin and agar are both naturally biocompatible, their co-hydrogels have garnered great study. Gelatin and agar are both polymers that occur naturally in food. Gelatin and agar are very inexpensive and readily available. Polysaccharide biopolymers such as agar whereas gelatin is a protein-based biopolymer [87]. Phase-separated hydrogels based on gelatin–polysaccharide have been widely described and investigated. Hydrogel phase separation is caused by high-concentration gelatin–polysaccharide solutions becoming thermodynamically unstable (solution), resulting in phase separation. Water-in-water emulsions arise when gelatin and polysaccharides are heated to room temperature and then cooled. Polysaccharide is concentrated in the internal phase of these emulsions, while gelatin is predominant in the exterior phase [88]. There are two types of emulsion gels: the emulgel and the emulsion gel. In a study, gelatin–starch phase separated emulgels were found to exist [89]. For example, they have found that the dispersed oil phase functions as active fillers and adds to emulgels' improved mechanical qualities [90] when the composition of the formulations is carefully chosen. The medication delivery applications of the developed formulations were confirmed to be safe and effective [91].

3 Companies Utilizing Seaweeds for Composites

See Tables 1 and 2.

Table 1 Start-up companies utilizing seaweeds for composites

Sr. No.	Companies	Products	Website link
1	Kelpi	Biotech company using bio-degradable plastics from seaweed for packaging purposes	https://www.kelpi.net/
2	Notpla	Company using seaweeds to make greaseproof and water proof bio-degradable cardboard coating, sachets, edible liquid packaging, pipette, papers	https://www.notpla.com/
3	B'zcos	Products of the company are food films and edible plastic free drinking straws to end the reliability on traditional plastics	https://www.bzeos.com/
4	Sway	The company uses colorful seaweeds to make poly retail bags	https://swaythefuture.com/
5	C-combinatory	Seaweeds are used to make fertilizers, vegan leather, bio-stimulants, and emulsifier	https://carbonwave.com/
6	Evoware	They make colorful disposable cup which are vegan friendly, seaweed sheet which can be turned into bags and sachets and are heat sealable	https://rethink-plastic.com/home/
7	SoluBlue	They make polymer sheets which performs like plastics and when heat sealed, they turn into bags Trays, containers, punnets, straws, and lids are rigid packaging material produce from seaweeds	https://solublue.com/

Table 2 Established companies utilizing seaweeds for composites

Sr. No.	Companies	Products	Website link
1	Acadian Seaplants Limited	The company uses seaweed for food, biochemical, agricultural and agricultural chemical products	https://www.acadianseaplants.com/
2	DuPont de Nemours, Inc.	The company deals in supplementary, nutrition, biosciences products The company uses carrageenan to produce beverages, frozen desserts, dairy, fruit, poultry products	https://www.dupont.com/
1	Seaweed and Co.	The company provide seaweed ingredients for food, beverage and nutrition market	https://www.seaweedandco.com/
2	Cargill, Incorporated	The company uses seaweed in food, beverage, nutrition and personal care products	https://www.cargill.co.in/ https://www.cargill.com/
3	Green Rise Agro Industries	Green seaweed liquid, seaweed extract powder for agricultural application as a fertilizer	https://www.greenriseagro.com/
4	VietDelta Ltd.	The company uses seaweed powder in fertilizer	http://vdelta.com.vn/
5	Ocean Rainforest	The company is engaged in growing, harvesting and processing several species of seaweed which can be used to make fresh, frozen, and dried products Also, the company uses seaweed in food and cosmetic products	https://www.oceanrainforest.com/
6	ALGA plus	The seaweeds are used in cosmetics and food products	https://www.algaplus.pt/en/
7	MYCSA Ag, Inc.	The company produces organic bulk fertilizers from seaweeds	https://www.mycsainc.com/en
8	Baoji Earay Bio Tech Co., Ltd.	The company uses seaweed products in pharmaceutical industries, food products, and beverages	http://www.earaybio.com/

4 Conclusion

Seaweed's medicinal capabilities have been employed by people since the beginning of civilisation. Food additives, emulsifiers, gelling agents and stabilisers have all been used for a long time in the food industry with great success. Seaweed polysaccharides such as agar, alginate, and carrageenan are the most widely used. Alginate, on the other hand, is the most studied seaweed polysaccharide due to its widespread use in food and pharmaceuticals and its ability to react with di- and trivalent cations to produce sodium or calcium alginate. When extracting seaweed polymers, the extraction process has a significant impact on their functional qualities. Compound materials that have high mechanical strength and barrier qualities have been created by mixing seaweed-based polymers with essential oils, other biopolymers, as well as nanoparticles. Seaweed composites can be made using a variety of methods, including solvent casting, intercalation of nano clays, and emulsification of essential oils, depending on their intended use. In addition to food packaging and coating, seaweed-based composites can be used to produce innovative drug delivery systems, biomedical scaffolds, and cell encapsulations, thanks to recent breakthroughs in the sector.

References

1. Gade, R., Tulasi, M. S., & Bhai, V. A. (2013). Seaweeds: A novel biomaterial. *International Journal of Pharmacy and Pharmaceutical Sciences, 5*(2), 975–1491.
2. Rao, P., et al. (2019). Seaweeds: distribution, production and uses. *Bioprospecting of Algae,* 59–78.
3. Khalil, H., et al. (2017). Biodegradable polymer films from seaweed polysaccharides: A review on cellulose as a reinforcement material. *Express Polymer Letters, 11*(4).
4. Salehi, B., et al. (2019). Current trends on seaweeds: Looking at chemical composition, phytopharmacology, and cosmetic applications. *Molecules, 24*(22), 4182.
5. Parveen, I., et al. (2019). Seaweed based bio polymeric film and their application: A review on hydrocolloid polysaccharides. *Scientific Review, 5*(5), 93–102.
6. Zollmann, M., et al. (2019). Green technology in green macroalgal biorefineries. *Phycologia, 58*(5), 516–534.
7. Goyanes, S. N., & D'Accorso, N. B. (2017). Industrial applications of renewable biomass products. *Past, Present, and Future* (Springer).
8. Saepudin, E., Sinurat, E., & Suryabrata, I. A. (2018). Depigmentation and characterization of fucoidan from brown seaweed Sargassum binderi Sonder. In *IOP Conference Series: Materials Science and Engineering*. IOP Publishing.
9. Hira, A., et al. (2018). Comparison in biodegradable films formulated from two red seaweeds of Karachi coast. *Pakistan Journal of Marine Sciences, 27*(2), 87–92.
10. Freile-Pelegrín, Y., & Madera-Santana, T. J. (2017). Biodegradable polymer blends and composites from seaweeds. *Handbook of Composites from Renewable Materials, Biodegradable Materials, 5,* 419.
11. Thiruchelvi, R., Das, A., & Sikdar, E. (2021). Bioplastics as better alternative to petro plastic. *Materials Today: Proceedings, 37,* 1634–1639.
12. Lim, C., et al. (2021). Bioplastic made from seaweed polysaccharides with green production methods. *Journal of Environmental Chemical Engineering,* 105895.

13. Xia, W., et al. (2017). Fabrication and characterization of biodegradable composites from poly (butylene succinate-co-butylene adipate) and Taihu Lake blue algae. *Advanced Composites Letters, 26*(5), 096369351702600501.
14. Gautam, S., & Mannan, M. A.-u. (2020). The role of algae in nutraceutical and pharmaceutical production. In *Bioactive natural products in drug discovery* (pp. 665–685). Springer.
15. Sanjeewa, K. K. A., et al. (2016). Anti-inflammatory and anti-cancer activities of sterol rich fraction of cultured marine microalga Nannochloropsis oculata. *Algae, 31*(3), 277–287.
16. Khalil, H. A., et al. (2017). Seaweed based sustainable films and composites for food and pharmaceutical applications: A review. *Renewable and Sustainable Energy Reviews, 77*, 353–362.
17. Shit, S. C., & Shah, P. M. (2014). Edible polymers: Challenges and opportunities. *Journal of Polymers, 2014.*
18. Joye, I. J., & McClements, D. J. (2014). Biopolymer-based nanoparticles and microparticles: Fabrication, characterization, and application. *Current Opinion in Colloid & Interface Science, 19*(5), 417–427.
19. The, D. P., et al. (2009). Biopolymer interactions affect the functional properties of edible films based on agar, cassava starch and arabinoxylan blends. *Journal of Food Engineering, 90*(4), 548–558.
20. Wang, S., et al. (2012). Fortification of dietary biopolymers-based packaging material with bioactive plant extracts. *Food Research International, 49*(1), 80–91.
21. Venkatesan, J., et al. (2016). Seaweed polysaccharide-based nanoparticles: Preparation and applications for drug delivery. *Polymers, 8*(2), 30.
22. Daemi, H., et al. (2016). A robust super-tough biodegradable elastomer engineered by supramolecular ionic interactions. *Biomaterials, 84*, 54–63.
23. Sousa, A. M., et al. (2010). Agar extraction from integrated multitrophic aquacultured Gracilaria vermiculophylla: Evaluation of a microwave-assisted process using response surface methodology. *Bioresource Technology, 101*(9), 3258–3267.
24. Huq, T., et al. (2012). Nanocrystalline cellulose (NCC) reinforced alginate based biodegradable nanocomposite film. *Carbohydrate Polymers, 90*(4), 1757–1763.
25. Emadian, S. M., Onay, T. T., & Demirel, B. (2017). Biodegradation of bioplastics in natural environments. *Waste Management, 59*, 526–536.
26. Jaramillo, C. M., et al. (2016). Biodegradability and plasticizing effect of yerba mate extract on cassava starch edible films. *Carbohydrate Polymers, 151*, 150–159.
27. Tavassoli-Kafrani, E., Shekarchizadeh, H., & Masoudpour-Behabadi, M. (2016). Development of edible films and coatings from alginates and carrageenans. *Carbohydrate Polymers, 137*, 360–374.
28. Lee, M. H., Kim, S. Y., & Park, H. J. (2018). Effect of halloysite nanoclay on the physical, mechanical, and antioxidant properties of chitosan films incorporated with clove essential oil. *Food Hydrocolloids, 84*, 58–67.
29. Blanco-Pascual, N., Montero, M., & Gómez-Guillén, M. (2014). Antioxidant film development from unrefined extracts of brown seaweeds Laminaria digitata and Ascophyllum nodosum. *Food Hydrocolloids, 37*, 100–110.
30. Gomaa, M., et al. (2018). Use of the brown seaweed Sargassum latifolium in the design of alginate-fucoidan based films with natural antioxidant properties and kinetic modeling of moisture sorption and polyphenolic release. *Food Hydrocolloids, 82*, 64–72.
31. Vijayan, S. R., et al. (2016). Seaweeds: A resource for marine bionanotechnology. *Enzyme and Microbial Technology, 95*, 45–57.
32. Mir, S. A., et al. (2018). Effect of plant extracts on the techno-functional properties of biodegradable packaging films. *Trends in Food Science & Technology, 80*, 141–154.
33. Siracusa, V., et al. (2008). Biodegradable polymers for food packaging: A review. *Trends in Food Science & Technology, 19*(12), 634–643.
34. Mihindukulasuriya, S., & Lim, L.-T. (2014). Nanotechnology development in food packaging: A review. *Trends in Food Science & Technology, 40*(2), 149–167.

35. Rinaudo, M. (2008). Main properties and current applications of some polysaccharides as biomaterials. *Polymer International, 57*(3), 397–430.
36. Chaichi, M., et al. (2017). Preparation and characterization of a novel bionanocomposite edible film based on pectin and crystalline nanocellulose. *Carbohydrate Polymers, 157*, 167–175.
37. Sung, S. H., Chang, Y., & Han, J. (2017). Development of polylactic acid nanocomposite films reinforced with cellulose nanocrystals derived from coffee silverskin. *Carbohydrate Polymers, 169*, 495–503.
38. George, J., & Sabapathi, S. (2015). Cellulose nanocrystals: Synthesis, functional properties, and applications. *Nanotechnology, Science and Applications, 8*, 45.
39. Wu, Y., et al. (2001). Moisture loss and lipid oxidation for precooked ground-beef patties packaged in edible starch-alginate-based composite films. *Journal of Food Science, 66*(3), 486–493.
40. Sorrentino, A., Gorrasi, G., & Vittoria, V. (2007). Potential perspectives of bio-nanocomposites for food packaging applications. *Trends in Food Science & Technology, 18*(2), 84–95.
41. Ruiz-Hitzky, E., et al. (2010). Hybrid materials based on clays for environmental and biomedical applications. *Journal of Materials Chemistry, 20*(42), 9306–9321.
42. Park, H.-M., et al. (2003). Environmentally friendly polymer hybrids Part I Mechanical, thermal, and barrier properties of thermoplastic starch/clay nanocomposites. *Journal of Materials Science, 38*(5), 909–915.
43. Ray, S. S., & Bousmina, M. (2005). Biodegradable polymers and their layered silicate nanocomposites: In greening the 21st century materials world. *Progress in Materials Science, 50*(8), 962–1079.
44. Hosseini, S. M., et al. (2013). Incorporation of essential oil in alginate microparticles by multiple emulsion/ionic gelation process. *International Journal of Biological Macromolecules, 62*, 582–588.
45. Nanaki, S., et al. (2010). Miscibility study of carrageenan blends and evaluation of their effectiveness as sustained release carriers. *Carbohydrate Polymers, 79*(4), 1157–1167.
46. Tirillini, B. (2000). Grapefruit: The last decade acquisitions. *Fitoterapia, 71*, S29–S37.
47. Kanmani, P., & Rhim, J.-W. (2014). Development and characterization of carrageenan/grapefruit seed extract composite films for active packaging. *International Journal of Biological Macromolecules, 68*, 258–266.
48. Lim, G.-O., Hong, Y.-H., & Song, K.-B. (2008). Incorporating grapefruit seed extract into Gelidium corneum-whey protein isolate blend packaging film increases the shelf life of fish paste. *Preventive Nutrition and Food Science, 13*(4), 370–374.
49. Soradech, S., et al. (2012). An approach for the enhancement of the mechanical properties and film coating efficiency of shellac by the formation of composite films based on shellac and gelatin. *Journal of Food Engineering, 108*(1), 94–102.
50. Rhim, J.-W. (2011). Effect of clay contents on mechanical and water vapor barrier properties of agar-based nanocomposite films. *Carbohydrate Polymers, 86*(2), 691–699.
51. Freile-Pelegrín, Y., et al. (2007). Degradation of agar films in a humid tropical climate: Thermal, mechanical, morphological and structural changes. *Polymer Degradation and Stability, 92*(2), 244–252.
52. Letendre, M., et al. (2002). Physicochemical properties and bacterial resistance of biodegradable milk protein films containing agar and pectin. *Journal of Agricultural and Food Chemistry, 50*(21), 6017–6022.
53. Wu, Y., et al. (2009). Effect of agar on the microstructure and performance of potato starch film. *Carbohydrate Polymers, 76*(2), 299–304.
54. Cabedo, L., et al. (2006). Optimization of biodegradable nanocomposites based on aPLA/PCL blends for food packaging applications. In *Macromolecular Symposia*. Wiley Online Library.
55. Pandey, J. K., et al. (2005). Recent advances in biodegradable nanocomposites. *Journal of Nanoscience and Nanotechnology, 5*(4), 497–526.
56. Pavlidou, S., & Papaspyrides, C. (2008). A review on polymer–layered silicate nanocomposites. *Progress in Polymer Science, 33*(12), 1119–1198.

57. Rhim, J.-W., & Ng, P. K. (2007). Natural biopolymer-based nanocomposite films for packaging applications. *Critical Reviews in Food Science and Nutrition, 47*(4), 411–433.
58. Brody, A. L. (2007). Nanocomposite technology in food packaging. *Food Technology.*
59. Mastromatteo, M., et al. (2011). Combined effect of active coating and MAP to prolong the shelf life of minimally processed kiwifruit (Actinidia deliciosa cv. Hayward). *Food Research International, 44*(5), 1224–1230.
60. Naik, M. I., et al. (2010). Antibacterial activity of lemongrass (Cymbopogon citratus) oil against some selected pathogenic bacterias. *Asian Pacific Journal of Tropical Medicine, 3*(7), 535–538.
61. Raybaudi-Massilia, R. M., Mosqueda-Melgar, J., & Martín-Belloso, O. (2008). Edible alginate-based coating as carrier of antimicrobials to improve shelf-life and safety of fresh-cut melon. *International Journal of Food Microbiology, 121*(3), 313–327.
62. Rojas-Graü, M. A., et al. (2007). Apple puree-alginate edible coating as carrier of antimicrobial agents to prolong shelf-life of fresh-cut apples. *Postharvest Biology and Technology, 45*(2), 254–264.
63. Azarakhsh, N., et al. (2014). Lemongrass essential oil incorporated into alginate-based edible coating for shelf-life extension and quality retention of fresh-cut pineapple. *Postharvest Biology and Technology, 88*, 1–7.
64. Atef, M., Rezaei, M., & Behrooz, R. (2015). Characterization of physical, mechanical, and antibacterial properties of agar-cellulose bionanocomposite films incorporated with savory essential oil. *Food Hydrocolloids, 45*, 150–157.
65. Malafaya, P. B., Silva, G. A., & Reis, R. L. (2007). Natural–origin polymers as carriers and scaffolds for biomolecules and cell delivery in tissue engineering applications. *Advanced Drug Delivery Reviews, 59*(4–5), 207–233.
66. Kurita, K. (2006). Chitin and chitosan: Functional biopolymers from marine crustaceans. *Marine Biotechnology, 8*(3), 203–226.
67. Singh, D. K., & Ray, A. R. (2000). Biomedical applications of chitin, chitosan, and their derivatives. *Journal of Macromolecular Science, Part C: Polymer Reviews, 40*(1), 69–83.
68. Grenha, A., et al. (2010). Development of new chitosan/carrageenan nanoparticles for drug delivery applications. *Journal of Biomedical Materials Research Part A: An Official Journal of The Society for Biomaterials, The Japanese Society for Biomaterials, and The Australian Society for Biomaterials and the Korean Society for Biomaterials, 92*(4), 1265–1272.
69. Frohlich, M., et al. (2008). Tissue engineered bone grafts: Biological requirements, tissue culture and clinical relevance. *Current Stem Cell Research & Therapy, 3*(4), 254–264.
70. Mohan, N., & Nair, P. D. (2005). Novel porous, polysaccharide scaffolds for tissue engineering applications. *Trends in Biomaterials and Artificial Organs, 18*(2), 219–225.
71. Ikada, Y. (2006). Challenges in tissue engineering. *Journal of the Royal Society Interface, 3*(10), 589–601.
72. Vögelin, E., et al. (2002). Prefabrication of bone by use of a vascularized periosteal flap and bone morphogenetic protein. *Plastic and Reconstructive Surgery, 109*(1), 190–198.
73. Chen, X.-G., & Park, H.-J. (2003). Chemical characteristics of O-carboxymethyl chitosans related to the preparation conditions. *Carbohydrate Polymers, 53*(4), 355–359.
74. Fernando, R., & Sérgio, P. (2004). Characteristics and properties of carboxymethyl chitosan related to the preparation conditions. *Carbohydrate Polymers, 75*, 214–221.
75. Soumya, S., et al. (2012). Development of a phytochemical scaffold for bone tissue engineering using Cissus quadrangularis extract. *Carbohydrate Polymers, 87*(2), 1787–1795.
76. Zheng, X., Huang, M., & Ding, C. (2000). Bond strength of plasma-sprayed hydroxyapatite/Ti composite coatings. *Biomaterials, 21*(8), 841–849.
77. Morozowich, N. L., Nichol, J. L., & Allcock, H. R. (2012). Investigation of apatite mineralization on antioxidant polyphosphazenes for bone tissue engineering. *Chemistry of Materials, 24*(17), 3500–3509.
78. Alves, N., et al. (2010). Designing biomaterials based on biomineralization of bone. *Journal of Materials Chemistry, 20*(15), 2911–2921.
79. Liao, S., Cui, F., & Zhu, Y. (2004). Osteoblasts adherence and migration through three-dimensional porous mineralized collagen based composite: NHAC/PLA. *Journal of Bioactive and Compatible Polymers, 19*(2), 117–130.

80. Liu, H., et al. (2014). Biomimetic and cell-mediated mineralization of hydroxyapatite by carrageenan functionalized graphene oxide. *ACS Applied Materials & Interfaces, 6*(5), 3132–3140.
81. Lahaye, M., & Robic, A. (2007). Structure and functional properties of ulvan, a polysaccharide from green seaweeds. *Biomacromolecules, 8*(6), 1765–1774.
82. Morelli, A., & Chiellini, F. (2010). Ulvan as a new type of biomaterial from renewable resources: Functionalization and hydrogel preparation. *Macromolecular Chemistry and Physics, 211*(7), 821–832.
83. Zhang, H.-J., et al. (2008). Chemical characteristics and anticoagulant activities of a sulfated polysaccharide and its fragments from Monostroma latissimum. *Carbohydrate Polymers, 71*(3), 428–434.
84. Muzzarelli, R. A. (2009). Chitins and chitosans for the repair of wounded skin, nerve, cartilage and bone. *Carbohydrate Polymers, 76*(2), 167–182.
85. Muzzarelli, R. A. (2011). Chitosan composites with inorganics, morphogenetic proteins and stem cells, for bone regeneration. *Carbohydrate Polymers, 83*(4), 1433–1445.
86. Toskas, G., et al. (2012). Ulvan and ulvan/chitosan polyelectrolyte nanofibrous membranes as a potential substrate material for the cultivation of osteoblasts. *Carbohydrate Polymers, 89*(3), 997–1002.
87. Ross, K., Pyrak-Nolte, L., & Campanella, O. (2006). The effect of mixing conditions on the material properties of an agar gel—microstructural and macrostructural considerations. *Food Hydrocolloids, 20*(1), 79–87.
88. Singh, V. K., et al. (2014). Gelatin-carbohydrate phase-separated hydrogels as bioactive carriers in vaginal delivery: Preparation and physical characterizations. *Journal of Applied Polymer Science, 131*(13).
89. Firoozmand, H., & Rousseau, D. (2013). Microstructure and elastic modulus of phase-separated gelatin–starch hydrogels containing dispersed oil droplets. *Food Hydrocolloids, 30*(1), 333–342.
90. Thakur, G., et al. (2012). Gelatin-based emulsion gels for diffusion-controlled release applications. *Journal of Biomaterials Science, Polymer Edition, 23*(5), 645–661.
91. Wakhet, S., et al. (2015). Characterization of gelatin–agar based phase separated hydrogel, emulgel and bigel: A comparative study. *Journal of Materials Science: Materials in Medicine, 26*(2), 118.

Sea Shell Extracted Chitosan Composites and Their Applications

Pragati Upadhayay, Preeti Pal, Dong Zhang, and Anjali Pal

1 Introduction

People love seafood, and as a result of modern seafood production practices membrane, brain, back end, carapace, lamina, spinal column, and other dilapidated production accumulates which can be used for the extraction of chitin. Chitin, a polysaccharide with exceptional intrinsic properties such as antimicrobial, biodegradability, antitumor, biocompatibility, biodegradability and antioxidant activities, is commonly found in seafood waste [1, 2]. Chitin is the most prevalent polysaccharide after cellulose. On a commercial scale, chitin is transformed to chitosan, a deacetylated derivative [3]. Because of their biological flexibility, chitin and chitosan have tremendous economic importance [4]. Living species in the ocean manufacture chitin every year [5]. Arthropods create 2.8×10^{10} kg in freshwater and 1.3×10^{12} kg in aquatic environments [6]. If industrial extraction techniques for

P. Upadhayay · P. Pal (✉)
Department of Biotechnology, Institute of Applied Sciences and Humanities, GLA University, Mathura 281406, India
e-mail: preeti.pal@gla.ac.in; preeti.p@acsipl.com; palpreiti@gmail.com

P. Upadhayay
e-mail: pragati.upadhayay_phd.bio20@gla.ac.in

P. Pal
Accelerated Cleaning Systems India Private Limited, Mumbai, India

D. Zhang
Department of Chemical, Biomolecular, and Corrosion Engineering, The University of Akron, Akron, OH 44325, USA
e-mail: dz39@uakron.edu

A. Pal
Civil Engineering Department, Indian Institute of Technology, Kharagpur, West Bengal 721302, India
e-mail: anjalipal@civil.iitkgp.ac.in

Sapuan S. M. and I. Ahmad (eds.), *Composites from the Aquatic Environment*,
Composites Science and Technology, https://doi.org/10.1007/978-981-19-5327-9_14

commercially competent polymers were established, vast amounts of chitin would be available as crude material [7]. Chitosan is a polysaccharide made up of (1,4) glycosidic linkages that connect deacetylated and acetylated D-glucosamine units. Deacetylation of chitin leads in the generation of acetate ions and a $-NH_2$ group via acetamide hydrolysis [8]. Chitosan, a low-cost polymer, has also been utilised in several experiments for heavy metal ion adsorption [9]. Biocompatibility and biodegradability have been demonstrated for this naturally occurring polymer [10]. Chitosan has long been known for its fascinating characteristics, and the polymer has been utilised in farming manufacturing, and medicine [11]. Chitosan has been recognized in agriculture as a plant antiviral and as a constituent in liquid multicomponent fertilizers [12], and an alloy-improving handler in agriculture industry. It has been widely recycled as a biomaterial due to its immunostimulatory properties [13], anticoagulant properties [14], antibacterial and antifungal action [15, 16], and in the realm of surgery, it can also act as a wound healing booster. Furthermore, Chitosan offers a number of exciting medicinal applications and is now being tested like a novel shipper material in medicine deliverance systems [17]. Depending on the process conditions, the degree of deacetylation has an impact on both chemical (exterior area, bending energy, thickness, conductivity, solubility, porosity, as well as flexibility) and biological properties (biodegradability, bioavailability, antioxidant, adsorption enhancer and biocompatibility) [18]. Chitin and chitosan's chemical structures are depicted in Fig. 1. Chitosan can be chemically modified to produce derivatives with regulated solubility, ionic properties, and hydrophilicity. Chemical modification of the polymeric backbone's hydroxyl, acetamido, and amine sites results in improved property profiles [19]. Abdelaal et al. [20] reported the production of different chitosan derivatives in neutral and slightly acidic environments utilizing succinic acid derivatives, phthalic anhydride, glycidyltrimethyl ammonium chloride, cellulose triacetate, and other polysaccharides. In Chitosan in its purest form is hydrophilic and has a lower scale of order. To make chitosan more hydrophobic, it is typically N-acylated with different fatty acid chlorides (C6–C16) [21].

2 Segregation of Chitosan

Due to the poor biodegradation rate, the seafood processing businesses generate a significant amount of waste [20]. When these food scraps are washed into the sea, they pollute coastal areas which create a lot of problems. This marine debris might be utilised as a processing substrate for sea crustaceans, the fungus Mucorrouxii, farmed mushroom fruit bodies, and other agricultural goods. The most essential stage in obtaining chitin is extracting it from natural sources. Several characteristics of distilled chitin are determined by the extraction settings and conditions, including molecular weight, polydispersity index, degree of purity, and deacetylation. Despite the fact that chemical extraction is an inefficient and ecologically unfriendly technique that alters the chemical as well as physical characteristics of chitin while removing minerals and proteins, it is still the very widely used technique of chitosan

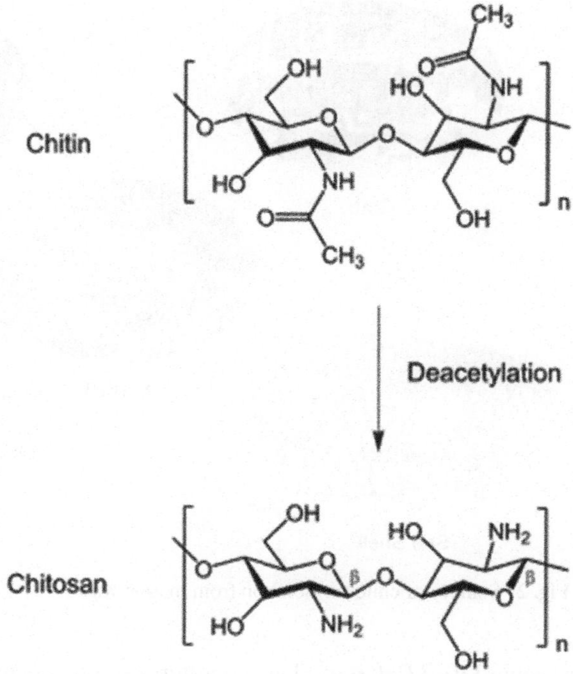

Fig. 1 Chitin and chitosan extracted from marine sources have different chemical structures

and chitin extraction [5, 22]. Chitin is extracted from exoskeletons of crab and shrimp shells, other producers are crustaceans, mollusks, insects, and some fungi [23]. It is available three allomorphic varieties α, β and γ forms. Chitin γ is a hybrid of the α and β forms of chitin. The mainly common kind is—chitin, which comes from crab exoskeletons, whereas α chitin comes from fungi and yeast. [24–27]. Alkaline conditions readily converts α-chitin to β-chitin following alkaline treatment, a water flush is performed. Chitin is also available on a commercial scale by various industries for different applications.

3 Applications of Chitosan and Their Composites

Chitin and chitosan have a ample variety of uses and profits, and the most appealing properties are their biodegradability and nontoxicity. Chitosan and different composites of chitosan can be effective heavy metal removal agent that is frequently employed in the purification of wastewater containing organic substances as a flocculating and coagulating agent [28]. It has the capacity to form water with a hydrocolloid and a gelifier, and it may be employed as a food additive because it acts in the food business as a water-storage agent for living organisms, coating agent and thickening agent [29]. Biocontrol agents like chitosan are also used in plants to manage diseases, and it is utilized to immobilize microbial cells and enzymes in medication delivery

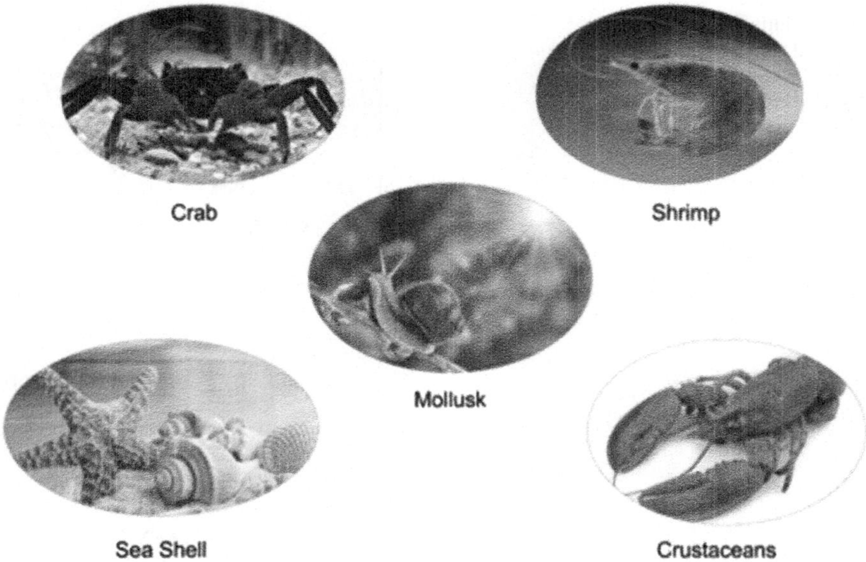

Fig. 2 Sources of chitosan isolation from marine waste

systems [30–32]. It is used in agriculture as a composting accelerator, in cosmetics to make cream, in biotechnology as porous particles for bioreactors and other skin-care products [33]. Chitosan has been shown to have wound-healing properties as well as to help improve plant defense systems by blocking the development of various bacteria. Chitosan has been used in agricultural systems for crop production and quality to minimize potential of disease. Chitosan against fungus and viruses serves as defensive mechanism for plant [34]. When applied to the soil, chitosan has been shown to reduce the occurrence and severity of plant diseases [35]. Figure 3 labels the various applications of chitosan.

4 Recent Advances in Chitosan Composites Preparation

A composite material (sometimes called a composition material or composite) is one that consists of two or more constituent components. These basic components have a large range of chemical and physical properties, and when combined, they produce a composite with characteristics that are distinct from the separate parts. Composites made of natural fibres are promising in applications of structural components [36]. Individual components stay distinct and separate throughout the completed construction, distinguishing composites from mixes and solid solutions. In addition to chitosan chemical alteration, many studies have been centered on the preparation of chitosan composites with a variety of materials. These composites have found use in a multiplicity of applications, including water and wastewater treatment. The

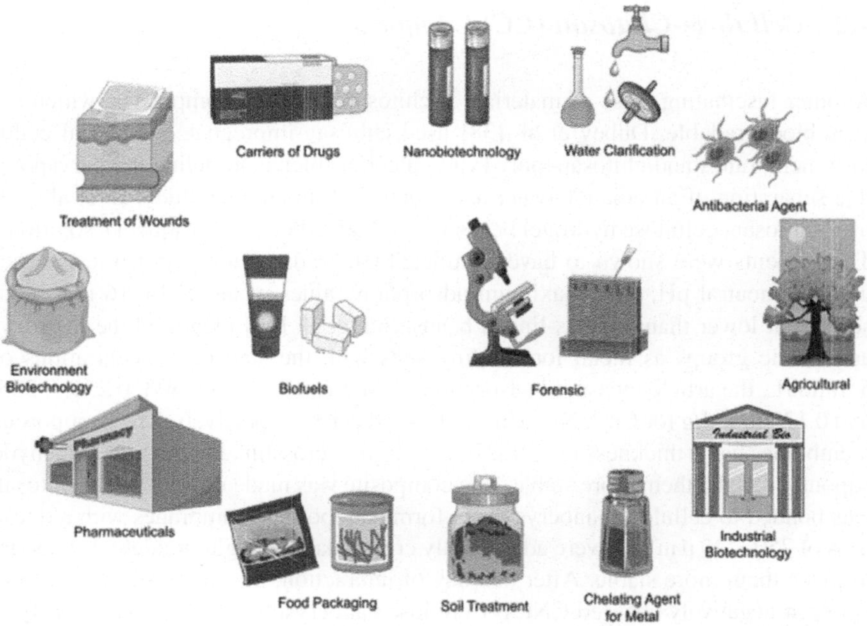

Fig. 3 Some major applications of chitosan

remainder of this chapter will be devoted to the various materials used to create chitosan composites.

4.1 *Chitosan of Palm Oil Ash Composites*

The palm oil ash-chitosan composites were reported by Nomanbhay and Palanisamy [37]. Its structure contains functional groups such as hydroxyl, carboxylic, and lactone, which give it a high affinity for metal ions, the simplest technique for palm oil may be obtained, which has good exchange/sorption properties. To make the composite biosorbent (up to about 21 wt%), On top of acid-treated oil palm shell charcoal, chitosan was applied, resulting in almost spherical particles with diameters varies from 100 to 150 μm. The sorbent's decisive capability was determined to be 154 mg/g [37].

4.2 Cellulose-Chitosan (CC) Composites

Another fascinating class of materials is chitosan cellulose composites, which are very biodegradable. Dubey et al. [38] used chitosan-impregnated bacterial cellulose membranes and chitosan–poly (vinyl alcohol) blends to define the pervaporative separation of an ethanol–water azeotrope [38]. In another study, Li et al. [39] used chitosan–cellulose hydrogel beads as Cu^{2+} adsorbents from aqueous solutions. The sorbents were shown to have significant Cu^{2+} adsorption capabilities, particularly at neutral pH, with maximum adsorption values of about 14–16 mg/g, but somewhat lower than uncross-linked beads. Sun et al. [40] identified the hydroxyl and amine groups as metal ion binding sites with the starting concentrations of 5 mmol/L, the article measured adsorption capacities of 0.417, 0.303, 0.251, 0.225, and 0.127 mmol/g for Cr^{+6}, Ni^{+2},Cu^{+2}, Pb^{+2} and Zn^{+2}, respectively. The nanoporous membranes had a thickness of 250–270 m and were crosslinked with gluteraldehyde vapours to make them more stable. The composite was made up of 10 wt% chitosan was bonded to cellulose nanocrystals to form nanoporous membranes with a thickness of 250–270 μm that were additionally crosslinked with gluteraldehyde vapours to make them more stable. After one day of interaction, the electrostatic attraction between negatively charged CNCs (Cellulose nanocrystals) and positively charged dyes resulted in the elimination of 70, 98, and 84% of Victoria Blue 2B, Methyl Violet 2B, and Rhodamine 6G, respectively [40].

4.3 Alginate-Chitosan (AC) Composites

For the removal of Ni^{2+} ions from aqueous solutions, Vijaya et al. [41] developed and utilised chitosan covered calcium alginate and chitosan covered silica (CCS) as an example of how calcium alginate-chitosan composites may be used to alter the characteristics of calcium alginate as well as chitosan. The researchers created chitosan-coated calcium alginate by spinning alginate beads in a 4% chitosan gel (chitosan in a 2% acetic acid (AA) solution) for 12 h (CCCA). The chitosan-coated beads were then washed and dried before being immersed in a 0.1 M NaOH solution for 4 h. The Langmuir adsorption isotherm determined the maximal monolayer Ni^{2+} adsorption potential of the CCCA to be 222.2 mg/g.

Ngah et al. [42] investigated Cu^{2+} adsorption by chitosan beads, chitosan–glutaraldehyde beads, and alginate-chitosan (CA) beads at various pHs, agitation durations, adsorbent doses, and starting concentrations. To make the alginate chitosan composites, chitosan flakes were dissolved in a 5% (v/v) acetic acid solution in distilled water, yielding a 5% (w/v) alginic acid solution. Before adding the alginic acid solution, the chitosan solution was heated and agitated for about 30 min at 60 °C Celsius, and the complete technique was combined for another 20 min at 60 °C Celsius after the solutions had been aged overnight. The CA beads were filtered and rinsed with distilled water to remove any leftover sodium hydroxide, dried, and sieved after being agitated in a 0.5 M NaOH solution. The CA beads, with an adsorption power of 67.66 mg/g, satisfactorily matched the non-linear Langmuir isotherm [42].

4.4 Polymer-Chitosan Composites

4.4.1 Polyurethane-Chitosan Composites

Polyurethane foams large amount of surface and open porous structure make them ideal materials for use as matrix materials for immobilizing various adsorbents. As a result, such composites have been produced and are being utilized to remove ions of heavy metals extracted from watery samples [43]. As a result, the notion of manufacturing chitosan and composites of polyurethane has aroused the interest of water sample treatment researchers [44]. The preparation procedure was formed by a series of reactions that happened during the polyurethane pre-polymer and water foaming phase. Carbon dioxide was produced as a forming agent, and the isocyanate groups on the polyurethane prepolymer react with the chitosan –NH2 groups, causing chitosan to be dispersed and immobilised on the polymeric matrix's surface. Conditions were established throughout the synthesis to produce composites with chitosan concentrations ranging from 5 to 20% by weight. The results also showed that using 0.25 wt% glutaraldehyde as the cross-linking agent resulted in the most immobilization [45]. In the realm of polyurethane-chitosan composites, Jayakumar and Sudha compared Pb^{2+} sorption of cross-linked glutaraldehyde-chitosan/Nylon 6/polyurethane foam blends to non-cross-linked chitosan/Nylon 6/polyurethane foam blends [46]. The procedure involved dissolving enough chitosan, 6 Nylon, and polyurethane foam in formic acid separately, then mixing them together in a 2:1:1 weight ratio with glutaraldehyde as a cross-linking agent. The cross-linking agent mixture was spun at room temperature for 1 h to disperse the solvent before being placed on a clean Petri plate and vacuum dried at 70 °C for 10 h. In the absence of the cross-linking agent, the mixture was dried following the method described before. Over a pH range of 2–8, the sorption activity of the two sorbents was examined, with the greatest adsorption occurring at pH 5. The percentages of Pb^{2+} removal for the non-cross-linked composite and cross-linked sorbents were found to be about 81 and 62%, respectively, for the non-cross-linked composite and cross-linked sorbents.

4.4.2 Polyvinyl Alcohol-Chitosan Composites

The first publication on the use of a polyvinyl alcohol-chitosan composite was published by Wan Ngah et al. [47], who looked at the equilibrium kinetics of Cu^{2+} sorption on chitosan polyvinyl alcohol beads. A known volume of chitosan flakes was dissolved in a 5% (v/v) acetic acid solution during the synthesis method. The solution was permitted to settle at room temperature for a day. Meanwhile, distilled water was dissolved in a volume of polyvinyl alcohol for 5 h at about 500 rpm at 70 °C on a magnetic stirrer, and the two solutions were combined for 2 h before being stored at room temperature for 2 days under the same stirring and temperature settings. After being sprayed into a 0.5 M NaOH precipitation and neutralisation vessel at 300 rpm, the chitosan coagulated and produced spherical beads. Before being utilised, the beads were filtered, thoroughly rinsed in DI water, air-dried, crushed in a jar mill, and sieved to a uniform size of <250 μm. During the adsorption investigations, the maximum adsorption of Cu^{2+} on the chitosan-PVA beads was determined to be 25 mg/g at pH 6.0. Another Cu^{2+} sorbent was created in a stereo-structural chitosan matrix utilising matrix scaffolds of evenly distributed ion exchange resins (H+ from Amberjet, ID 780 m). The chitosan (in a solution phase) was homogeneously cross-linked with aqueous polyvinyl alcohol (PVA) solutions during the synthesis process, and the particles suspended in the PVA cross-link chitosan solution were transferred into cylindrical aluminium containers to make PVA-chitosan using the freeze-gelation method. According to the findings, the composite was used to remove Cu^{2+} ions from water, with the optimum sorption pH of 6.0 and the highest adsorption capacity of 93.5 mg/g for copper solutions starting at 2500 mg/L. The desorption ratio of Cu^{2+} ions was found to be 97.9% at comparable low starting levels of 200 mg/L [48]. The production of nanoscale PVA/Chitosan particles and their use for the removal of Mn^{2+} species at an optimum pH of 6.0 were described in a recent study. PVA/Chitosan was purportedly created by grinding a 1:1 combination of PVA and chitosan solution for 8 h at 350 rpm at room temperature in a ball mill (Reech, PM400, Germany). The sorbent was shown to remove almost 95% of the Mn^{2+} content of a 20 mg/L solution at a pH of 5.5, with an adsorption capacity of roughly 9.2 mg/g, at an initial Mn^{2+} concentration of 100 mg/L [49].

4.5 *Hydroxyapatite*

During the last two decades, chitosan, a natural polymer produced from chitin, a fundamental component of crustacean exoskeleton, has played a vital role in bone tissue engineering. Chitosan composite materials have gotten a lot of attention in recent years due to their minimal foreign body reactions, biodegradability, intrinsic antibacterial nature, biocompatibility, and ability to be moulded into various geometries and forms such as porous structures, suitable for cell in growth and osteoconduction. Because of its biodegradability and biocompatibility, the chitosan-hydroxyapatite composite is frequently utilized [50]. Composite materials are increasingly being used as scaffolds in bone tissue engineering. CTS, as previously noted and elsewhere, has a number of advantages in orthopaedic applications, making it a suitable bone transplant alternative [51]. Because CTS scaffolds are not capable to sustain load bearing bone implants, their mechanical characteristics are inferior to those of actual bone. CTS scaffolds are unable to replicate all of the characteristics of natural bone on their own. CTS/HAp composite materials may imitate both the organic and inorganic components of genuine bone, whereas calcium phosphate materials are osteoconductive and can duplicate both the organic and inorganic components of natural bone [52]. In bone tissue engineering, calcium phosphate compounds are very important. One of the most stable forms of calcium phosphate is hydroxyapatite [$Ca_{10}(PO_4)_6(OH)_2$], is a significant component of bone (60–65%) [53].

4.6 Chitosan/Gelatin

Chitosan/gelatin composite sponges have been studied as a wound dressing alternative. Because of its better mechanical characteristics and thermostability, this two-polymer combination was chosen. Chitosan has antibacterial and hemostatic properties, but gelatin is good at absorbing water and forming films. Both of these polymers are biocompatible, which means they won't cause allergic reactions, and biodegradable, which means they won't leave any scars on the healing process. The surface roughness of gelatin-based membranes is rough, whereas the surface roughness of chitosan membranes is smooth [54]. The most appropriate porous scaffold for use as a dermal equivalent is made up of composites of these polymers in the correct composition. In cytotoxic tests on L929 cell lines, chitosan/gelatin sponge biocompatibility was established. After 72 h of chitosan/gelatin sponge incubation, it was discovered that about 140% of L929 cells were alive, and it was deemed to be superior than merely or gelatin or chitosan produced sponge. When the sponge was composed completely of chitosan or gelatin, its mechanical properties improved considerably, with a significantly greater Young's modulus and tensile strength. The water absorption capacity and retention duration of the chitosan/gelatin sponge are significantly higher than those of chitosan or gelatin alone. Furthermore, in terms of speeding up the healing process, a chitosan/gelatin sponge containing growth factors outperformed a sponge without growth factors. All of these findings contribute to a better understanding of what constitutes a suitable skin substitute for wound dressings [55].

4.7 Chitosan/Collagen Biocomposite

A wound-healing skin replacement made of chitosan and collagen has also been proposed. Collagen (Types I, IV, and VII) is a significant component of skin protein, accounting for around 70% of dry skin mass. It is chosen for skin replacement manufacturing because of its stronger cellular affinity [56]. Collagen has the drawback of degrading fast and having weak mechanical properties when used alone [57]. Biocompatibility, biodegradability, and mechanical characteristics are all improved when chitosan and collagen are combined. Mixing chitosan and collagen in a composite enhances collagen biostability without the use of synthetic cross-linkers, according to in vitro biodegradation tests. The medicated chitosan/collagen composite left no scar on albino rats following 28 days of wound healing in a full thickness wound model. In both medicated and nonmedicated wounds treated with chitosan/collagen scaffold, no pus developed, suggesting that chitosan is antibacterial (Table 1) [58].

5 Applications of Chitosan Composites in Different Fields

5.1 Pharmaceutical and Biomedical Applications

When employed for certain purposes, non-toxicity, sustainable origin, non-allergenicity, biocompatibility and biodegradability in the body are some of the main qualities of chitin and chitosan. Furthermore, due to their biological behavior such as immunoadjuvant, anticholesteremic agent, antifungal, antibacterial, anti-cancer, antithrombogenic, and bioadhesivity, its attractiveness has increased. They are commonly used as hydration boosters and absorption boosters, moreover in film making and injury curing. Chitosan has been utilised as a wound healing accelerator in several research [58–60]. It encourages the growth of inflammatory cells like macrophages as well as osteopontin and leukotriene 4, polymorphonuclear leukocytes, platelet-derived growth factor, transforming growth factor b1, and fibroblasts [61, 62]. Depending on the intended use, chitin and, more simply, chitosan may be converted into a variety of forms such as films, capsules, fibres, powders, solutions, sponges, beads and gels. Chitin promotes wound healing in sprays, gels, and gauze [62–64]. Considering low toxicity, hydrophilic character, protein affinity, biodegradability, physiological inertness, antibacterial characteristics, gel-forming characteristics, and mucoadhesivity, it is used to aid treatments or to regulate drug release [8, 65]. Antithrombogenic materials for drug encapsulation, drug encapsulation, enzyme and cell immobilisation, and gene carriers include chitosan (the lone pseudo-natural polycationic molecule) and its electrostatic complexes with synthetic or natural polymers (such as alginate) [66]. Biodegradability, antimicrobial activity, hydrophilicity, and the presence of polar groups capable of secondary interaction with other polymers are all advantages of chitosan-based goods (Hydrogen bonds are formed by the –OH and $-NH_2$ groups, whereas hydrophobic contacts are formed by the N-acetyl groups). Chitosan films, like many other polysaccharide-based films, are good at blocking fat diffusion and selective gas permeability, but not so good at blocking water and water vapour transfer [67]. This is due to their hydrophilic nature, which means they have a strong attraction to water [62]. To solve this issue, chitosan-based bioactive and stable coatings are created via polymer mixing, biocomposites, and multilayer systems. The chitosan-calcium phosphate cement has a one-of-a-kind application. Chitosan glycerophosphate, also known as chitosan, was combined with

Table 1 Composites of chitosan and their roles

Composites	Purpose	Types of conditions	References
Palm oil ash-chitosan composites	Exchange/sorption capabilities	The sorbent's decisive capability was determined to be 154 mg/g	[38]
Cellulose-chitosan (CC) composites	Utilised chitosan–cellulose hydrogel beads as Cu^{2+} adsorbents from aqueous solutions	The sorbents were shown to have considerable Cu^{2+} adsorption capacities, notably at neutral pH values, with maximum adsorption values of around 14–16 mg/g	[39]
Alginate-chitosan (AC) composites	Elimination of Ni^{2+} ions from aqueous solutions	The Langmuir adsorption isotherm determined the maximal monolayer Ni^{2+} adsorption potential of the CCCA to be 222.2 mg/g	[41]
Polyurethane-chitosan composites	Removal of Heavy metals (Pb^{+2})	The percentages of Pb^{2+} removal for the non-cross-linked composite and the cross-linked sorbents were found to be around 81 and 62% respectively	[44]
Polyvinyl alcohol-chitosan composites	Examining the equilibrium kinetics of Cu^{2+} sorption on chitosan polyvinyl alcohol beads	The maximum adsorption of Cu^{2+} on the chitosan-PVA beads was determined to be 25 mg/g at pH 6.0	[48]
Hydroxyapatite	Bone tissue engineering	Calcium phosphate is hydroxyapatite $[Ca_{10}(PO_4)_6(OH)_2]$ is a significant component of bone (60–65%)	[50]
Chitosan/gelatin	Wound dressing replacement	Both of these polymers are biocompatible, which means they won't cause allergic reactions, and biodegradable, which means they won't leave any scars on the healing process	[54]
Chitosan/collagen biocomposite	Possible skin replacement for wound healing	Collagen (Types I, IV, and VII) is a major component of skin protein, accounting for around 70% of dry skin mass	[57]

calcium phosphate and citric acid to create an aesthetically appealing injectable self-hardening device for bone repair and filling [58]. Depending on the device and the desired function, chitosan can be converted more effectively into sponge, capsule, or nanoparticle forms than chitin.

5.2 Agriculture

Since the 1990s, chitosan products have been used to protect agricultural crops from plant pathogenic bacteria that cause decay and damage throughout the growing season and after harvest [62]. They are bacteriostatic and/or bactericidal (kill bacteria) (hindering the development of bacteria). The most well-known approach is to use chitosan's polycationic feature, which allow it to relate with negatively charged organisms (bacterium cell membrane). Chitosan is a powerful antifungal agent due to its chelating properties [57, 58]. Plant defence reactions are triggered by the presence of chitosan. It is often used as a strong elicitor in plant disease control. Chitosan products have been proposed as a way of limiting the discharge of agro-chemicals (fertilizers and pesticides. Only a few examples are plant diseases (bacteria and fungal management), plant growth stimulation, insects and pests, seed covering, and post-harvest storage [65, 67, 68]. Plant viruses and viroid are also inhibited by chitosan. It has a great deal of promise as a biopesticide. It's a seed soaker, root application agent, and spray agent. Plant disease control and stress tolerance are heavily influenced by these methods. To increase its delayed pesticide release characteristics, chitosan can be blended with new ingredients such as gum, starch, and alginate. The usage of chitosan products will provide protection against over 60 diseases on a wide range of plants. Their antibacterial and plant innate immunity induced activity has a significant impact on plant disease control. Sporulation, mycelia, spore viability and germination, and the production of fungal virulence factors were all suppressed, as were other aspects of fungal development. Seed coating (for cucumber, rice, cotton, soybean, and wheat), soil improvement (for potato, spinach and soybean lettuce), foliar spraying (for cabbage, soybean, maize, peanut, rice, and cotton), hydroponic supplementation (for rice, wheat, and peanut), and plant tissue culture medium supplementation are all possible applications for chitosan products (chrysanthemum, limonium, carrot). In plants including soybean sprouts, ornamental plants, maize, wheat, lentil, rice, and peanuts, they enhance germination rate, seedling development parameters, and yield. Bioactivities that are significant in agricultural applications include antifungal behaviour, crop production improvement, activation of plant defence systems, and plant growth stimulation. The degree of acetylation, molecular weight, chitosan concentration, solution pH, viscosity, and the target microorganism all impact chitosan bioactivities [69, 70]. During the last two decades, the antibacterial activity of non-modified chitosan against a wide spectrum of microorganisms, including bacteria, yeasts, fungi, and viruses, has sparked interest. According to research, bioactivity is influenced by a range of variables, including molecular weight, which is most likely the most significant feature determining efficacy, but the evidence is often inconsistent. In general, antibacterial effectiveness reduces as molecular weight decreases; however, no direct association exists between molecular weight and bioactivity [71].

5.3 Biotechnology

Chitosan can be used to immobilise a variety of enzymes (e.g., Amylases, Urease, *Escherichia coli* cells, lysozyme). They become entrapped and swallowed by macromolecule chains. Chitosan is used in biochemistry as an enzyme support, notably in cross-linking activities. Chitosan and its derivatives have been used in biotechnology as biosensors and biodevices. As recently studied by Grifoll-Romero et al. [18] chitinases and deacetylases depolymerize and de-N-acetylate chitin, resulting in a variety of derivatives such as chitooligosaccharides, which have a variety of applications in biotechnology [18].

5.4 Food Industry

The US Food and Drug Administration has approved the use of chitosan on the market (FDA) as a normally recognised safe foodstuff ingredient, nutritional fibre (hypocholesterolemic impact), and efficient component. Since the 1990s, chitosan has been utilised as a food ingredient in Japan and Korea. Because of its bioactive activity and cationic nature, chitosan is utilised as a therapeutic element (food additives, functional food), antibacterial and antioxidant agent (food protection), antimicrobial coatings for fruits and vegetables, anticholesterolemic dairy products, and nutraceuticals [72–74]. The antibacterial properties of chitosan in powders, solutions, coatings, and edible films were emphasised by Friedman and Juneja [75]. The lowest molecular weight chitosan gave the best outcomes. New chitosan derivatives and oligomers are now being developed as antibacterial agents against bacteria found in food by researchers. These compounds look to be promising, particularly for use in nutraceuticals [75]. According to Kardas et al. [74] chitosan and its derivatives have a wide range of uses in the food sector, including food preservation against microbial destruction, shelf-life extension, biodegradable film production, and food packaging [74]. The products can be produced into fibres, films, gels, beads, or nanoparticles for use as packaging or coating materials. Van Broek et al. [76] found a similar outcome while researching the production of chitosan films and blends as packaging materials. Chitosan's antibacterial and film-forming characteristics have been explored as a viable alternative to non-biodegradable and non-renewable polymers as a food additive or coating material, reducing the widespread usage of harmful pesticides in food safety. In fact, their films have selective gas permeability [77]. Furthermore, the chitosan films demonstrated exceptional results and excellent mechanical qualities, as well as the potential to incorporate useful ingredients such as vitamins and antibacterial agents as carriers. Chitosan films, on the other hand, are extremely permeable to water vapour in packaging applications, and demonstrate fat diffusion resistance and selective gas permeability due to their hydrophilic nature. As food packaging films, chitosan-containing blends, composites, and multilayer systems have been proposed, and they appear promising. The food and nutrition industries are the biggest users of chitosan [77–79], with the biggest markets in Asia (Korea, and China and Japan), North America, and Europe. In 2018, the food and beverage industry in the United States is expected to reach 2288 metric tonnes. Chitosan is in high demand, especially for its potential use in nutraceutical components and feedstocks. Functional foods, nutritional supplements, and herbal/natural goods are all part of the nutraceutical industry. In addition to its role as a dietary fibre, chitosan has antioxidant, antibacterial, antiulcer bioactivities, anti-inflammatory and anti-carcinogenic. In reality, the United States has the world's largest nutraceutical market, followed by Europe (Sweden, Germany, France and the Netherlands). By 2030, China is anticipated to be the world's largest user of nutraceuticals. It appears to have promise for the treatment of obesity and weight loss. The effectiveness of chitosan in reducing fat absorption then again, remains questionable. The viscosity of chitosan is quite high and is highly water soluble, yet it is indigestible in the upper gastrointestinal system [78]. As a dietary fibre, it has the capacity to lower cholesterol by inhibiting the absorption of dietary fat and cholesterol. Chitosan and its derivatives have been proven to help with weight loss and body fat reduction, as well as lower systolic and diastolic blood pressure [80]. As an essential prebiotic, chitosan can help enhance intestinal health. Animal feed additives are another use for chitosan products, which is a growing industry [79, 81].

5.5 Aquaculture

A requirement for larger industrial use of chitin is the development of low-cost production methods and the expansion of cost-effective procedures for recovering chitin and by-products like as proteins and colours. The recovery of chitinous objects from garbage is widely known to be a source of additional cash. Crustacean shells contain substantial levels of carotenoids, which have yet to be synthesised and are offered in aquaculture as a fish food ingredient, particularly for salmon. Chitosan and its derivatives are used in aquaculture, according to Alishahi and Ader [82]. It may be utilised as a pollutant remover in water and wastewater, as well as a functional food, a nutritional supplement (synbiotics), a carrier for bioactive chemicals, medication liberate pathogen encapsulation, or nucleic acid encapsulation. Furthermore, there is a continuing requirement for competent vaccinations and delivery techniques to avoid and monitor developing and re-emerging infectious illnesses in aquaculture. It has proven challenging to develop effective vaccinations for a number of infectious illnesses. The failure is mostly due to a lack of capacity to create vaccines that trigger sufficient protected responses. The development of vaccine delivery methods that are both effective and antigen stable, as well as adjuvant-like, has been aided by the introduction of chitosan-based nanoparticles. Many nanoparticles can reach antigen-presenting cells through a variety of ways and evoke appropriate immune responses. Vinay et al. [83] investigated the utilize of chitosan for vaccine administration and weighed the pros and cons of several delivery methods for the development of novel fish disease vaccines.

5.6 Environmental Chemistry

This natural polymer has a variety of characteristics that make it a good choice for environmental applications. Its usage is supported by four major benefits: (1) flexibility, (2) inexpensive cost in compared to commercial activated carbon or organic resins, (3) possible biodegradability after usage and (4) superior pollutant-binding capabilities and high selectivity. Indeed, chitosan's remarkable ability to bind a wide range of pollutants is one of its most essential applications. Chitosan can be used in aqueous solutions to remove colours, heavy metals, and other pollutants [66]. The effects of a variety of parameters on biosorption, including chitosan characteristics, activation settings, process variables, dye chemistry, and experimental settings used in batch systems, were examined. The writers also went through the numerous adsorption methods that are now in use. They came to the conclusion that biosorbents were effective in removing pollutants while being inexpensive, nontoxic, and biocompatible. For decolorization, metal, and organic removal, biosorption onto chitosan is a viable alternative to conformist adsorbents, according to Kyzas et al. [84], Desbrières and Guibal [85], and Pakdel and Peighambardoust [86]. Flexible chitosan materials are frequently used as coagulating and flocculating agents wastewater treatment and water, as well as clarity and water purification. They have the potential to be a more ecologically friendly alternative to metallic salts and synthetic polyelectrolytes for the removal of both particulate and dissolved contaminants in water treatment [78]. Despite obvious progress and a great quantity of study, laboratory-scale investigations on the use of chitosan for pollutant resurrection procedures are still the norm. Indeed, these chitosan research disciplines have had difficulty finding industrial-scale applications. Because simultaneous flocculating and adsorbing agents are less costly, real-world applications are still uncommon, such as Pennofloc™ for water clarifying and ChitoVan™ for biofiltration. Even though chitosan is more effective in pollution removal, conventional products are sufficient to fulfil present regulatory requirements.

5.7 Paper and Pulp Industry

In 1936, chitosan was first employed in the papermaking business [87]. The main application was to increase paper's wet strength. Chitosan may also cooperate with cellulose pulp during paper manufacture to generate films that give cohesive rupture resistance. Non-toxic, biodegradable, and environmentally friendly, this biopolymer makes it easy to comply with environmental standards. Chitosan is also utilised as a chelating and complexing agent in the purification of pulp and paper wastewaters to remove lignin, colour, and unwanted impurities, as well as to minimise total organic carbon and chemical oxygen demand.

5.8 Tissue Bone Engineering

For orthopaedic therapy, chitosan scaffolds (CTS) may be easily changed into numerous forms such as films, fibres, beads, sponges, and more complicated structures [65]. CTS's cationic nature attracts a variety of negatively charged proteoglycans. In the process of bone implantation, porous materials play an essential role. CTS acetic acid solutions can be frozen and lyophilized in appropriate moulds to create porous CTS structures [88].

HAp, alginate, hyaluronic acid, calcium phosphate, poly (methyl methacrylate), poly-L-lactic acid, and growth hormones have all been utilised in combination with CTS for use in orthopaedics. HAp can be utilised in orthopaedics, dentistry, and maxillofacial surgery, among other fields. As a result, HAp has recently become a key ingredient in the production of artificial bone. It is utilised in the treatment of dental implants and orthopaedic replacements, particularly bone regeneration. Because of its weak mechanical characteristics, HAp cannot be utilised in load-bearing bone structures. To get better Hap's mechanical properties (compressive potency, Young's modulus, and fracture toughness), polymers were used. CTS in combination with HAp may be able to mimic normal bone function [89].

6 Conclusions

Based on a vast number of relevant published references, the goal of this chapter is to offer an overview of the state of the art in chitosan, sources of chitosan, applications, and chitosan composites. The relevance of chitosan in a variety of disciplines has been addressed in this chapter. The properties of and chitosan and chitin are detailed in this review. Chitosan is a highly effective, biodegradable, and environmentally acceptable polymer for environmental remediation. Heavy metals, radioactive metals, colours, and oil and grease wastes may all be successfully removed from polluted resources using chitosan. Chitosan may be extracted from a variety of sources, including crabs, shrimp, mollusks, and seashells. When harsh chemicals are used in extraction operations, polymers with variable characteristics such as charge, molecular weight, size, and degree of acetylation/deacetylation are commonly produced. As a result, greener and cleaner techniques, such as biological extractions, must be developed to extract chitin while decreasing the number of irregular by-products. Composite materials are utilised in a variety of applications, including water and agriculture, pharmaceuticals and biomedicine, followed by biotechnology. The rest of this chapter is dedicated to the various materials used to create chitosan composites.

Acknowledgements Authors would like to thank Department of Biotechnology, GLA University Mathura and Accelerated Cleaning Systems India Private Limited, Mumbai, India for providing necessary support for carrying out this activity.

References

1. Yadav, M., Goswami, P., Paritosh, K., Kumar, M., Pareek, N., & Vivekanand, V. (2019). Seafood waste: A source for preparation of commercially employable chitin/chitosan materials. *Bioresources and Bioprocessing*, *6*(1). https://doi.org/10.1186/s40643-019-0243-y
2. Perinelli, D., Fagioli, L., Campana, R., Lam, J., Baffone, W., Palmieri, G., et al. (2018). Chitosan-based nanosystems and their exploited antimicrobial activity. *European Journal of Pharmaceutical Sciences, 117*, 8–20. https://doi.org/10.1016/j.ejps.2018.01.046
3. Malerba, M., & Cerana, R. (2019). Recent applications of chitin- and chitosan-based polymers in plants. *Polymers, 11*(5), 839. https://doi.org/10.3390/polym11050839
4. Aam, B., Heggset, E., Norberg, A., Sørlie, M., Vårum, K., & Eijsink, V. (2010). Production of chitooligosaccharides and their potential applications in medicine. *Marine Drugs, 8*(5), 1482–1517. https://doi.org/10.3390/md8051482
5. Dhillon, G., Kaur, S., Brar, S., & Verma, M. (2012). Green synthesis approach: Extraction of chitosan from fungus mycelia. *Critical Reviews in Biotechnology, 33*(4), 379–403. https://doi.org/10.3109/07388551.2012.717217
6. Cauchie, H.-M. (2002). Chitin production by arthropods in the hydrosphere. *Hydrobiologia, 470*(1/3), 63–95. Springer Science and Business Media LLC. https://doi.org/10.1023/a:1015615819301
7. Xu, Y., Bajaj, M., Schneider, R., Grage, S., Ulrich, A., Winter, J., & Gallert, C. (2013). Transformation of the matrix structure of shrimp shells during bacterial deproteination and demineralization. *Microbial Cell Factories, 12*(1), 90. https://doi.org/10.1186/1475-2859-12-90
8. Salomon, C., Goycoolea, F., & Moerschbacher, B. (2017). Recent trends in the development of chitosan-based drug delivery systems. *An Official Journal of the American Association of Pharmaceutical Scientists, 18*(4), 933–935. https://doi.org/10.1208/s12249-017-0764-7
9. Upadhyay, U., Sreedhar, I., Singh, S., Patel, C., & Anitha, K. (2021). Recent advances in heavy metal removal by chitosan based adsorbents. *Carbohydrate Polymers, 251*, 117000. https://doi.org/10.1016/j.carbpol.2020.117000
10. Viarsagh, M. S., Janmaleki, M., Falahatpisheh, H. R., & Masoumi, J. (2010). Preparing chitosan from Persian Gulf shrimp shells and studying the impact of time on the degree of deacetylation. *Journal of Paramedical Science, 1*, 3.
11. Hirano, S. (1999). *Biotechnology and bioactive polymers*.https://doi.org/10.1007/978-1-4757-9519-6
12. Struszczyk, H., Pospieszny, H., & Kotlinski, S. (1989). In G. Skjk-Brk, T. Anthonsen, & P. Sandford (Eds.), *Chitin and chitosan* (pp. 733–742). Elsevier Science.
13. Nishimura, K., Nishimura, S., Seo, H., Nishi, N., Tokura, S., & Azuma, I. (1987). Effect of multiporous microspheres derived from chitin and partially deacetylated chitin on the activation of mouse peritoneal macrophages. *Vaccine, 5*(2), 136–140. https://doi.org/10.1016/0264-410x(87)90061-2

14. Finch, C. (1992). Advances in chitin and chitosan. In C. J. Brine, P. A. Sandford, & J. P. Zikakis (Eds.), Elsevier Science Publishers, London (pp. xxii + 685). ISBN 1-85166-899-3. *Polymer International, 31*(4), 404–404. https://doi.org/10.1002/pi.4990310419
15. H. Seo, K. Mitsuhashi, & H. Tanibe. (1992). Advances in chitin and chitosan. In C. J. Brine, P. A. Eldridge & J. P. Zikakis (Eds.), Elsevier Science (pp. 34–40).
16. Badawy, M., Rabea, E., Taktak, N., & El-Nouby, M. (2016). The antibacterial activity of chitosan products blended with monoterpenes and their biofilms against plant pathogenic bacteria. *Scientifica, 2016*, 1–10. https://doi.org/10.1155/2016/1796256
17. Muzzarelli, R., Biagini, G., Pugnaloni, A., Filippini, O., Baldassarre, V., Castaldini, C., & Rizzoli, C. (1989). Reconstruction of parodontal tissue with chitosan. *Biomaterials, 10*(9), 598–603. https://doi.org/10.1016/0142-9612(89)901130
18. Divya, K., & Jisha, M. (2017). Chitosan nanoparticles preparation and applications. *Environmental Chemistry Letters, 16*(1), 101–112. https://doi.org/10.1007/s10311-017-0670-y
19. Benhabiles, M., Salah, R., Lounici, H., Drouiche, N., Goosen, M., & Mameri, N. (2012). Antibacterial activity of chitin, chitosan and its oligomers prepared from shrimp shell waste. *Food Hydrocolloids, 29*(1), 48–56. https://doi.org/10.1016/j.foodhyd.2012.02.013
20. Abdelaal, M., Sobahi, T., & Al-Shareef, H. (2013). Modification of chitosan derivatives of environmental and biological interest: A green chemistry approach. *International Journal of Biological Macromolecules, 55*, 231–239. https://doi.org/10.1016/j.ijbiomac.2013.01.013
21. El Knidri, H., Belaabed, R., Addaou, A., Laajeb, A., & Lahsini, A. (2018). Extraction, chemical modification and characterization of chitin and chitosan. *International Journal of Biological Macromolecules, 120*, 1181–1189. https://doi.org/10.1016/j.ijbiomac.2018.08.139
22. Abdou, E., Nagy, K., & Elsabee, M. (2008). Extraction and characterization of chitin and chitosan from local sources. *Bioresource Technology, 99*(5), 1359–1367. https://doi.org/10.1016/j.biortech.2007.01.051
23. Yen, M., Yang, J., & Mau, J. (2009). Physicochemical characterization of chitin and chitosan from crab shells. *Carbohydrate Polymers, 75*(1), 15–21. https://doi.org/10.1016/j.carbpol.2008.06.006
24. Philibert, T., Lee, B., & Fabien, N. (2016). Current status and new perspectives on chitin and chitosan as functional biopolymers. *Applied Biochemistry and Biotechnology, 181*(4), 1314–1337. https://doi.org/10.1007/s12010-016-2286-2
25. Rinaudo, M. (2006). Chitin and chitosan: Properties and applications. *Progress in Polymer Science, 31*(7), 603–632. https://doi.org/10.1016/j.progpolymsci.2006.06.001
26. Anitha, A., Sowmya, S., Kumar, P., Deepthi, S., Chennazhi, K., Ehrlich, H., et al. (2014). Chitin and chitosan in selected biomedical applications. *Progress in Polymer Science, 39*(9), 1644–1667. https://doi.org/10.1016/j.progpolymsci.2014.02.008
27. Aranaz, I., Mengibar, M., Harris, R., Panos, I., Miralles, B., Acosta, N., et al. (2009). Functional characterization of chitin and chitosan. *Current Chemical Biology, 3*(2), 203–230. https://doi.org/10.2174/2212796810903020203
28. Yang, R., Li, H., Huang, M., Yang, H., & Li, A. (2016). A review on chitosan-based flocculants and their applications in water treatment. *Water Research, 95*, 59–89. https://doi.org/10.1016/j.watres.2016.02.068
29. Osifo, P., Neomagus, H., Everson, R., Webster, A., &vd Gun, M. (2009). The adsorption of copper in a packed-bed of chitosan beads: Modeling, multiple adsorption and regeneration. *Journal of Hazardous Materials, 167*(1–3), 1242–1245.https://doi.org/10.1016/j.jhazmat.2009.01.109
30. Jobin, G., Couture, G., Goyer, C., Brzezinski, R., & Beaulieu, C. (2004). Streptomycete spores entrapped in chitosan beads as a novel biocontrol tool against common scab of potato. *Applied Microbiology and Biotechnology, 68*(1), 104–110. https://doi.org/10.1007/s00253-004-1843-8
31. Ravi Kumar, M. (2000). A review of chitin and chitosan applications. *Reactive and Functional Polymers, 46*(1), 1–27. https://doi.org/10.1016/s1381-5148(00)00038-9
32. Zhang, H., Tachikawa, H., Gao, X., & Nakanishi, H. (2014). Applied usage of yeast spores as chitosan beads. *Applied and Environmental Microbiology, 80*(16), 5098–5105. https://doi.org/10.1128/aem.00677-14

33. Jia, Y., & Fiedler, B. (2020). Tensile creep behaviour of unidirectional flax fibre reinforced bio-based epoxy composites. *Composites Communications, 18*, 5–12. https://doi.org/10.1016/j.coco.2019.12.010
34. Sharp, R. (2013). A review of the applications of chitin and its derivatives in agriculture to modify plant-microbial interactions and improve crop yields. *Agronomy, 3*(4), 757–793. https://doi.org/10.3390/agronomy3040757
35. Beauséjour, J., Clermont, N., & Beaulieu, C. (2003). Effect of Streptomyces melanosporofaciens strain EF-76 and of chitosan on common scab of potato. *Plant and Soil, 256*(2), 463–468. https://doi.org/10.1023/a:1026177714855
36. Fazeli, M., Florez, J., & Simão, R. (2019). Improvement in adhesion of cellulose fibers to the thermoplastic starch matrix by plasma treatment modification. *Composites Part B: Engineering, 163*, 207–216. https://doi.org/10.1016/j.compositesb.2018.11.048
37. Nomanbhay, S., & Palanisamy, K. (2005). Removal of heavy metal from industrial wastewater using chitosan coated oil palm shell charcoal. *Electronic Journal of Biotechnology, 8*(1). https://doi.org/10.2225/vol8-issue1-fulltext-7
38. Dubey, V., Pandey, l., & Saxena, C. (2005). Pervaporative separation of ethanol/water azeotrope using a novel chitosan-impregnated bacterial cellulose membrane and chitosan? poly(vinyl alcohol) blends. *Journal of Membrane Science, 251*(1–2), 131–136.https://doi.org/10.1016/j.memsci.2004.11.009
39. Li, N., & Bai, R. (2005). Copper adsorption on chitosan–cellulose hydrogel beads: Behaviors and mechanisms. *Separation and Purification Technology, 42*(3), 237–247. https://doi.org/10.1016/j.seppur.2004.08.002
40. Sun, X., Peng, B., Ji, Y., Chen, J., & Li, D. (2009). Chitosan(chitin)/cellulose composite biosorbents prepared using ionic liquid for heavy metal ions adsorption. *Aiche Journal, 55*(8), 2062–2069. https://doi.org/10.1002/aic.11797
41. Vijaya, Y., Popuri, S., Boddu, V., & Krishnaiah, A. (2008). Modified chitosan and calcium alginate biopolymer sorbents for removal of nickel (II) through adsorption. *Carbohydrate Polymers, 72*(2), 261–271. https://doi.org/10.1016/j.carbpol.2007.08.010
42. Ngah, W., & Fatinathan, S. (2008). Adsorption of Cu(II) ions in aqueous solution using chitosan beads, chitosan–GLA beads and chitosan–alginate beads. *Chemical Engineering Journal, 143*(1–3), 62–72. https://doi.org/10.1016/j.cej.2007.12.006
43. Alhakawati, M., & Banks, C. (2004). Removal of copper from aqueous solution by Ascophyllum nodosum immobilised in hydrophilic polyurethane foam. *Journal of Environmental Management, 72*(4), 195–204. https://doi.org/10.1016/j.jenvman.2004.04.010
44. Jang, S., Min, B., Jeong, Y., Lyoo, W., & Lee, S. (2008). Removal of lead ions in aqueous solution by hydroxyapatite/polyurethane composite foams. *Journal of Hazardous Materials, 152*(3), 1285–1292. https://doi.org/10.1016/j.jhazmat.2007.08.003
45. Jayakumar, S., Gomathi, T., & Sudha, P. (2013). Sorption studies of lead (II) onto crosslinked and non crosslinked biopolymeric blends. *International Journal of Biological Macromolecules, 59*, 165–169. https://doi.org/10.1016/j.ijbiomac.2013.04.031
46. Jayakumar, S., & Sudha, P. (2013). Physicochemical characterization of chitosan/nylon6/polyurethane foam chemically cross-linked ternary blends. *Spectrochimica Acta Part A: Molecular and Biomolecular Spectroscopy, 105*, 20–23. https://doi.org/10.1016/j.saa.2012.12.012
47. Wan Ngah, W., Kamari, A., & Koay, Y. (2004). Equilibrium and kinetics studies of adsorption of copper (II) on chitosan and chitosan/PVA beads. *International Journal of Biological Macromolecules, 34*(3), 155–161. https://doi.org/10.1016/j.ijbiomac.2004.03.001
48. Hsien, T., Liu, Y., Huang, C., Lin, P., & Wang, D. (2011). Synthesis of novel PVA crosslink mixed matrix scaffolds and adsorption of copper ions from waste water. *Desalination and Water Treatment, 34*(1–3), 354–360. https://doi.org/10.5004/dwt.2011.2800
49. Abdeen, Z., Mohammad, S., & Mahmoud, M. (2015). Adsorption of Mn (II) ion on polyvinyl alcohol/chitosan dry blending from aqueous solution. *Environmental Nanotechnology, Monitoring & Management, 3*, 1–9. https://doi.org/10.1016/j.enmm.2014.10.001

50. Venkatesan, J., & Kim, S.-K. (2010). Chitosan composites for bone tissue engineering—an overview. *Marine Drugs, 8*(8), 2252–2266. https://doi.org/10.3390/md8082252
51. Kumirska, J., Weinhold, M., Thöming, J., & Stepnowski, P. (2011). Biomedical activity of chitin/chitosan based materials—Influence of physicochemical properties apart from molecular weight and degree of N-Acetylation. *Polymers, 3*(4), 1875–1901. https://doi.org/10.3390/polym3041875
52. Kumar, P., Saini, M., Dehiya, B., Umar, A., Sindhu, A., Mohammed, H., et al. (2020). Fabrication and in-vitro biocompatibility of freeze-dried CTS-nHA and CTS-nBG scaffolds for bone regeneration applications. *International Journal Of Biological Macromolecules, 149*, 1–10. https://doi.org/10.1016/j.ijbiomac.2020.01.035
53. Hu, Q., Li, B., Wang, M., & Shen, J. (2004). Preparation and characterization of biodegradable chitosan/hydroxyapatite nanocomposite rods via in situ hybridization: A potential material as internal fixation of bone fracture. *Biomaterials, 25*(5), 779–785. https://doi.org/10.1016/s0142-9612(03)00582-9
54. Lu, B., Wang, T., Li, Z., Dai, F., Lv, L., Tang, F., et al. (2016). Healing of skin wounds with a chitosan–gelatin sponge loaded with tannins and platelet-rich plasma. *International Journal of Biological Macromolecules, 82*, 884–891. https://doi.org/10.1016/j.ijbiomac.2015.11.009
55. Parvez, S., Rahman, M., Khan, M., Khan, M., Islam, J., Ahmed, M., et al. (2012). Preparation and characterization of artificial skin using chitosan and gelatin composites for potential biomedical application. *Polymer Bulletin, 69*(6), 715–731. https://doi.org/10.1007/s00289-012-0761-7
56. Debels, H., Hamdi, M., Abberton, K., & Morrison, W. (2015). Dermal matrices and bioengineered skin substitutes. *Plastic and Reconstructive Surgery Global Open, 3*(1), e284. https://doi.org/10.1097/gox.0000000000000219
57. Rothamel, D., Schwarz, F., Sager, M., Herten, M., Sculean, A., & Becker, J. (2005). Biodegradation of differently cross-linked collagen membranes: An experimental study in the rat. *Clinical Oral Implants Research, 16*(3), 369–378. https://doi.org/10.1111/j.1600-0501.2005.01108.x
58. Mahmoud, A., & Salama, A. (2016). Norfloxacin-loaded collagen/chitosan scaffolds for skin reconstruction: Preparation, evaluation and in-vivo wound healing assessment. *European Journal of Pharmaceutical Sciences, 83*, 155–165. https://doi.org/10.1016/j.ejps.2015.12.026
59. Venter, J., Kotze, A., Auzelyvelty, R., & Rinaudo, M. (2006). Synthesis and evaluation of the mucoadhesivity of a CD-chitosan derivative. *International Journal of Pharmaceutics, 313*(1–2), 36–42. https://doi.org/10.1016/j.ijpharm.2006.01.016
60. Rinaudo, M. (2008). Main properties and current applications of some polysaccharides as biomaterials. *Polymer International, 57*(3), 397–430. https://doi.org/10.1002/pi.2378
61. Ishihara, M., Nakanishi, K., Ono, K., Sato, M., Kikuchi, M., Saito, Y., et al. (2002). Photocrosslinkable chitosan as a dressing for wound occlusion and accelerator in healing process. *Biomaterials, 23*(3), 833–840. https://doi.org/10.1016/s0142-9612(01)00189-2
62. Alemdaroğlu, C., Değim, Z., Çelebi, N., Zor, F., Öztürk, S., & Erdoğan, D. (2006). An investigation on burn wound healing in rats with chitosan gel formulation containing epidermal growth factor. *Burns, 32*(3), 319–327. https://doi.org/10.1016/j.burns.2005.10.015
63. Sung, J., Hwang, M., Kim, J., Lee, J., Kim, Y., Kim, J., et al. (2010). Gel characterisation and in vivo evaluation of minocycline-loaded wound dressing with enhanced wound healing using polyvinyl alcohol and chitosan. *International Journal of Pharmaceutics, 392*(1–2), 232–240. https://doi.org/10.1016/j.ijpharm.2010.03.024
64. Yang, C., Xu, L., Zhou, Y., Zhang, X., Huang, X., Wang, M., et al. (2010). A green fabrication approach of gelatin/CM-chitosan hybrid hydrogel for wound healing. *Carbohydrate Polymers, 82*(4), 1297–1305. https://doi.org/10.1016/j.carbpol.2010.07.013
65. Li, X., Chen, S., Zhang, B., Li, M., Diao, K., Zhang, Z., et al. (2012). In situ injectable nano-composite hydrogel composed of curcumin, N, O-carboxymethyl chitosan and oxidized alginate for wound healing application. *International Journal of Pharmaceutics, 437*(1–2), 110–119. https://doi.org/10.1016/j.ijpharm.2012.08.001
66. Pal, P., Pal, A., Nakashima, K., & Yadav, B. (2021). Applications of chitosan in environmental remediation: A review. *Chemosphere, 266*, 128934. https://doi.org/10.1016/j.chemosphere.2020.128934

67. Liu, J., Zhu, L. (2009). Method for preparing chitin composite artificial skin that can be used as woundplast. *Faming Zhuanli Shenqing CN 101411897 A 20090422*.
68. Bordenave, N., Grelier, S., Pichavant, F., & Coma, V. (2007). Water and moisture susceptibility of chitosan and paper-based materials: Structure-property relationships. *Journal of Agricultural and Food Chemistry, 55*(23), 9479–9488. https://doi.org/10.1021/jf070595i
69. Yin, H., & Du, Y. (2011). Mechanism and plant defence applications of chitin/chitosan and their derivatives. In S. K. Kim (Ed.), *Chitin, chitosan, oligosaccharides, and their derivatives: biological activities and applications* (Chapter 41, pp. 605–618). Taylor & Francis Group LLC, CRC Press. ISBN 9781439816035.
70. Munoz-Bonilla, M. Cerrada, & Fernández-Garca, M. (Eds.). (2014). *Chitosan's antimicrobial role in food, agriculture, and biomedicine. RSC polymer chemistry series* (Vol. 10, Chapter 2, pp. 22–53). RSC Publishing.https://doi.org/10.1039/9781782624998-00022
71. Grifoll-Romero, L., Pascual, S., Aragunde, H., Biarnés, X., & Planas, A. (2018). Chitin deacetylases: Structures, specificities, and biotech applications. *Polymers, 10*(4), 352. https://doi.org/10.3390/polym10040352
72. Sharif, R., Mujtaba, M., Rahman, M., Shalmani, A., Ahmad, H., Anwar, T., Deng, T. C., & Wang, X. P. (2018). A analysis of the multifunctional role of chitosan in horticultural crops. *Molecules, 23*, 872. https://doi.org/10.3390/molecules230408
73. Kardas, I., Struszczyk, M. H., Kucharska, M., van den Broek, L. A. M., van Dam, J. E. G., & Ciechaska, D. (2012). Chitin and chitosan as industrial biopolymers with functional properties. In *The European polysaccharide network of excellence* (Vol. 11, pp. 329–373). https://doi.org/10.1007/978-3-7091-0421-7
74. Van den Broek, L., Knoop, R., Kappen, F., & Boeriu, C. (2015). Chitosan films and blends for packaging material. *Carbohydrate Polymers, 116*, 237–242. https://doi.org/10.1016/j.carbpol.2014.07.039
75. Fernandez-Saiz, P. (2010). The use of chitosan in antimicrobial films for food protection. *CAB Reviews: Perspectives in Agriculture, Veterinary Science, Nutrition and Natural Resources, 5*(024). https://doi.org/10.1079/pavsnnr20105024
76. Friedman, M., & Juneja, V. (2010). Review of antimicrobial and antioxidative activities of chitosans in food. *Journal of Food Protection, 73*(9), 1737–1761. https://doi.org/10.4315/0362-028x-73.9.1737
77. Elsabee, M., & Abdou, E. (2013). Chitosan based edible films and coatings: A review. *Materials Science and Engineering: C, 33*(4), 1819–1841. https://doi.org/10.1016/j.msec.2013.01.010
78. Wang, H., Qian, J., & Ding, F. (2018). Emerging chitosan-based films for food packaging applications. *Journal of Agricultural and Food Chemistry, 66*(2), 395–413. https://doi.org/10.1021/acs.jafc.7b04528
79. Muzzarelli, R. A. A., & de Vincenzi, M. (1997). Chitosan as dietary food additives. In M. F. A. Goosen (Ed.), *Applications of chitin and chitosan* (Chapter 7, pp. 115–128). CRC Press LLC.
80. Alishahi, A., & Aïder, M. (2012). Applications of chitosan in the seafood industry and aquaculture: A review. *Food and Bioprocess Technology, 5*(3), 817–830. https://doi.org/10.1007/s11947-011-0664-x
81. Roller, S., & Valley, T. (2003). Chitosan: New food preservative or laboratory curiosity? In S. Roller (Ed.), *Natural antimicrobials for the minimal processing of foods* (Chapter 8, pp. 158–175). CRC Press, Woodhead Publishing in Food Science and Technology.
82. Vinay, T., Bhat, S., GonChoudhury, T., Paria, A., Jung, M., ShivaniKallappa, G., & Jung, S. (2017). Recent advances in application of nanoparticles in fish vaccine delivery. *Reviews in Fisheries Science & Aquaculture, 26*(1), 29–41. https://doi.org/10.1080/23308249.2017.1334625
83. Kyzas, G., Bikiaris, D., & Mitropoulos, A. (2017). Chitosan adsorbents for dye removal: A review. *Polymer International, 66*(12), 1800–1811. https://doi.org/10.1002/pi.5467
84. Desbrières, J., & Guibal, E. (2018). Chitosan for wastewater treatment. *Polymer International, 67*(1), 7–14. https://doi.org/10.1002/pi.5464
85. MohammadzadehPakdel, P., & Peighambardoust, S. (2018). Review on recent progress in chitosan-based hydrogels for wastewater treatment application. *Carbohydrate Polymers, 201*, 264–279. https://doi.org/10.1016/j.carbpol.2018.08.070

86. Struszczyk, M. (2002). Chitin and Chitosan. Part II. Applications of chitosan. *Polimery, 47*(06), 396–403. https://doi.org/10.14314/polimery.2002.396
87. Song, Z., & Li, G. (2018). Application of chitin/chitosan and their derivatives in the papermaking industry. *Polymers, 10*(4), 389. https://doi.org/10.3390/polym10040389
88. Rodríguez-Vázquez, M., Vega-Ruiz, B., Ramos-Zúñiga, R., Saldaña-Koppel, D. A., & Quiñones-Olvera, L. F. (2015). Chitosan and Its Potential Use as a Scaffold for Tissue Engineering in Regenerative Medicine. *BioMed Research International, 2015*, 821279. https://doi.org/10.1155/2015/821279
89. Grande-Tovar, C., Chaves-Lopez, C., Serio, A., Rossi, C., & Paparella, A. (2018). Chitosan coatings enriched with essential oils: Effects on fungi involved in fruit decay and mechanisms of action. *Trends in Food Science & Technology, 78*, 61–71. https://doi.org/10.1016/j.tifs.2018.05.019

A Review of Seaweed Based Composites

M. H. M. Rizalludin, S. M. Sapuan, M. N. M. Rodzi, M. S. Ibrahim, and S. F. K. Sherwani

1 Introduction

The extensive acknowledgement of natural fibres and natural polymers as green materials are being driven by the rapid exhaustion of petroleum resources, as well as a growing attention of global environmental problems associated with the usage of conventional plastics. Natural fibres and biopolymers have piqued the interest of scientists and industry because of their environmentally favourable and long-lasting properties [1].

Seaweed polysaccharide-derived biopolymers have a lot of potential because they are renewable, biodegradable, biocompatible, and environmentally friendly [2]. Many naturally existing polymers generated from seaweeds, such as alginates, carrageenan, and agar, have been proposed for application in various goods, while others have remained unused such as fucoidan and ulvan [3]. Seaweed polymers offer a number of features that make them a possible material of choice. These polysaccharides are readily accesible in big numbers and at a reasonable cost from renewable and agricultural feedstock [4].

M. H. M. Rizalludin · S. M. Sapuan (✉) · M. N. M. Rodzi · M. S. Ibrahim · S. F. K. Sherwani
Advanced Engineering Materials and Composites Research Centre (AEMC), Department of Mechanical and Manufacturing Engineering, Universiti Putra Malaysia, 43400 UPM Serdang, Selangor, Malaysia
e-mail: sapuan@upm.edu.my

S. M. Sapuan
Laboratory of Biocomposite Technology, Institute of Tropical Forestry and Forest Products (INTROP), Universiti Putra Malaysia, 43400 UPM Serdang, Selangor, Malaysia

Sapuan S. M. and I. Ahmad (eds.), *Composites from the Aquatic Environment*, Composites Science and Technology, https://doi.org/10.1007/978-981-19-5327-9_15

1.1 Definition of Seaweeds

Seaweeds are a macroalgae that normally populate along the coastal areas. Usually, In the algal group, seaweeds are considered as one the members where they have various sizes. Their sizes can be as small as microscopic single cells to as large as marine algae which can grow up to 60 m in length [5, 6]. There are about 10,000 species of macroalgae and it provides more or less 10% of the productivity in the total world of marine life. There are three families that seaweed belongs to: Chlorophyceae which is green in colour, Rhodophyceae which is red in colour, and also Phaeophyceae which is brown in colour [7]. Seaweeds are macroalgae that generally live by attaching to rock or any other substrate in the coastal areas. Figure 1 shows the types of algae such as Chlorophyceae (Green), Rhodophyceae (Red), and Phaeophyceae (Brown).

(a) (b) (c)

Fig. 1 Types of Algae: (**a**) Chlorophyceae (Green), (**b**) Rhodophyceae (Red), and (**c**) Phaeophyceae (Brown) [8]

1.2 Benefits of Seaweeds

Nowadays, people need to get rid of the idea that seaweeds are one of the plants that are 'weak' and only can be used as food materials. Actually, seaweed has many advantages that not everyone knows about. Currently, many researchers in the world are trying their best to find the potential candidates to decrease the use of petroleum as one of the materials in the composite industries. One of the great candidates for the production of bioplastic or composites are seaweed [9]. One of the advantages of seaweeds is that they are 'flexible' plants. Seaweeds are capable of growing in many different environments in the sea. This facilitates their growth in the natural environment [10]. Using seaweeds for Other than that, the bad effects on the food chain in the marine ecosystem can be lowered when seaweeds are used as an alternative for bioplastics production [9, 11]. This is because seaweed based composites can be decomposed in a shorter amount of time than synthetic composite. Notpla, which is one the startups in London, stated that their natural plastic-like casings which are made from seaweed can be biodegradable between four to six weeks compared to the synthetic plastic that takes several hundred years [12]. This can reduce the water pollution happening in the sea if the composites are not disposed correctly. This pollution can destroy the food chain of the organism in the sea and eventually the food web will be unstable. Moreover, seaweed has potential as an alternative for petroleum because it can form film [13]. Seaweeds are types of plants that are easy to find, are economical, do not need fertilizer and pesticides and easy to nurture. Seaweed has high biomass and rich in polysaccharides, agarose, ulvan and fucoidal [14]. Lastly, the most important factor why seaweed is chosen as one of the materials to replace petroleum is because it is a renewable source and can be lasting for a long time. Seaweeds can be planted again and again after being taken to process in the industries unlike petroleum, where the source can be depleted in the future.

1.3 Classification of Seaweed

In general, seaweeds can be classified as red (Rhodophyta), green (Chlorophyta), and brown (Phaeophyta), Red and brown seaweeds are solely found in the ocean, whereas green algae can be found in freshwater environments such as rivers and lakes, as well as on land [15]. Usually, chlorophyta is also known as green algae. The reason it is called green algae is because it comprises the two pigments which are chlorophyll a and b. Examples of the genus of the green algae are Chlamydomonas, Spirogyra, and Chara. While for Phaeophyta, it is known as brown algae. Normally, brown algae is mostly available in marine plants. Unlike green algae which has only two pigments, brown algae consists of four pigments. The pigments that are present in the phaeophyta are chlorophyll A, C, carotenoids, and xanthophyll. Examples of the genus of phaeophyta are Dictyota, Laminaria, and Sargassum. Lasly, Rhodophyta which is known as red algae because it has red pigment. The name of the red pigment is r-phycoerythrin. Examples of the genus of Rhodophyta are Porphyra, Gracilaria, and Gelidium [16]. In term of reproduction, all algae which are green algae, brown algae and red algae capable of undergo sexually and asexually reprodution. But there are slight differences in how they undergo the process. For chlorophyta, the asexual reproduction is by the fusion of male and female gametes. Other than that, it can also be performed by using other parts of an organism [17]. For its sexual reproduction, it requires reciprocation of two nuclei through conjuction tubes. For brown algae, most of the them except the Fucales order, undergo sexual reproduction by sporic meiosis process. As for some orders in Phaephyta, they undergoe asexual reproduction by motile zoospores process. Lastly, for red algae, the sexual reproduction is same as green algae which is via fusion of gametes. In spite of that, the male gamete of Rhodophyta is not moved because of the absence of flagellum. Due to this, it depends on water current to move to the female gamete. For asexual reproduction, it has three ways which are via spore production, fragmentation, and propagules production [18].

2 Seaweed Derivatives

Seaweeds are high in vitamins and minerals and they (mostly red and brown seaweeds) are commonly employed as both herbal medicine and a human food source. Although all seaweed species have a similar chemical composition, the percentage of protein content varies depending on the species. Table 1 shows the chemical composition of seaweeds.

Seaweeds have been used as the source of hydrocolloids in diverse areas of biotechnology, microbiology, food technology and plastics industry [20, 21]. A hydrocolloid is a non-crystalline material with big molecules that dissolves in water to produce a viscous (thickened) solution [22]. There are three main products (hydrocolloid) derived from seaweeds which are alginate, carrageenan and agar. Table 2 shows the source of polysaccharides in brown, red and green seaweeds.

Table 1 The chemical composition of seaweeds [19]

Components	Composition
Water	80–90%
Carbohydrates	50% dry weight
Lipids	1–3% dry weight
Minerals	7–38% dry weight
Proteins	Brown seaweed: 3 to 15% dry weight Red or Green seaweeds: 10 to 47% dry weight

Table 2 The source of hydrocolloids in red, brown and green seaweeds [21, 23]

Hydrocolloids	Red seaweed	Brown seaweed	Green seaweed
Agar	/	x	x
Alginate	x	/	x
Carrageenan	/	x	x
Cellulose	/	/	/
Floridean Starch (α-1,4-binding glucan)	/	x	x
Fucoidan (sulfated fucose)	x	/	x
Laminarin (β-1, 3 glucan)	x	/	x
Porphyran	/	x	x
Mannan	/	x	x
Mannitol	x	/	x
Sulphuric acid polysaccharides	x	x	/
Sargassan	x	/	x
Sulphatedgalactans	/	x	/
Xylans	/	x	/

2.1 *Alginate*

Alginates which exist extensively in brown SW are biopolymers containing of linear copolymers of β-(1–4) linked d-mannuronic acid and β-(1–4)-linked l-guluronic acid units [24]. Figure 2 shows the chemical structure of mannuronic acid and guluronic acid. Thickening, emulsion-stabilizing, film-forming, stabilizing, and gel-producing agents are all popular uses for alginate [25]. Alginate was discovered in 1881, but alginate fibres were first developed in the 1940s, and they began to be used in wound dressing in the 1980s due to their outstanding gel-forming qualities as well as superior hemostatic and absorbent properties [26, 27].

2.2 *Carrageenan*

Sulphated polysaccharides derived from red seaweed are called as carrageenan. It's a natural anionic linear polysaccharide made up of galactose and anhydrogalactose units [28]. Carrageenan is basically a polymer that soluble in water that has a linear chain of partially sulfonated galactans [15]. There are three main groups of carrageenan, which are iota-carrageenan (i-carrageenan), lambda-carrageenan (λ-carrageenan) and kappa-carrageenan (κ-carrageenan) [29]. Carrageenans have been used as natural ingredients in a broad range of food applications especially in the elaboration of gels and as thickeners in the past decades. Figure 3 shows the chemical structure of the three groups of carrageenan.

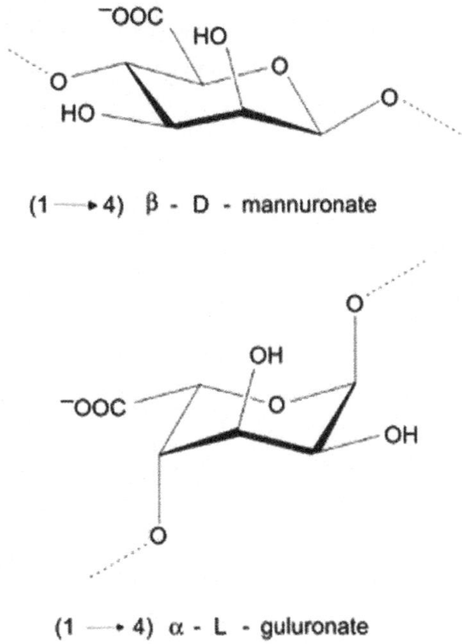

Fig. 2 The chemical structure of mannuronic acid and guluronic acid [24]

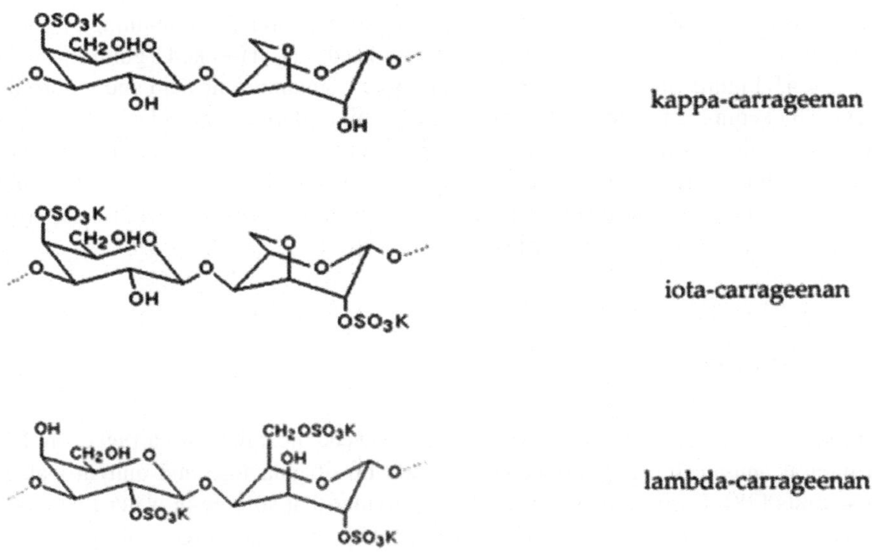

Fig. 3 The chemical structure of the three groups of carrageenan [30]

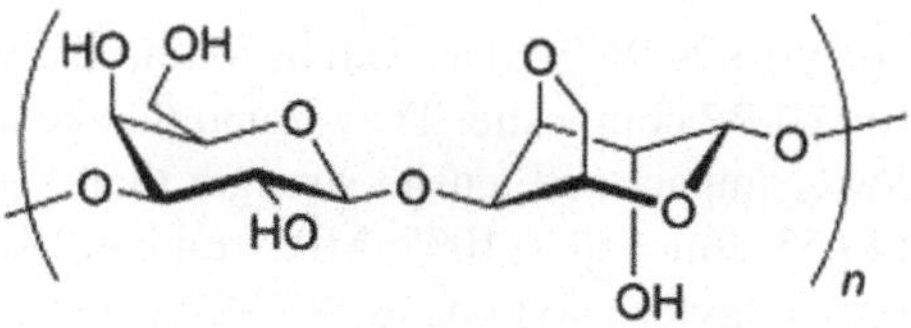

Fig. 4 The chemical structure of agarose [31]

2.3 Agar

Agar has been widely utilized in the food industry as a gelling agent because of its hydrophilic colloidal qualities [15]. It is made up primarily of two polysaccharides which are agaropectin and agarose. Agarose consists of (1–4)-linked 3,6-anhydro-α-l-galactopyranose and (1–3)-linked β-d-galactose units that alternate. Agaropectin and agarose basically have the same backbone but agaropectin has a lot of anionic groups like sulphate, pyruvate, and glucuronate [31]. Figure 4 shows chemical structure of agarose.

3 Seaweed Based Composite

A composite is a material formed by combining at least two or more materials, frequently having chemical and physical properties that differ [32]. A biocomposite, on the other hand, is a material made up of two or more unique constituent materials (one of which is naturally sourced) that are combined to create a new material that outperforms the constituent materials individually [33]. Composites and biocomposites have attracted a lot of attention in recent years as a result of rising environmental alertness, concerns about fossil fuel depletion and a push for more sustainable technology. There are several studies on seaweed based composites with the reinforcement of synthetic polymer and natural polymer.

3.1 Seaweed Reinforced Synthetic Polymer Composites

Synthetic polymers or also called as man-made polymers are polymers that are made synthetically in laboratories [34]. Polyethylene (PE), polypropylene (PP), polyamides (nylon), polystyrene (PS), poly(vinyl chloride) (PVC), teflon, epoxy and synthetic rubber, are examples of synthetic polymers. There are several investigations on seaweed reinforced synthetic polymer composites especially with polypropylene.

Saad et al. [35] had prepared the kenaf/seaweed reinforced polypropylene composite by using hot pressing and extrusion techniques. The goal of their study was to examine the possibility of kenaf/seaweed to be transformed into wood plastic composites (WPCs) as well as to find the various properties of kenaf/seaweed reinforced PP composite. They found that kenaf/seaweed reinforced PP composite had lower impact and tensile strength than kenaf reinforced polypropylene composite (49.53 J/m and 0.1098 MPa respectively) but greater than seaweed reinforced polypropylene composite (40.79 J/m and 0.0501 MPa respectively). This is because kenaf has high cellulose content which contributes to the kenaf reinforced polypropylene composite's strong tensile and impact strength [36]. When seaweed is incorporated into kenaf, however, the tensile and impact strength begins to reduce. Furthermore, they also discovered that kenaf/SW reinforced polypropylene composite has the second-highest percentage of water absorption, whereas kenaf polypropylene composite having the highest and SW/polypropylene composite having the lowest water absorption.

Abdullah, Salim and Roslan [37] had investigated the effect of alkaline treatment on seaweed reinforced propylene blend composites. They concluded that the seaweed and propylene that undergo alkaline treatment have the good prospects to be utilized into a composite as a result of their research. Alkaline treatment or also known as mercerization is the steps of exposing a plant fibre to the action of a sufficiently concentrated aqueous solution of a strong base to produce significant swelling, resulting changes in the fine structure, size, morphology, and mechanical performance [38]. Alkaline treatment involves immersing fibres in a sodium hydroxide (NaOH) solution for a length of time, which increases surface roughness and develops better mechanical interlocking, as well as increase the amount of cellulose exposed on the fibre surface (the number of potential reaction sites increased) [39].

The untreated seaweed/propylene composites have a greater melt flow rate (MFR) value (2.513 g/10 min with loads of 1.2 kg) compared to treated seaweed/propylene composites (1.870 g/10 min with loads of 1.2 kg) because of the waxy and cellulose constituents retained together in the composites and easier to circulate through the melt indexer. The observation from the test on water absorption revealed that the treated SW/PP composites had a reduced percentage of water absorption because unwanted materials from the seaweed were removed, causing the composites to withstand absorbing water into the films. The untreated seaweed reinforced polypropylene composites have a low tensile strength. The mechanical interlocking of the composites materials causes the interfacial link between the PP matrix and SW fibres to be weaker in untreated seaweed reinforced polypropylene composites than in treated SW/PP composites, which is the main cause for the drop in strength. The treated seaweed reinforced polypropylene composites performed well in Izod impact tests, indicating that they are robust. Due to the weak interfacial bonding between polar hydrophilic SW and non-polar hydrophobic PP, the alkaline treatment on the SW resulted in a considerable improvement in impact strength.

3.2 *Seaweed Reinforced Biopolymer Composites*

Biopolymers, also known as natural polymers are compounds that exist naturally or can be taken from plants or animals. Biopolymers are appealing substitutes for nonbiodegradable petroleum-based plastics because they are both biodegradable and biocompatible [40]. There are numerous studies on seaweed reinforced biopolymer composites since these types of composites can produce a product with good mechanical properties [41].

3.2.1 Seaweed (SW)/Sugar Palm Fibre (SPF) Reinforced Sugar Palm Starch (SPS)/Agar Hybrid Composite

Jumaidin et al. [42] had prepared a hybridized seaweed/sugar palm fibre filler at three different weight ratio of 25:75, 50:50 and 75:25 using thermoplastic SPS/agar (TPSA) as a matrix. The special properties of thermoplastic starch allow it to melt and stiffen repeatedly, making it ideal for a variety of typical plastic fabrication procedures [43]. The composite was fabricated by using melt mixing and hot pressing techniques. Seaweed and sugar palm fibre were shown to be compatible and their hybridization in composites resulted in an increase in intermolecular hydrogen bonding of the composite.

Mechanical tests of the hybrid composites showed that the incorporation of SPF enhanced the tensile and flexural properties while lowering the impact resistance. Figure 5 shows the scanning electron micrograph (SEM) of seaweed composites, seaweed and SPF composites and sugar palm fibre composites. SEM analysis of tensile fractures revealed better fibre-matrix adhesion and effective stress transfer from matrix to fibre (fibre breakage). The hybrid composites improved their water resistance in all tests, including water solubility, water absorption, and thickness swelling.

Due to the enhanced hydrophobicity of the fibre, soil burial studies revealed that hybridising SW composites with SPF resulted in a slower biodegradation process. Overall, combining SW with SPF in TPSA improves the properties (thermal, mechanical and physical) of the composites and can expanding the range of applications for this biodegradable material.

3.2.2 Microcrystalline Cellulose (MCC) Based Seaweed Composites

Cellulose is a key component of all fibre sources, consisting of linear carbohydrate polymer chains made up of β-D,1,4 glucose units linked together by glycosidic linkage. Cellulose is the most plentiful natural polymer in the world because it can be found in numerous sources such as from algae, cottons, wood and bacteria [44]. Nanocrystalline cellulose (NCC) and microcrystalline cellulose (MCC) are crystalline phases of cellulose that can be isolated from purified cellulosic fibres

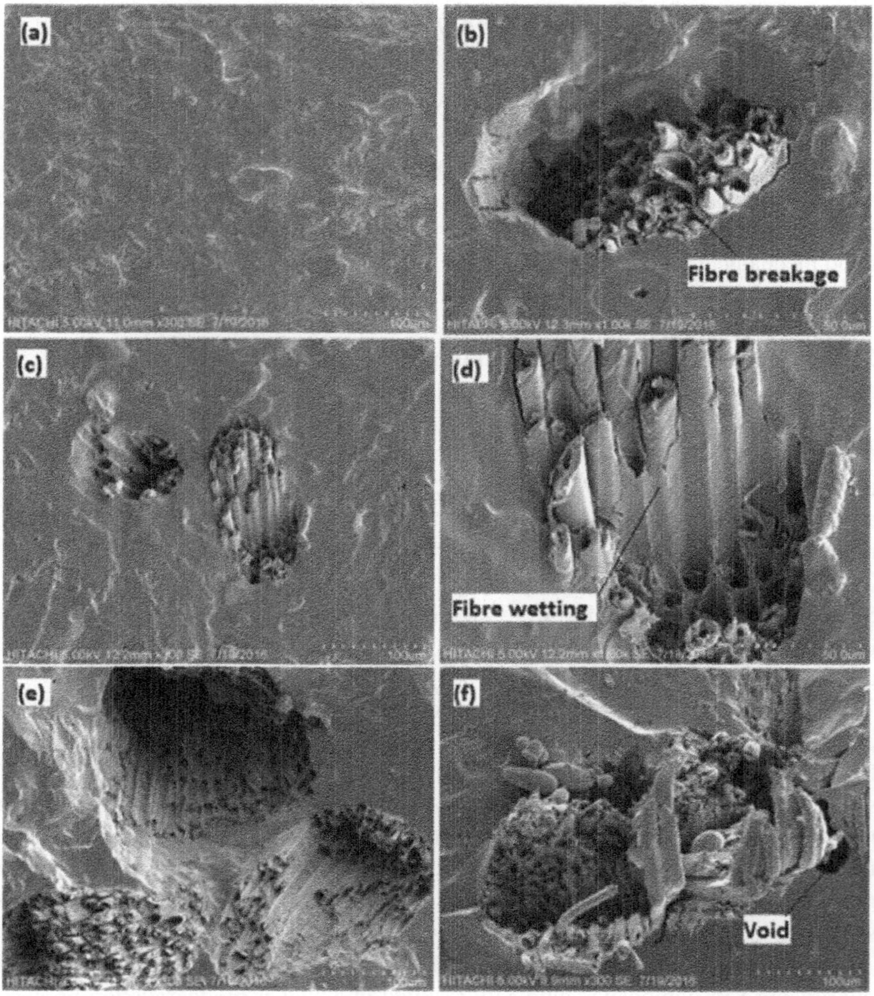

Fig. 5 The scanning electron micrograph images of (**a**) seaweed composites, (**b**) 75% seaweed + 25% SPF composites, (**c**) 50% seaweed +50% SPF composites, (**d**) enlarged image of 50% seaweed +50% SPF composites, (**e**) 25% seaweed +75% SPF composites, (**f**) sugar palm fibre composites [42]

(pure cellulose) using acid hydrolysis. MCC had numerous advantages such as good mechanical capabilities, low density, less or non-abrasive behaviour, high reactivity, biodegradability and renewability when compared to silica, glass fibres, and carbon black [45]. Bamboo, one of the most important natural fibre plants due to its rapid growth rate and adaptability, might be an excellent source for MCC production [46].

Hasan et al. [44] developed and studied seaweed films that are biodegradable with various proportions of MCC taken from two different types of bamboo which are from *Lemang Bamboo* or *L. Bamboo* (*Schizostachyum brachycladum*) and *Semantan*

Bamboo or *S. Bamboo* (*Gigantochloa scortechinii*). They discovered that when different amounts and categories of MCC were added to pure seaweed films, the film morphological properties become more rougher and displayed ranged waves. This resulted in reduced tensile strength in pure seaweed films compared to MCC reinforced seaweed composite films. Figure 6 shows the morphology of SW composite fractured films reinforced with 3% *S. Bamboo* MCC, 5% *L. Bamboo* MCC and 7% commercial MCC particles.

Figure 7 displays the images of seaweed reinforced MCC composites films before and after the soil burial test for a duration of one month. Each set of the seaweed reinforced MCC composite film was categorized by a different colour, specifically red (pure seaweed films), black (seaweed/commercial MCC), yellow (seaweed/*L. Bamboo* MCC) and blue (seaweed/*S. Bamboo* MCC). They discovered that adding different loadings and categories of MCC particles to the seaweed films had no significant effect because all produced seaweed/MCC composite films tend to degrade quickly. Seaweed and MCC are vulnerable to microbial attack in the soil burial test because both are biodegradable polymers.

Khalil et al. [47] have used the MCC from another different bamboo which is Sacred Bali Bamboo (*Schizostachyum brachycladum*) as the reinforcement in SW based composite film. They found that MCC produced from Sacred Bali bamboo demonstrated strong mechanical reinforcement effects in the SW-based film. The

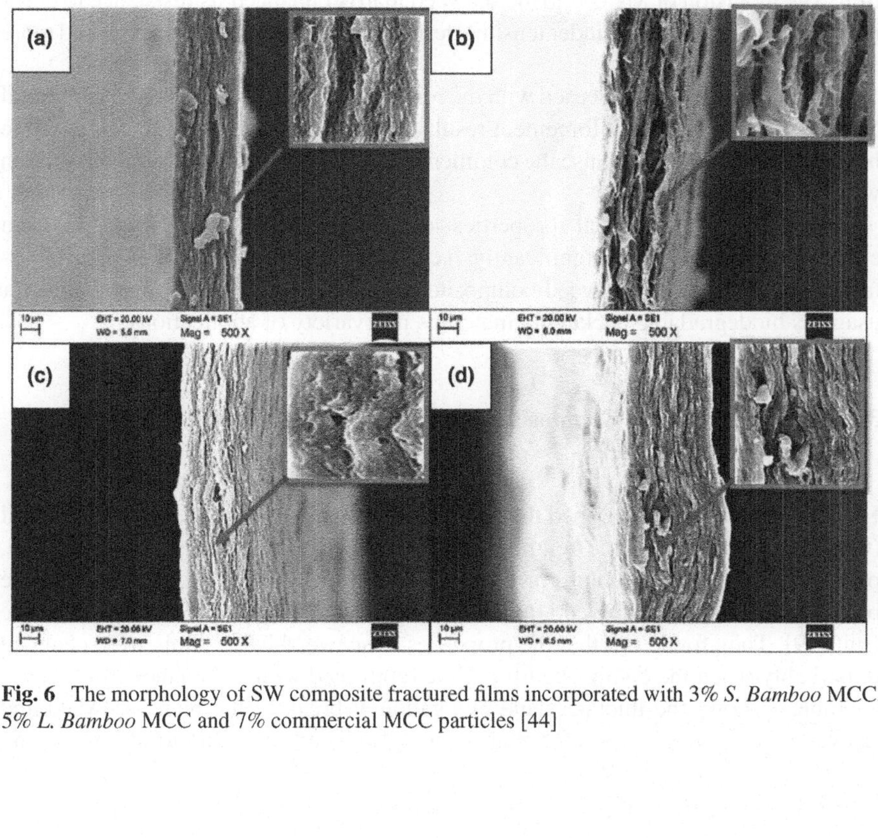

Fig. 6 The morphology of SW composite fractured films incorporated with 3% *S. Bamboo* MCC, 5% *L. Bamboo* MCC and 7% commercial MCC particles [44]

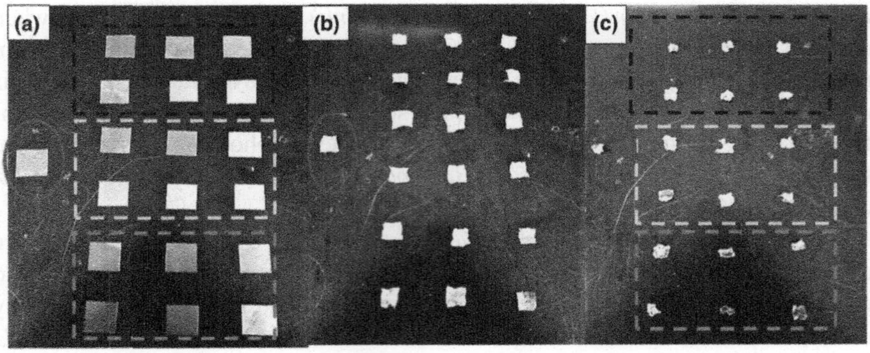

Fig. 7 Digital photos of MCC reinforced seaweed composite films' soil burial biodegradability test (**a**) prior to the test, (**b**) after 2 weeks of testing, and (**c**) after one month of testing [44]

tensile strength increased with the reinforcement of bamboo MCC but the commercial MCC portrayed better reinforcement results and better tensile strength compared to bamboo MCC. This is because the commercial MCC has good interfacial interaction with the seaweed matrix. Figure 8 shows the SEM fractures surfaces of SW composite films at magnification of 500× and 1000× . From the observation, the tensile strength of seaweed composite film reinforced MCC fillers was higher than pure seaweed film. The inclusion of MCC into the seaweed matrix enhanced its resistance to crack initiation and propagation under tensile stresses, and crack propagation required more energy.

The tensile strength increased with the reinforcement of MCC but the commercial MCC portrayed better reinforcement results and better tensile strength compared to bamboo MCC. This is because the commercial MCC has good interfacial interaction with seaweed matrix.

Overall, the mechanical properties of all produced MCC-based seaweed composite films improved significantly. It can be concluded that bamboo micro crystalline cellulose based seaweed composite films have a great deal of potential for usage as biodegradable packaging materials in a variety of applications.

3.2.3 Seaweed Based Composite Film Reinforced Oil Palm Shell Nanofiller (Done)

Khalil et al. [48] had developed the seaweed-composite films that use oil palm shell (OPS) nanoparticles as the reinforcement. They did the analysis on mechanical, physical and surface structure properties of the film. OPS is a by-product of the palm oil mill that is produced after the oil has been removed or extracted from the fruit [49]. They found out that the properties of seaweed based films were changed remarkably when the composite films were reinforced with OPS nanoparticles.

Table 3 shows the thickness and elongation at break of the blank seaweed and seaweed incorporated with OPS nanofiller composite films. Due to the increasing

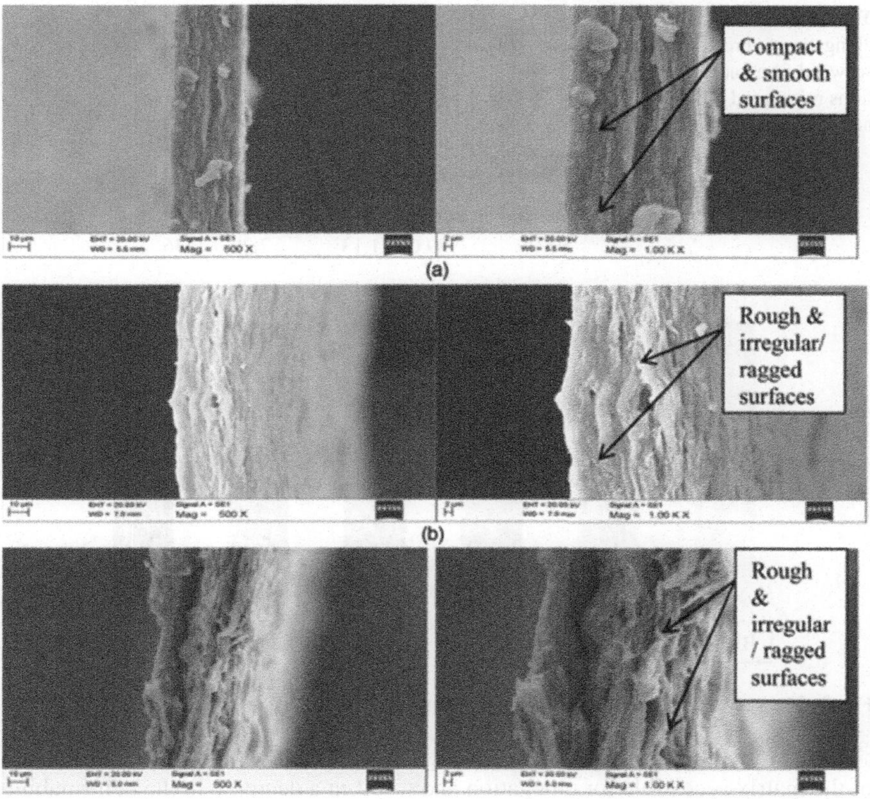

Fig. 8 SEM fractures surfaces of SW composite films (**a**) seaweed (pure) (**b**) seaweed +5% Sacred Bali Bamboo MCC and (**c**) seaweed +5% commercial MCC at magnification of 500 × and 1000 × [47]

solids content, the thickness of the films rises moderately with increased OPS nanofiller content. Furthermore, because the thickness of a film can alter its transparency, the opacity of the films increased as the concentration of OPS nanofillers increased [50]. As a result, darker films are linked to the insertion of a greater number of nanofillers into the SW matrix. Figure 9 shows the seaweed film and seaweed films incorporated with OPS nanofillers.

The elongation at break of the blank seaweed film is the highest which is around 3.30% but for the seaweed films that incorporated with OPS nanofillers, the elongation at break was reduced when the OPS nanofiller content was increased. The ratio of the changed length to the original length of the test specimen (films) after it breaks is known as elongation at break (fracture strain) [51]. According to this research, The addition of OPS nanofillers decreased the flexibility of the composite films while increasing their brittleness. Because the OPS nanofillers were stiffer than the seaweed matrix, the increase in nanofiller content would restrict the chain mobility

Table 3 The thickness and elongation at break of blank seaweed film and seaweed films reinforced OPS nanofillers [48]

Film	Thickness	Elongation at break (%)
Blank	79.1 ± 0.47	3.30 ± 0.35
1%	82.3 ± 0.32	2.72 ± 0.45
5%	83.2 ± 0.19	2.45 ± 0.40
10%	83.8 ± 0.11	2.20 ± 0.49
20%	84.1 ± 0.12	2.10 ± 0.33
30%	89.0 ± 1.13	2.08 ± 0.46

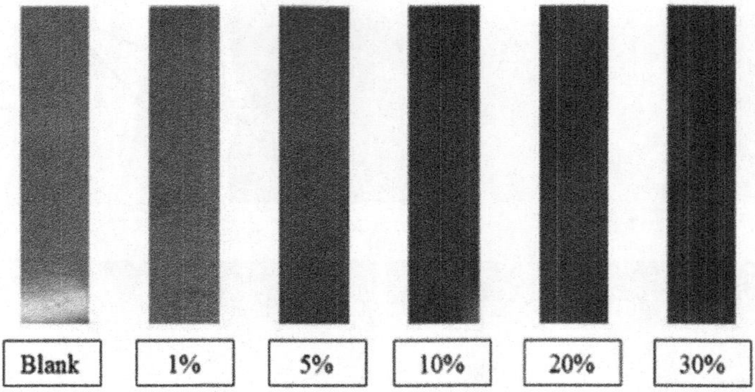

Fig. 9 The seaweed film and seaweed films incorporated with OPS nanofillers [48]

of the matrix accessible for elongation and result in a decrease in the deformability of the interface between the filler and the matrix [52].

3.2.4 Seaweed Based Films Reinforced Cellulosic Pulp Fibre

Khalil et al. [53] had fabricated seaweed biocomposite films reinforced with empty fruit bunch (EFB) pulp fibres. EFB fibre is primarily generated from the oil extraction mills process and it is categorized as a palm oil industry waste [54]. The physical, mechanical and morphological properties of seaweed biocomposite films are noticeably changed when the EFB pulp fibres were used as the reinforcement. The tensile strength of seaweed-EFB composite films improved due to the good compatibility between both seaweed and EFB pulp fibre. In addition, when the EFB pulp fibres content increased in the seaweed-EFB composite films, the contact angle decreased. The main aim of contact angle analysis is to determine the surface hydrophobicity and wettability characteristics of the materials [55]. Figure 10 shows the stationary drops for water contact angle of the SW incorporated EFB pulp films with changing EFB pulp content. Seaweed composite film reinforced EFB pulp fibre has a bright future to use as packaging material because it has better mechanical properties and acceptable hydrophilicity.

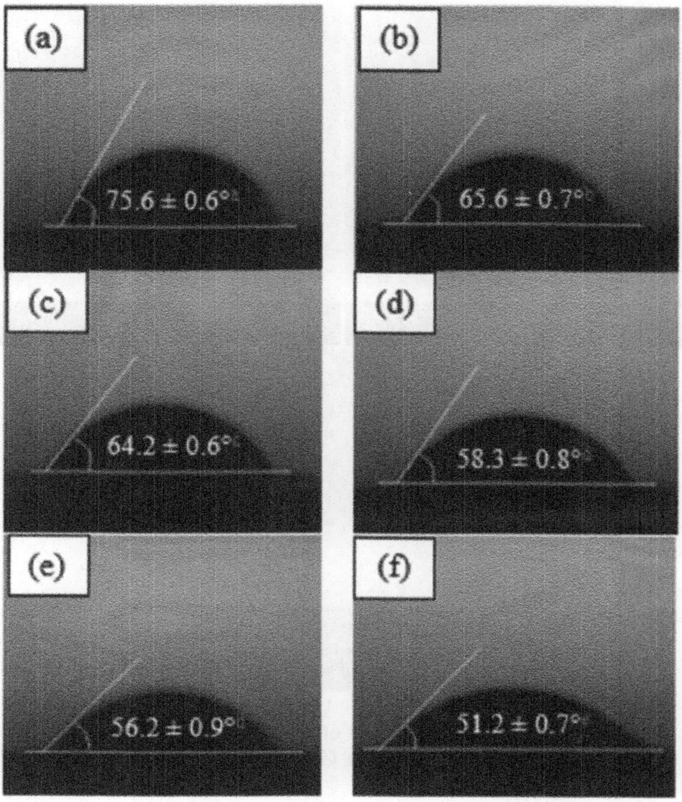

Fig. 10 Stationary drops for water contact angle of the SW incorporated EFB pulp films with changing EFB pulp content: (**a**) 0%, (**b**) 10%, (**c**) 20%, (**d**) 30%, (**e**) 40%, and (**f**) 50% [53]

3.2.5 Seaweed Based Film Incorporated Grapefruit Seed Extract

Kanmani & Rhim [56] had prepared the antimicrobial films from carrageenan incorporated with grapefruit seed extract (GSE). GSE is mainly extracted from the pulps, germs and peel of the grapefruit. The active composite films were created by mixing GSE with carrageenan as an antibacterial agent and the fabricated composite films were interpreted physically and mechanically. Figure 11 shows the scanning electron micrographs photos of cross-section of the carrageenan control film and common carrageenan reinforced GSE composite films with addition of two different levels of GSE. By observation, there are noticeable differences between the control and composite films.

The inclusion of GSE changed the characteristics of carrageenan-based composite films dramatically. The existence of polyphenolic chemicals in the GSE improved the UV barrier characteristics of films. After adding hydrophilic GSE, the films' water contact angle reduced but their water vapour permeability increased marginally. The

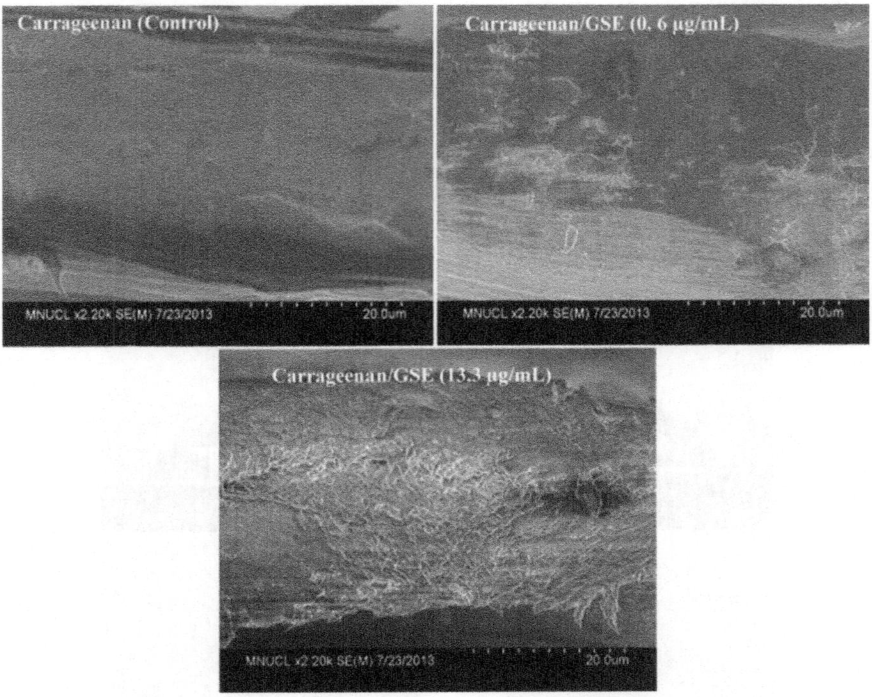

Fig. 11 Scanning electron micrographs photos of cross section of (**a**) carrageenan control film, (**b**) carrageenan reinforced with 0.6 μg/mL of GSE film, and (**c**) carrageenan reinforced with 13.3 μg/mL of GSE film [56]

addition of GSE had no significant effect on the thermal stability of the carrageenan film. Furthermore, the carrageenan/GSE composite films have a significant potential for application as UV screening films for packaging UV-sensitive foods due to their high transparency and excellent UV barrier characteristic. However, further research is required to evaluate potential improved performance for industrialized application of the film. However, more research into enhancing the film's mechanical qualities is required for commercial food packaging applications.

4 Application of Seaweed

Seaweeds based polysaccharide are one of nature most abundant biopolymers that can be used as dispersant, scaffold, stabilising, packaging, thickening agent and coating in the food, biomedical and biomass industries due there, excellent film characteristics, high water retaining capacity, biodegradability and biocompatibility.

4.1 Food Packaging

Application of the seaweed polysaccharides in the food industry are mostly on the capacity to stabilise, emulsify, and produce gels. It's frequently used as a culinary additive to improve and solidify the structure of jams, jellies, ice creams, and dairy goods. Food packaging also makes use of seaweed-based composite sheets and coatings. Active packaging is a system of the product. To improve the shelf life and safety also maintain the quality product by interacting the product, package and environment together. It has components that allow substances to be exchanged to the packaging from the product, start at the environment to the product, or from the product into the environment. Single layer or multilayer active packaging is possible [57]. The active compound is integrated into the polymer in the single layer system, whereas the active compound is placed between a layer of polymer in the multilayer system to control its release. Temperature control and chemical addition are two tactics used in this procedure.

Antimicrobial active packaging is the one of the active food packaging systems [58]. By coating antimicrobial chemicals as a food packaging, antimicrobial packaging decreases, impedes, or slows the growth of microorganisms. Antimicrobial packaging is divided into four categories: pads, antimicrobial coating, naturally antimicrobial polymer and antimicrobial sachets or direct integration in polymer [59]. The ingredients in antimicrobials are contained in a sachet and added to the packaging in antimicrobial sachets or pads. The direct integration of an antimicrobial agent into the polymer, one of the other types, releases the chemical into the packing headspace or onto the food surface. Coating the packaging with a matrix acting is a third category as a transporter for the antimicrobial substance that evaporates into the empty air area or moves into the food through diffusion. The third approach for producing antibacterial properties is to employ naturally antimicrobial polymers [60].

Antimicrobials like ethanol, which is found in baked goods, fish, and cheese is exhaled as a vapor from sachets, are just a few examples. Chlorine dioxide is another substance that is effective against fungi, bacteria, and viruses. When moisture from the product chlorine dioxide contacts the hydrophobic phase, the chlorine dioxide will be released. Acid precursors and Sodium chlorite are saved in a hydrophilic and hydrophobic copolymer [59]. Nisin-loaded chitosan/poly L-lactic acid is one of example for antimicrobial polymers, that have possibly become the new active food packaging film because of its antibacterial habit that fight Staphylococcus aureus [61]. Table 4 shows the application of antimicrobial packaging.

Table 4 The application of antimicrobial packaging [62]

Principle/Reagent	Application
Bacterial cellulose (treated by polypyrrole—Zinc oxide nanocomposite)	All these mesophilic and psychrophilic bacteria grow slowly in chicken thigh [63]
Polyvinyl alcohol/cinnamon essential oil/b cyclodextrin	Antimicrobial activity extends the shelf life of strawberries [64]
Phenolic acids/extracts, e.g. thymol, gallic acid, carvacrol, vanillic acid, cinnamic acid	Applications of fruit and seedling coatings to reduce bacterial contamination/spoilage [65]

Table 5 Biomedical and pharmaceutical applications of SW-based polysaccharides [62]

Type of composites	Application
Carrageenan/Graphene oxide	Bone regeneration and implantation [66]
Carrageenan/Gelatin hydrogels	Drug delivery applications [67]
Oxidized alginate/Gelatin	Hydrogel wound dressing [68]

4.2 Pharmaceutical

Seaweed-based composites might be employed in a variety of applications, including wound dressing, materials, tablet dispersants, bone tissue engineering, cell encapsulation and scaffolds [66, 67]. Polysaccharides generated from seaweed are often used in medicine delivery systems because they gel quickly. The gel formation kinetics has been demonstrated to have an important impact on a number of functional aspects, including stability, biodegradability, immunological properties, and biocompatibility, in several investigations. The active component loss owing to leaching through the pores of the beads during manufacture is the main disadvantage of using SW polymer in drug delivery systems [48]. Many seaweed-based composites have been developed and tested to solve the problem of drug delivery applications that are faced. Table 5 shows the biomedical and pharmaceuticals applications of SW-based polysaccharides.

4.3 Biomass in Fuel

Seaweed is a large and diverse group of aquatic plants (or macroalgae). Sugar kelp and other common species have the potential to become useful biofuel sources. Seaweed is a great contender for biofuels. Anaerobic digestion for biogas and fermentation for ethanol are suitable techniques for seaweed in biofuel production, since it contains between 85 and 90% water. Sugar kelp is one of the example seaweed species that have low lignin content and high carbohydrate that are suitable candidates for bioethanol synthesis. Seaweed is one of species with high effectiveness in

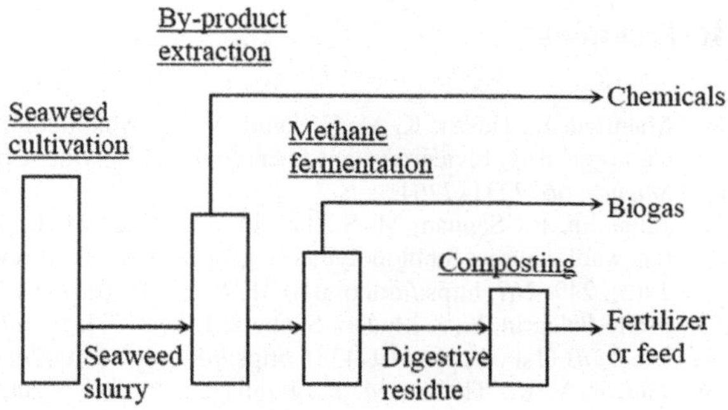

Fig. 12 Seaweed production process for energy and chemicals schematic [69]

absorbing nutrients like nitrogen and phosphate. Because seaweed grows so rapidly, it can absorb high CO_2, up to 66 tonnes CO_2 per hectare, which can help counteract ocean acidification. CO_2 emissions from SW biofuel, for example, are immediately absorbed by new growth due to fast growth. Seaweed is also a high-yielding crop, with a dry weight yield of roughly 26 tonnes per hectare, compared to 2.3 tonnes for soya and 5.1 tonnes for maize [69]. Figure 12 shows the schematic of energy and chemicals production process from seaweed.

5 Conclusion

Seaweed has a lot of potential as a sustainable biopolymer source. These biopolymers have shown to be effective for a number of applications due to their unique film-forming capabilities and superior mechanical properties. Seaweed, it may be concluded, is a highly promising renewable resource for the development of biocompatible and environmentally acceptable materials. Humans have employed seaweed for its therapeutic benefits. It's also used in food as an additive, emulsifier, gelling agent, and stabilizer. Alginate, agar and carrageenan are the most utilized polysaccharides from seaweed. Synthetic polymers and natural polymers are basically having their own advantages when reinforced into seaweed-based composites. Synthetic polymer such as polypropylene and natural polymers such as thermoplastic starch/agar (TPSA) and microcrystalline cellulose (MCC) can be used as the reinforcement in the seaweed-based composites. The tensile strength, impact strength and other mechanical properties were improved and the composites can be used in more various applications. Recent advancements in the field of seaweed-based composites have allowed them to be used in food packaging, pharmaceuticals and biomass in fuel.

References

1. Mahmud, S., Hasan, K. M. F., Jahid, M. A., Mohiuddin, K., Zhang, R., & Zhu, J. (2021). Comprehensive review on plant-fiber reinforced polymeric biocomposites. *Journal of Materials Science, 56*, 7231–7264.
2. Jumaidin, R., Sapuan, M. S., Jawaid, M., Ishak, M. R., & Sahari, J. (2018). Seaweeds as renewable sources for biopolymers and its composites: A review. *Current Analytical Chemistry, 14*(3), 249–267. https://doi.org/10.2174/1573411013666171009164355
3. Freile-Pelegrin, Y., & Madera-Santana, T. J. (2017). *Handbook of composites from renewable materials* (1st ed., pp. 419–438). https://doi.org/10.1002/9781119441632.ch98
4. Thakur, V. K., Thakur, M. K., Raghavan, P., & Kessler, M. R. (2014). Progress in green polymer composites from lignin for multifunctional applications: A review. *ACS Sustainable Chemistry & Engineering, 2*, 1072.
5. Lee Thomas, F. (1986). *The seaweed handbook: An illustrated guide to seaweeds from North Carolina to the Arctic* (pp. 1–42). Dover Publications Inc.
6. Thomas, D. N. (2002). *Seaweeds*. Smithsonian Books.
7. Suthin Raj, T., Nishanthi, P., Graff, K., & Suji, H. (2018). Seaweed extract as a biostimulant and a pathogen controlling agent in plants. *International Journal Of Tropical Agriculture*.
8. Samanthi. (2021). Difference between Chlorophyceae Phaeophyceae and Rhodophyceae. In *Compare the Difference Between Similar Terms*. Retrieved January 17, 2022, from https://www.differencebetween.com/difference-between-chlorophyceae-phaeophyceae-and-rhodophyceae/
9. Rajendran, N., Puppala, S., Sneha, R. M., Ruth, A. B., & Rajam, C. (2012). Seaweeds can be a new source for bioplastics. *Journal of Pharmacy Research, 5*(3), 1476–1479.
10. Alaswad, M., Dassisti, T., Prescott, A. G., & Olabi. (2015). Technologies and developments of third generation biofuel production. *Renewable Sustainable Energy Rev, 51*, 1446–1460. https://doi.org/10.1016/j.rser.2015.07.058
11. Castilho, L. R., Mitchell, D. A., & Freire, D. M. G. (2009). Production of polyhydroxyalkanoates (PHAs) from waste materials and by-products by submerged and solid-state fermentation. *Bioresource Technology, 100*(23), 5996–6009. https://doi.org/10.1016/j.biortech.2009.03.088
12. Price, C. (2020). Seaweed may be the solution to our plastic crisis. A London startup is making edible packaging out of it. *Business Insider*. Retrieved January 17, 2022, from https://www.businessinsider.com/notpla-plastic-waste-seaweed-packaging-2020-10
13. Lim, C., Yusoff, S., Ng, C. G., Lim, P. E., & Ching, Y. C. (2021). Bioplastic made from seaweed polysaccharides with green production methods. *Journal of Environmental Chemical Engineering, 9*(5), 105895. https://doi.org/10.1016/j.jece.2021.105895
14. Shravya, S. C., Vybhava, L. N., Pooja, P., Kishore, K. C. M., & Sadashiva, M. B. M. (2021). Seaweed a sustainable source for bioplastic: a review. *International Research Journal of Modernization in Engineering Technology and Science, 3*(7).
15. Khalil, H. P. S. A., Tye, Y. Y., Saurabh, C. K., Leh, C. P., Lai, T. K., Chong, E. W. N., Fazita, M. R. N., Hafiidz, J. M., Banerjee, A., & Syakir, M. I. (2016). Biodegradable polymer films from seaweed polysaccharides: A review on cellulose as a reinforcement material. *Express Polymer Letters, 11*(4), 244–265. https://doi.org/10.3144/expresspolymlett.2017.26
16. Kennedy, J. (2019). What are the 3 types of Seaweed (marine algae)? ThoughtCo. Retrieved January 18, 2022, from https://www.thoughtco.com/types-of-marine-algae-2291975
17. Dery, B. (n.d.). Reproduction of green algae. Accueil—Dictionnaire Visuel. Retrieved January 19, 2022, from https://infovisual.info/en/biology-vegetal/reproduction-of-green-algae
18. Biology Online. (2021). Chlorophyta definition and examples—biology online dictionary. biology articles, tutorials & dictionary online. https://www.biologyonline.com/dictionary/chlorophyta#Classification
19. García-Casal, M. N., Pereira, A. C., Leets, I., Ramírez, J., & Quiroga, M. F. (2007). High iron content and bioavailability in humans from four species of marine algae. *Journal of Nutrition, 137*, 2691–2695.

20. Gade, R., Tulasi, M. S., & Bhai, V. A. (2014). Seaweeds: A novel biomaterial. *International Journal of Pharmacy and Pharmaceutical Sciences, 5*, 40–44.
21. El-Said, G. F., & El-Sikaily, A. (2013). Chemical composition of some seaweed from Mediterranean Sea coast Egypt. *Environmental Monitoring and Assessment, 185*, 6089–6099.
22. Kılınç, B., Cirik, S., Turan, G., Tekogul, H., & Koru, E. (2013). *Food industry* (1st ed., pp. 735–748). https://doi.org/10.5772/53172
23. Kraan, S. (2012). Algal polysaccharides, novel applications and outlook. In C. F. Chang (Ed.), *Carbohydrates—comprehensive studies on glycobiology and glycotechnology'*. InTech. https://doi.org/10.5772/51572
24. Alihosseini, F. (2016). *Antimicrobial textiles* (1st ed., pp. 155–195). Woodhead Publishing Series in Textiles. https://doi.org/10.1016/B978-0-08-100576-7.00010-9
25. Rhim, J. W. (2004). Physical and mechanical properties of water resistant sodium alginate films. *LWT-Food Science and Technology, 37*(3), 323–330. https://doi.org/10.1016/j.lwt.2003.09.008
26. Qin, Y. (2004). Gel swelling properties of alginate fibers. *Journal of Applied Polymer Science, 91*, 1641–1645.
27. Goh, C. H., Heng, P. W. S., & Chan, L. W. (2012). Alginates as a useful natural polymer for microencapsulation and therapeutic applications. *Carbohydrate Polymers, 88*, 1–12. https://doi.org/10.1016/j.carbpol.2011.11.012
28. Cheng, H., Zhang, X., Chui, Z., & Mao, S. (2021). Advances and challenges in pharmaceutical technology. *Materials, Process Development and Drug Deliveries Strategies, 75–129,*. https://doi.org/10.1016/B978-0-12-820043-8.00010-4
29. Chee, B. S., & Nugent, M. (2019). *Natural polysaccharides in drug delivery and biomedical applications* (pp. 589–615). https://doi.org/10.1016/B978-0-12-817055-7.00026-1
30. Andrea, G. Z., Miguel, A. P. L., Cecilia, J. L., Juan, C. M., & Jesus, S. G. (2019). The potential of seaweeds as a source of functional ingredients of prebiotic and antioxidant value. *Antioxidants, 8*, 406. https://doi.org/10.3390/antiox8090406
31. Song, E. H., Shang, J., & Ratner, D.M. (2012). *Polymer Science: A Comprehensive Review* (1st ed.), *9*, 137–155. https://doi.org/10.1016/B978-0-444-53349-4.00246-6
32. Amir, S. M. M., Sultan, M. T. H., Jawaid, M., Ariffin, A. H., Mohd, S., Salleh, K. A. M., Ishak, M. R., & Shah, A. U. M. (2019). *Durability and life prediction in biocomposites, fibre-reinforced composites and hybrid composites* (1st ed.). Woodhead Publishing Series in Composites Science and Engineering. https://doi.org/10.1016/B978-0-08-102290-0.00016-7
33. Okonkwo, P. C., Collins, E., & Okonkwo, E. (2017). *Biopolymer composites in electronics* (1st ed.). https://doi.org/10.1016/B978-0-12-809261-3.00018-8
34. Shrivastava, A. (2018). *Introduction to plastic engineering* (1st ed.). Plastic Design Library. https://doi.org/10.1016/B978-0-323-39500-7.00001-0
35. Saad, F. U. S. M., Salim, N., & Roslan, R. (2021). Physical and mechanical properties of kenaf/seaweed reinforced polypropylene composite. *Materials Today: Proceedings*. https://doi.org/10.1016/j.matpr.2021.11.411
36. Muzakir, S., Salim, N., Roslan, R., Abdullah, Y., & Samsudin, A. S. (2021). Investigation of hydrophobic properties and mechanical stability of hydrophobic compressed oil palm trunk (OPT) panel. *Materials Today: Proceedings, 49*(8), 3205–3209. https://doi.org/10.1016/j.matpr.2020.11.981
37. Abdullah, N. S. E., Salim, N., & Roslan, R. (2021). Investigation on the effect of alkaline treatment on seaweed/polypropylene (SW/PP) blend composites. *Materials Today: Proceedings*. https://doi.org/10.1016/j.matpr.2021.11.407
38. Chand, N., & Fahim, M. (2021). *Tribology of Natural Fiber Polymer Composites* (2nd ed., pp. 1–59). Woodhead Publishing Series in Composites Science and Engineering. https://doi.org/10.1016/B978-0-12-818983-2.00001-3
39. Valadez-Gonzalez, A., Cervantes-Uc, J. M., Olayo, R., & Herrera-Franco, P. J. (1999). Effect of fiber surface treatment on the fiber-matrix bond strength of natural fiber reinforced composites. *Composites Part B: Engineering, 30*, 309–320. https://doi.org/10.1016/S1359-8368(98)00054-7

40. Das, O., Kim, N. K., Kalamkarov, A. L., Sarmah, A. K., & Bhattacharyya, D. (2017). Biochar to the rescue: Balancing the fire performance and mechanical properties of polypropylene composites. *Polymer Degradation and Stability, 144*, 485–496. https://doi.org/10.1016/j.polymdegradstab.2017.09.006
41. Aaliya, B., Sunooj, K., & Lackner, M. (2021). Biopolymer composites: A review. *International Journal of Biobased Plastics, 3*(1), 40–84. https://doi.org/10.1080/24759651.2021.1881214
42. Jumaidin, R., Sapuan, S. M., Jawaid, M., Ishak, M. R., & Sahari, J. (2017). Thermal, mechanical, and physical properties of seaweed/sugar palm fibre reinforced thermoplastic sugar palm starch/agar hybrid composites. *International Journal of Biological Macromolecules, 97*, 606–615. https://doi.org/10.1016/j.ijbiomac.2017.01.079
43. Jumaidin, R., Zainel, S. N. M., & Sapuan, S. M. (2020). *Advanced processing, properties and applications of starch and other bio-based polymers* (1st ed., pp. 11–19). https://doi.org/10.1016/B978-0-12-819661-8.00002-0
44. Hasan, M., Lai, T. K., Gopakumar, D. A., Jawaid, M., Owolabi, F. A. T., Mistar, E. M., Alfatah, T., Noriman, N. Z., Haafiz, M. K. M., & Khalil, H. P. S. A. (2019). Micro crystalline bamboo cellulose based seaweed biodegradable composite films for sustainable packaging material. *Journal of Polymers and the Environment, 27*, 1602–1612. https://doi.org/10.1007/s10924-019-01457-4
45. Ashori, A., & Nourbakhsh, A. (2010). Performance properties of microcrystalline cellulose as a reinforcing agent in wood plastic composites. *Composites Part B: Engineering, 41*(7), 578–581. https://doi.org/10.1016/j.compositesb.2010.05.004
46. Rasheed, M., Jawaid, M., & Chuah, A. L. (2020). Morphological, physiochemical and thermal properties of microcrystalline cellulose (MCC) extracted from bamboo fiber. *Molecules, 25*(12), 2824. https://doi.org/10.3390/molecules25122824
47. Khalil, H. P. S. A., Lai, T. K., Tye, Y. Y., Paridah, M. T., Fazita, M. R. N., Azniwati, A. A., Dungani, R., & Rizal, S. (2018). Preparation and characterization of microcrystalline cellulose from sacred bali bamboo as reinforcing filler in seaweed-based composite film. *Fibers and Polymers, 19*(2), 423–434. https://doi.org/10.1007/s12221-018-7672-7
48. Khalil, H. P. S. A., Tye, Y. Y., Ismail, Z., Leong, J. Y., Saurabh, C. K., Lai, T. K., Chong, E. W. N., Aditiawati, P., Paridah, M. T., & Dungani, R. (2017). Oil palm shell nanofiller in seaweed-based composite film: mechanical, physical and morphological properties. *Bioresources, 12*(3), 5996–6010. https://doi.org/10.15376/BIORES.12.3.5996-6010
49. Muthusamy, K., Budiea, A. M. A., Azhar, N. W., Jaafar, M. S., Mohsin, S. M. S., Arifin, N. F., & Yahaya, F. M. (2021). Durability properties of oil palm shell lightweight aggregate concrete containing fly ash as partial cement replacement. *Materials Today: Proceedings, 41*(1), 56–60. https://doi.org/10.1016/j.matpr.2020.10.1003
50. Atef, M., Rezaei, M., & Behrooz, R. (2015). Characterization of physical, mechanical and antibacterial properties of agar-cellulose bionanocomposite films incorporated with savory essential oil. *Food Hydrocolloids, 45*. 150–157. https://doi.org/10.1016/j.foodhyd.2014.09.037
51. Petroudy, S. R. D. (2017). *Advanced High Strength Natural Fibre Composites in Construction*, pp. 59–83. https://doi.org/10.1016/B978-0-08-100411-1.00003-0
52. Rosamah, E., Hossain, M. S., Abdul Khalil, H. P. S., Wan Nadirah, W. O., Dungani, R., Nur Amiranajwa, A. S., Suraya, N. L. M., Fizree, H. M., & Mohd Omar, A. K. (2016). Properties enhancement using oil palm shell nanoparticles of fibers reinforced polyester hybrid composites. *Advanced Composite Materials., 1–14*,. https://doi.org/10.1080/09243046.2016.1145875
53. Khalil, H. P. S. A., Tye, Y. Y., Chow, S. T., Saurabh, C. K., Paridah, M. T., Dungani, R., & Syakir, M. I. (2017). Cellulosic pulp fiber as reinforcement materials in seaweed-based film. *Bioresources, 12*(1), 29–42. https://doi.org/10.15376/biores.12.1.29-42
54. Faizi, M. K., Shahriman, A. B., Abdul Majid, M. S., Shamsul, B. M. T., Ng, Y. G., Basah, S. N., Cheng, E. M., Afendi, M., Zuradzman, M. R., Khairunizam, W., & Hazry, D. (2017). An overview of the Oil Palm Empty Fruit Bunch (OPEFB) potential as reinforcing fibre in polymer composite for energy absorption applications. *Matec Web Of Conferences, 90*. https://doi.org/10.1051/matecconf/20179001064

55. Rane, L. R., Savadekar, N. R., Kadam, P. G., & Mhaske, S. T. (2014). Preparation and characterization of K-carrageenan/nanosilica biocomposite film. *Journal of Materials*. https://doi.org/10.1155/2014/736271
56. Kanmani, P., & Rhim, J.W. (2014). Development and characterization of carrageenan/grapefruit seed extract composite films for active packaging. *International Journal of Biological Macromolecules, 68*, 258–266. https://doi.org/10.1016/j.ijbiomac.2014.05.011
57. Domínguez, R., Barba, F. J., G´omez, B., Putnik, P., Kovaˇcevi´c, D. B., Pateiro, M., Santos, E. M., & Lorenzo, J. M. (2018). Active packaging films with natural antioxidants to be used in meat industry: A review. *Food Research International, 113*, 93–101.
58. Sharma, S., Barkauskaite, S., Jaiswal, A. K., & Jaiswal, S. (2020). Essential oils as additives in active food packaging. *Food Chemistry, 343*,. https://doi.org/10.1016/j.foodchem.2020.128403
59. Biji, K. B., Ravishankar, C. N., Mohan, C. O., & Gopal, T. S. (2015). Smart packaging systems for food applications: A review. *Journal of Food Science and Technology, 52*(10), 6125–6135. https://doi.org/10.1007/s13197-015-1766-7
60. Fang, Z., Zhao, Y., Warner, R. D., & Johnson, S. K. (2017). Active and intelligent packaging in meat industry. *Trends in Food Science & Technology, 61*, 60–71. https://doi.org/10.1016/j.tifs.2017.01.002
61. Wang, H., Liu, H., Chu, C., She, Y., Jiang, S., Zhai, L., Jiang, S., & Li, X. (2015). Diffusion and antibacterial properties of nisin-loaded chitosan/poly (L-Lactic acid) towards development of active food packaging film. *Food and Bioprocess Technology, 8*, 1657–1667.
62. Carina, D., Sharma, S., Jaiswal, A. K., & Jaiswal, S. (2021). Seaweed polysaccharides in active food packaging: A review. *Trends in Food Science & Technology, 110*, 559–572.
63. Pirsa, S., & Shamusi, T. (2019). Intelligent and active packaging of chicken thigh meat by conducting nano structure cellulose-polypyrrole-ZnO film. *Material Science and Engineering, 102*(24), 798–809.
64. Wen, P., Zhu, D. H., Feng, K., Liu, F. J., Lou, W. Y., Li, N., Zhong, M. H., & Wu, H. (2016). Fabrication of electrospun polylactic acid nanofilm incorporating cinnamon essential oil/B-cyclodextrin inclusion complex for antimicrobial packaging. *Food Chemistry, 196*, 996–1004.
65. Alkan, D., & Yemenicioglu, A. (2016). Potential application of natural phenolic antimicrobials and edible film technology against bacterial plant pathogens. *Food Hydrocolloids, 55*, 1–10.
66. Liu, H., Cheng, J., Chen, F., Hou, F., Bai, D., Xi, P., & Zeng, Z. (2014). Biomimetic and cell-mediated mineralization of hydroxyapatite by carrageenan functionalized graphene oxide. *ACS Applied Materials & Interfaces, 6*(5), 3132–3140.
67. Balakrishnan, B., Mohanty, M., Umashankar, P. R., & Jayakrishnan, A. (2005). Evaluation of an in situ forming hydrogel wound dressing based on oxidized alginate and gelatin. *Biomaterials, 26*(32), 6335–6342.
68. Varghese, J. S., Chellappa, N., & Fathima, N. N. (2014). Gelatin–carrageenan hydrogels: Role of pore size distribution on drug delivery process. *Colloids and Surfaces B: Biointerfaces, 113*, 346–351.
69. Bellona Europa, Factsheet, Pros and Cons Seaweed For biofuel.

Smart and Sustainable Product Development from Environmentally Polluted Water Hyacinth (*Eichhornia Crassipes*) Plant

A. Ajithram, J. T. Winowlin Jappes, and S. Vignesh

1 Introduction

The water hyacinth is an aquatic plant that is found mostly in tropical and subtropical regions such as the Amazon River and South Africa [1]. These plants can be found in many of the lakes and ponds in Tamil Nadu, including some of the top surfaces of water bodies. This plant grows to about 5 to 8 cm in height in water bodies. Hyacinth plant growth is extremely fast in comparison to other natural fibres and plants and can withstand exposure to the water surface for up to 28 years. Oxygen and nutrients in water bodies are depleted by these plants. It is the main cause of the death of other plants and animals [2]. Hydrophilic properties were achieved by these plants. As a result, they absorb 70% of the water content. Normally, water hyacinth plants are used to produce many applications like ornamental items by separating the parts. Long stalks are characteristic of these plants. Two or three daughter plants can be produced from one water hyacinth plant. Seeds from daughter plants survive longer [3]. There are varieties of water hyacinths that grow from 2 to 5 m tall in a particular region, and there are varieties that grow from 5 to 8 m tall in other regions. The growth characteristics of plants depend on temperature ranges. Studies report that temperatures below 18 °C are not favourable for the growth of that plant. The optimal temperature range is reported as 25–30 °C and for high growth levels, the temperature range is 33 to 35 °C. Further, it is reported the non-growth of these plants is when exposed to the greatest amount of heat. Petioles of these plants contain the bacteria azotobacter [4]. Nitrogen is fixed by this bacterium. Scientists found that

J. T. W. Jappes (✉) · S. Vignesh
Department of Mechanical Engineering, Centre for Composite Materials, Kalasalingam Academy of Research and Education, Virudhunagar, Tamilnadu, India
e-mail: winowlin@klu.ac.in

A. Ajithram
Department of Mechanical Engineering, Karpaga Vinayaga College of Engineering and Technology, Chengalpattu, Tamilnadu, India

Sapuan S. M. and I. Ahmad (eds.), *Composites from the Aquatic Environment*, Composites Science and Technology, https://doi.org/10.1007/978-981-19-5327-9_16

these plants created deficiencies like itchiness and triterpenoids in their experiments. North America, Europe, Africa, and a number of new lands are originally home to these water hyacinth plants. Water hyacinths cover all lakes and ponds once they are unable to be controlled, resulting in the detrimental flow rate of the water in the water bodies. Plants affect water pH and salinity levels at the same time. This unhealthy condition will cause an imbalance in water body properties and alter the whole system's physical and chemical composition [5]. Water bodies are harmed by these characteristics as fishes, other animals, and other water plants are decimated [6]. These plants were used by the German Kaiser submarine mission during the First World War. This infestation of water hyacinth plants was spread to the British by the entire Japanese aircraft during the landing. Based on known facts, water hyacinths are mostly found in amazon basins, a region whose distribution has been greatly affected by wind and birds.

A mat-like structure is the real nature of the plant, used for occupying gaps of other plants and to cover surfaces of water bodies. Water hyacinth plant bulbous blocks the sun rays from reaching deep within water bodies, which affects the life of aquatic plants, animals, and destroys the water content that would otherwise be usable by humans [7, 8]. By blocking the canal system, hyacinth plants directly affect the water, tourism, and irrigation departments. It is possible to cover a large number of water bodies (canals and rivers) within a short period of time, negatively impacting the aquatic environment. Hyacinth plants are considered to be a disaster and waste by the government, so a huge amount of money is spent to remove them from water bodies. It is mostly mechanical methods that are used to remove these plants. As a result of other methods like chemical, manual and biological methods, toxic substances are created, which harm our water bodies. Fibre extraction is higher in stems, as they contain a higher amount of celluloses. In recent years, plant properties such as mechanical, thermal, and vibrational properties were studied to produce products that are strong and durable in commercial applications. Natural fibres are biodegradable, strong, recyclable, and are more readily available.

An effective alternative is to mix materials in a way that produces variations depending on individual applications. In terms of matrix materials, there are thermoplastic resins and thermosetting resins. A thermoset process is able to provide polymers that are cross-linked and have low-temperature resistance with a high modulus (stiffness). A heavy cross-link is created during the polymerization process with a mixture of hardeners. This material is stronger than thermoplastics with non-cross-linked properties [9]. Thermosetting resins are always amorphous at the end of the process. In addition to having high thermal stability, rigidity, high dimensional stability, and weight reduction, thermoset products are also highly insulating and lightweight. Water hyacinth plant composites have attracted many researchers, however, there is a need for significant development in producing powder and ash composites. Various contaminated waters can be cleaned up using water hyacinth. Among the industries in which wastewater can be treated are dairies, tanneries, sugar factories, pulp and paper mills, palm oil mills, distilleries, etc. Heavy metals from the water column are absorbed into the plants' tissues, and the plants grow well in water polluted with organic contaminants and high levels of nutrients [10]. Several small cottage industries have used water hyacinth for paper, rope, baskets, mats, shoes, sandals, bags, wallets, vases, etc. Water hyacinth is a readily available resource in the Philippines, Indonesia, and India. Water hyacinth populations are weakly affected by these products, and their market share is far too small to have any effect on infestations."Under its own brand name "Ban-Aoy", the group has produced 1000 different items ranging in price from 10 baht to 10,000 baht, such as coasters, slippers, baskets, hats, fruit trays, and furniture. In spite of this, the most popular product is handbags and lady purses, more than 80% of which are exported to nations like Japan, the USA, Hong Kong, Singapore, Italy, and more [11].

This research aims to develop an efficient way to utilize the aquatic wastewater plant (Eichhornia crassipes) power, ash, and fibres and to study the use of powder, ash, and fibre material from water hyacinth plants in combination with an epoxy resin matrix to create sustainable products. Hyacinth plant powder and ash are extracted through the traditional process. A new mechanical way of fabricated extracting machines is used to extract the fibre from the water hyacinth plant. Based on the previous kinds of literature reports and mechanical testing results the hyacinth fibre reinforced composites are strongly recommended for commercial particleboard production. From the initial stage of this research work, hyacinth plants are characterized to identify the physical and chemical composition then by utilizing the compression moulding technique composite plats are produced.

2 Materials and Methods

2.1 Materials

The water hyacinth plants are collected from the nearby water bodies located in thiruchirapalli mukkombu river. After the collection of plants, the hyacinth plant parts are removed respectively stem, petiole, bulbous, root and leafs. From the composite production process the secondary matrix materials epoxy and hardeners were purchased at Seenu and company located in Coimbatore. From the composite production process the 2 kg of hyacinth powder and ash particles, and 1 kg of water hyacinth fibres are extracted from the parent water hyacinth plants.

2.2 Water Hyacinth Powder Extraction Process

From the earlier stage of identification of water hyacinth plant, the plants are collected from the tropical and subtropical regions. In India, Tamil Nadu most of the southern district water bodies fully covered by this water hyacinth weeds. Especially Trichy, Madurai, Virudhunagar, Thiruvarur, Thanjavur district water bodies are fully covered by these plants. Once the plants are collected then it is removed by its parts like plant

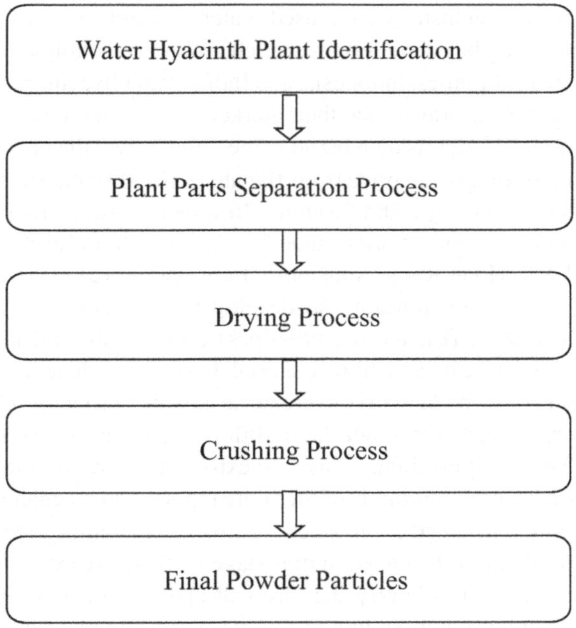

Fig. 1 Methodology for Water Hyacinth Powder Extraction

leafs, roots, petioles, stem, bulbous, and flowers. These plant parts are fully cleaned by using distilled water. Then these parts are dried to the open sunlight air for two weeks except stem. Because, the hyacinth plant stem contains 65 to 70% of water molecules. So, it needs huge time to remove the moisture content from the stem part. Water hyacinth plant stem part alone requires 5 to 6 weeks to get fully dried with the help of room temperature [12]. Once the plant parts are dried, it was crushed by crushing process. With the help of the crushing machine the dried parts of hyacinth plants are fully crushed and the powder particles are finally produced. After the filtration process, the smooth water hyacinth powder particles are derived from the parent plants (Figs. 1 and 2).

2.3 *Water Hyacinth Ash Extraction Process*

The water hyacinth weed completely covers the southern districts of Tamil Nadu. Upon collecting the plants, they are separated by their parts such as leaves, roots, petioles, stems, bulbs, and flowers. These parts are then placed in direct sunlight for five to six weeks [13]. Then the dried hyacinth plant parts are burned with open sunlight air condition. Once the burning process is completed, hyacinth ash particles are produced by utilizing the appropriate filtration process. In order to produce fine ash particles, the extracted ash is filtered again by the filtration machine and then the hyacinth plant fine ash particles are produced (Figs. 3, 4).

Fig. 2 Water Hyacinth Powder Extraction Process

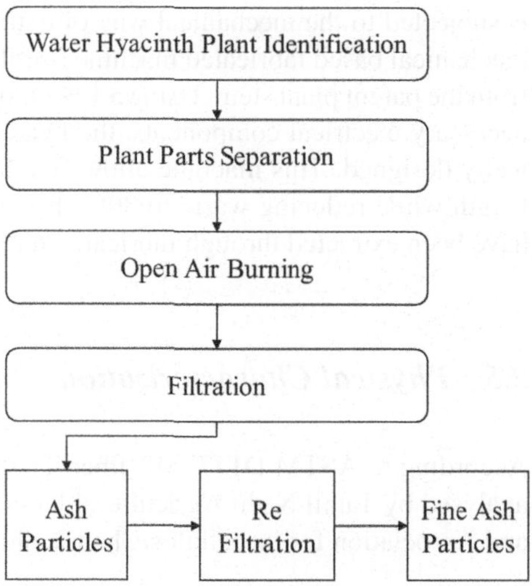

Fig. 3 Methodology for Water Hyacinth Ash Extraction

2.4 Water Hyacinth Fibre Extraction Process

From the beginning stage the hyacinth plant stem is collected from the parent plant. Then, the different types of fibre extraction process are utilized to extract the hyacinth plant fibre. Initially, the plant stem is cleaned by using distilled water. Then the plant stem is subjected to the conventional retting process. At the end of the process, only minimum amount of the fibres are derived from the parent plant. Then the plant stem is subjected to the hot water boiling method and chemical way of extraction method. From these methods, a minimum quantity of the fibres are derived which is similar that of the manual fibre extraction process. Finally, the hyacinth plant stem

Fig. 4 Hyacinth Plant Ash Particle Extraction Process

is subjected to the mechanical way of extraction process [14, 15]. In this work the mechanical based fabricated machine is utilized to remove hyacinth fibre effectively from the parent plant stem. Using a 1 HP motor, mono block bearings, couplings, and necessary electrical components, the hyacinth fibre extraction machine is mechanically designed. This machine allows the hyacinth fibres to maintain their original length while reducing waste to 80%. It is the first time fibres from water hyacinth have been extracted through fabricated mechanical machine (Figs. 5, 6).

2.5 Physical Characterization

According to ASTM D1577-07, fiber length is measured. Water hyacinth fiber was analysed by Tamil Nadu Agricultural University and South Indian Textile Research and Association for its cellulose, hemicellulose, lignin, ash, and pectin content.

3 Result and Discussion

3.1 Physical Characterization Studies

The diameter and density of the hyacinth powder, ash particles are identified at SITRA (South Indian Textile Research Association), India. Cellulose, Hemi cellulose, and essential functional groups of hyacinth fibres are determined at Tamil Nadu Agricultural University, India. The hyacinth single fibre is resulted with a tensile strength of 1.02 MPa. The hyacinth powder particle has 751.8 mm diameter, and ash

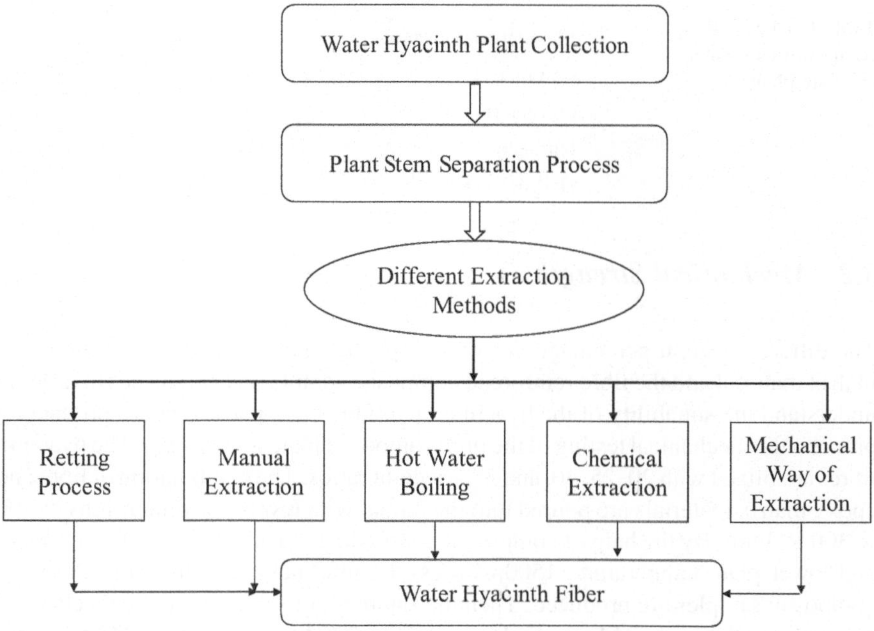

Fig. 5 Water Hyacinth Fibre Extraction Process Flowchart

Fig. 6 Mechanical Way of Hyacinth Fibre Extraction Process

particle has 1332.5 mm diameter. Water hyacinth powder is 1.14 g/cc density, ash is 0.38 g/cc, and water hyacinth fibre is 1.36 g/cc density [16] Table 1.

Table 1 Chemical composition of water hyacinth plant

Element composition	Percentage (%)
Carbon (C)	40–42.5
Oxygen (O)	27–29
Nitrogen (N)	1.2–4.6
Hydrogen (H)	5.2–6.5

3.2 Mechanical Strength

The different weight percentage of hyacinth plant fibres are reinforced with epoxy matrix material and the fibre reinforced composite samples are produced. In order to understand the suitability of the hyacinth plant fibres in a particle board production process, the mechanical testing of the fibre composite plate is done [17]. The hyacinth fibres are mixed with 20, 25, 30, and 35% weight ratios. The combination of fibre and epoxy matrix materials are poured into the mould with respective dimensions of 300 × 300 × 3mm. By the help of compression moulding machine with 100 °C of upper and lower plate temperature, 1500psi pressure, the final hyacinth fibre reinforced composite samples are produced. Then the samples are subjected to the mechanical test with respective ASTM standards like ASTM D3039 for tensile, ASTM D790 for flexural, and ASTM D256 for impact test. The fibre composite tensile strength varied from 17.2 MPa, 19.6 MPa, 23.5 MPa, and 20.8 MPa. The hyacinth fibre composite flexural strength varied from 24.33 MPa, 28.27 MPa, 29.66 MPa, and 20.47 MPa. The impact strength of the hyacinth composite varied from 0.10 J, 0.10 J, 0.5 J, 0.10 J. The hardness values of hyacinth fibre based composite samples are evaluated by using SHORE D hardness tester. From each sample, five different zone hardness values are taken. The hardness values of fibre composite varied from 72, 78, 89, and 81 shore D hardness. Compared to the hardness values of hyacinth composites to the other traditional fibre composites this hyacinth based composites have achieved higher hardness values. Based on the mechanical strength and hardness test results 30% of the hyacinth fibre reinforced composite samples are strongly recommended for particle board production process [18] (Fig. 7).

3.3 Commercial Sustainable Products from Water Hyacinth Plants

Water hyacinth fibre products have the advantage of being tough and durable, as they can last up to three to five years. Forms can be changed according to market trends and tailored to fit any form. At present, water hyacinth fibre is used in various products, including coasters, shoes, hats, baskets, furniture, and women's purses. Water hyacinth stems, which are the main raw material for producing water hyacinth fibre, are known to be tough yet flexible. Water hyacinth fibres have such properties

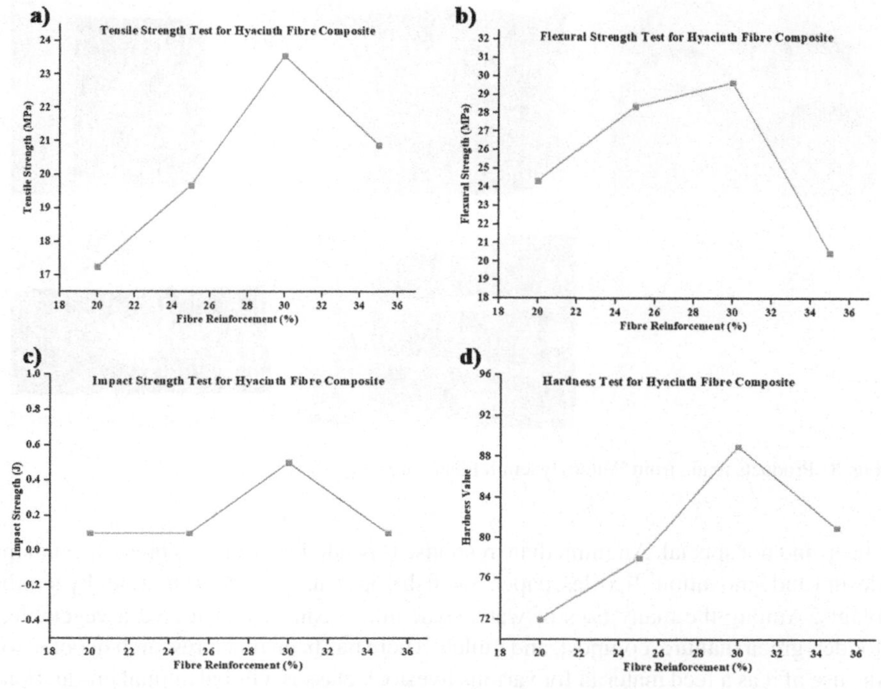

Fig. 7 Mechanical Strength of Water Hyacinth Fibre Composite

by which they can be woven into any form imaginable through 3D weaving. Despite the golden brown colour of natural water hyacinth fibre, natural and chemical dyes can be added to create artefacts that are even more vibrant in colour. The fibres of water hyacinth alone are beautiful and can be used for crafting products. However, other materials such as fabric, leather, and jewellery may be used to enhance the pieces and create only the most delicate pieces. Fibres from water hyacinths are biodegradable as well. Products made from water hyacinth could last up to 5 years with proper coating and care. However, the fibres of water hyacinth can also naturally decompose over time [19] (Fig. 8).

3.4 Development of Hyacinth Products-Present Trend

Handicraft products made from water hyacinth have become more popular on the market from a variety of sources [20]. Having more natural materials in supply to satisfy demand might sound good, but replicas or clones are not ideal since they are repetitive. Competitors have been doing the same thing since they just copy the top selling products of others and offer them for less. Everyone in the system has been affected by this inappropriate action. Products made from water hyacinths now seem

Fig. 8 Products made from Water Hyacinth Plant parts

cheap and not special. An immediate response is needed, as well as a breakthrough in design and innovation. Textiles, paper, and fish traps can be made from water hyacinth plants. Among the many uses of water hyacinth, it can be considered a vegetable, fodder, green manure, compost, and mulch. There has been much research devoted to the use of it as a feed material for various livestock classes. Global animal production has been increased by developing integrated fish, pig, and water hyacinth farming systems in South-East Asia. In fish ponds, the growth of water hyacinths can provide high nutritional value (high protein content) to fish and pigs in various forms.

4 Conclusion

Water hyacinth powder, ash, and fibre particles are successfully extracted and converted to the composite materials in this work. Convert biological waste of aquatic water hyacinth plant into zero waste concept is achieved. Initially, the hyacinth plant fibre is derived, then the wastage of fibre production process is converted into powder and ash form of particles. Sample with 30% of fibre attained highest mechanical strength (23.5 tensile, 29.6 flexural, and 0.5impact) compare to the other weight percentage samples. Based on the mechanical, hardness test results this hyacinth fibre reinforced composite sample is highly recommended for sustainable commercial particle board production process.

References

1. Akash, Sreenivasa Rao K. V., Venkatesha Gupta N. S., Arun Kumar D. S. (2016) Mechanical properties of sisal/coir fiber reinforced hybrid composites fabricated by cold pressing method. In: *IOP Conference Series: Materials Science and Engineering, 149*(1).
2. Rakotoarisoa, T. F., Richter, T., Rakotondramanana, H., & Mantilla-Contreras, J. (2016). Turning a problem into profit: Using water hyacinth (eichhornia crassipes) for making handicrafts at lake alaotra. *Madagascar Economic Botany, 70*(4), 365–379.
3. Sumrith, N., Techawinyutham, L., Sanjay, M. R., Dangtungee, R., & Siengchin, S. (2020). Characterization of alkaline and silane treated fibers of 'water hyacinth plants' and reinforcement of 'water hyacinth fibers' with bioepoxy to develop fully biobased sustainable ecofriendly composites. *Journal of Polymers and the Environment, 28*(10), 2749–2760.
4. Wauton, I., William-Ebi, D. (2019). Characterization of water hyacinth (eichhornia crassipes) for the production of thermochemical fuels. *Journal of Multidisciplinary Engineering Science Studies, 5*(7):2458–925. Available from: www.jmess.org.
5. Guna, V., Ilangovan, M., Anantha Prasad, M. G., Reddy, N. (2017). Water hyacinth: a unique source for sustainable materials and products. *ACS Sustainable Chemistry & Engineering*, Jun 5, *5*(6):4478-90.
6. Gao, Y, Chen, H. P., Wang, J., Shi, T., Yang, H. P., Wang, X. H., (2011) Characterization of products from hydrothermal liquefaction and carbonation of biomass model compounds and real biomass. Ranliao Huaxue Xuebao/Journal Fuel Chem Technol. *39*(12):893–900. Available from: https://doi.org/10.1016/S1872-5813(12)60001-2.
7. Rahman, M. A., (2018). Pyrolysis of water hyacinth in a fixed bed reactor: Parametric effects on product distribution, characterization and syngas evolutionary behavior. *Waste Manag, 80*:310–8. Available from: https://doi.org/10.1016/j.wasman.2018.09.028.
8. Silva, R. P., de Melo, M. M. R., Silvestre, A. J. D., Silva, C. M., (2015). Polar and lipophilic extracts characterization of roots, stalks, leaves and flowers of water hyacinth (Eichhornia crassipes), and insights for its future valorization. *Industrial Crops and Products, 76*:1033–8. Available from: https://doi.org/10.1016/j.indcrop.2015.07.055.
9. Zhang, C., Ma, X., Huang, T., Zhou, Y., Tian, Y. (2020) Co-hydrothermal carbonization of water hyacinth and polyvinyl chloride: Optimization of process parameters and characterization of hydrochar. Bioresour Technol, *314*:123676. Available from: https://doi.org/10.1016/j.biortech.2020.123676.
10. Wu, J. S., Ho, T. C., Chien, H. C., Wu, Y. J., Lin, S. M., & Juang, R. H. (2008). Characterization of the high molecular weight Cd-binding complex in water hyacinth (Eichhornia crassipes) when exposed to Cd. *Journal of Agriculture and Food Chemistry, 56*(14), 5806–5812.
11. Hajar, U., Wardoyo, A. Y. P., Masruroh (2019). Developing and characterization of an ultrafine filter made of banana leaf and water hyacinth to reduce motorcycle emission. *Applied Ecology and Environmental Research, 17*(2):1959–65.
12. Asrofi, M., Abral, H., Kasim, A., Pratoto, A (2017). Characterization of the microfibrillated cellulose from water hyacinth pulp after alkali treatment and wet blending. *IOP Conference Series: Materials Science and Engineering, 204*(1).
13. Supatata, N., Buates, J., & Hariyanont, P. (2013). Characterization of fuel briquettes made from sewage sludge mixed with water hyacinth and sewage sludge mixed with sedge. *International Journal of Environmental Science and Development, 4*(2), 179–181.
14. Asrofi, M., Abral, H., Kasim, A., Pratoto, A., Mahardika, M., & Hafizulhaq, F. (2018). Characterization of the sonicated yam bean starch bionanocomposites reinforced by nanocellulose water hyacinth fiber (WHF): The effect of various fiber loading. *Journal of Engineering Science and Technology, 13*(9), 2700–2715.
15. Barua, V. B., Kalamdhad, A. S., (2017). Effect of various types of thermal pretreatment techniques on the hydrolysis, compositional analysis and characterization of water hyacinth. *Bioresour Technol, 227*:147–54. Available from: https://doi.org/10.1016/j.biortech.2016.12.036.

16. Asrofi, M., Abral, H., Kasim, A., Pratoto, A., Mahardika, M., Park, J. W., et al. (2018). Isolation of Nanocellulose from Water Hyacinth Fiber (WHF) Produced via Digester-Sonication and Its Characterization. *Fibers and Polymers, 19*(8), 1618–1625.
17. Venkatesan, J., Qian, Z. J., Ryu, B., Ashok Kumar, N., Kim, S. K., (2011). Preparation and characterization of carbon nanotube-grafted-chitosan—Natural hydroxyapatite composite for bone tissue engineering. *Carbohydr Polym, 83*(2):569–77. Available from: https://doi.org/10.1016/j.carbpol.2010.08.019.
18. Deepak Joel Johnson, R., Arumugaprabu, V., Nagarajan, R., Souza, F. G., Shanmugam, V. (2021). Temperature-Dependent dynamic mechanical properties and static mechanical properties of sansevieria cylindrica reinforced biochar-tailored vinyl ester composite. *Mechanical and Dynamic Properties of Biocomposites*, Jun 21:235–53.
19. Sundari, M. T., & Ramesh, A. (2012). Isolation and characterization of cellulose nanofibers from the aquatic weed water hyacinth—Eichhornia crassipes. *Carbohydrate Polymers, 87*(2), 1701–1705.
20. Sanjay, M. R., Madhu, P., Jawaid, M., Senthamaraikannan, P., Senthil, S., Pradeep, S. (2018) Characterization and properties of natural fiber polymer composites: A comprehensive review. *Journal of Cleaner Production*, Elsevier B.V. *172*, 566–581 p.

The manufacturer's authorised representative in the EU is Springer Nature Customer Service Centre GmbH, Europaplatz 3, 69115 Heidelberg, Germany. If you have any concerns regarding our products, please contact ProductSafety@springernature.com

Printed and bound by CPI Group (UK) Ltd, Croydon, CR0 4YY
30/09/2026
02225237-0011